Ecological Studies, Vol. 109

Analysis and Synthesis

Edited by

O.L. Lange, Würzburg, FRG
H. A. Mooney, Stanford, USA
H. Remmert †, Marburg, FRG

Ecological Studies

Volumes published since 1989 are listed at the end of this book.

G.W. Davis D.M. Richardson (Eds.)

Mediterranean-Type Ecosystems

The Function of Biodiversity

With 67 Figures and 21 Tables

Springer

Dr. GEORGE W. DAVIS
National Botanical Institute
Stress Ecology Research Unit
Private Bag X7
7735 Claremont
South Africa

Dr. DAVID M. RICHARDSON
University of Cape Town
Institute for Plant Conservation
Botany Department
7700 Rondebosch
South Africa

Front cover: For millennia human societies in the Mediterranean Basin have been exploring the balance between Nature's ability to provide, and Humanity's need to consume. The cover illustration is a design by Peithinos (6th century B.C.), and depicts the marine deity Thetis, granddaugther of the earth goddess Gaea, enploying water, fire, serpents and wild beasts to resist advances by the mortal Peleus. Capitulation and consummation in this encounter resulted in the birth of the hero Achilles, a paradoxical symbol of both human strength and frailty.

Library of Congress Cataloging-in-Publication Data. Mediterranean-type ecosystems: the function of biodiversity/ G.W. Davis, D.M, Richardson, eds. p. cm. — (Ecological studies; vol. 109) Includes bibliographical references and index.
1. Mediterranean-type ecosystems. 2. Biological
diversity. I. Davis, G.W. (George W.), 1947-. II. Richardson, D.M-. (David M.), 1958-. III. Series: Ecological studies; v. 109. QH541.5.M44M43 1994 574.5' 262—dc20 94-34429

ISBN-13: 978-3-642-78883-3 e-ISBN-13: 978-3-642-78881-9
DOI: 10.1007/978-3-642-78881-9

Softcover reprint of the hardcover 1st edition 1995

Typesetting: Camera ready by author
31/3130-5 4 3 2 1 0 - Printed on acid-free paper

Foreword

The threat, and reality, of global change has caused us to examine more closely the controls on the abundances and distributions of organisms. As a result, the study of biogeography and palaeoecology is receiving renewed attention. New insights are being attained that are providing the tools to make predictions of how the world's biotic systems will be reconfigured due to changing land use, a changing atmosphere and a changing climate.

An additional element of global change is the biotic rearrangements that are occurring on Earth due to biological invasions and biotic extinctions. We have ample evidence of the massive intercontinental biotic exchanges that are occurring and of the consequences of these invasions. We also have indications of the great losses in biotic diversity that are occurring at all levels, particularly now in the tropics. For some reason more attention has been given to documenting these losses than in understanding the ecological consequences of the new biotic world in which we live.

To address this issue, SCOPE (Scientific Committee on Problems of the Environment) initiated a program to examine the "ecosystem function of biodiversity". Specifically, the program addresses two basic questions: "How do species affect ecosystem processes in the both the short and long term and in the face of global change?" and, "How is system stability and resistance affected by species diversity and how will global change affect these relationships?". These questions are species-centered, as have been most of the issues related to biodiversity. However, as the program has developed, additional attention has been given to the other changing dimensions of biodiversity, such as populations, communities, and even landscapes.

The basic approach of the SCOPE program has been to examine the functioning of the major ecosystems of the Earth to assess how they have been impacted by species losses and additions as well as by habitat fragmentation and disturbance. This review provides us with the first global assessment of the nature of the problem. The next phase of the program, to be developed by the International Geosphere-Biosphere Program (IGBP), will refine the assessment by launching a specific program to provide new information from experiments specifically designed to address the diversity/function problem rather than relying on the imperfect information derived from inadvertent "experiments" instigated by human activities or from natural environmental fluctuations.

This Volume is one of the first products of the SCOPE global assessment of the ecosystem consequences of changes of biodiversity. Not surprisingly, the first group to achieve this synthesis are the scientists of the mediterranean regions of the world. As documented in the introductory chapter, these scientists have had a long history of international collaboration and of ecosystem comparative analysis. Further, they live and work in natural ecosystems of extraordinarily rich biodiversity, and yet ones that have been heavily impacted by human activities. These circumstances have led to this pathbreaking volume on the relationship between structure and function of natural ecosystems.

Stanford, Autumn 1994 Harold Mooney

Preface

This book is set within the broad research framework of a project, under the auspices of IUBS, SCOPE and UNESCO, which is aimed at providing insights into ecological complexity and the way in which ecosystems will respond to global change (see *Foreword*). That is the top-down perspective. From the bottom up, there have been a series of events which have gathered momentum over several years, and of which this Volume is a staging point. The process began when the mediterranean-type ecosystems (MTEs), which occur in the Mediterranean Basin, California, Central Chile, southern and southwestern Australia, and the southern and southwestern Cape Province of South Africa, were targeted as a focal biome-type for the project. Brian Huntley, member of the SCOPE Committee and Chief Director of the National Botanical Institute in South Africa, was assigned to coordinate this project.

Regional synthesis meetings were arranged in the five regions. The proceedings of the Australian meeting were published as a book (Hobbs 1992), and selected contributions from the South African synthesis meeting appeared in a special issue of the *South African Journal of Science* (Vol. 90 part 6) in 1994. Meetings and consultations in the other regions were directed at producing the framework for the synthesis chapters that appear in this Volume. An international meeting was convened in Cape Town in September 1992 to coincide with the South African synthesis conference. At this meeting, delegates from Australia, California, Chile and the Mediterranean Basin presented draft summaries of their regional syntheses. These, the South African contributions, and other related issues were debated at a 3-day workshop at Springfield on the Agulhas Plain (Davis et al. 1993; Richardson and Cowling 1993). The constructive interaction between delegates at the workshop led to far-reaching discussion on the subject, and the chapters in this book emerged from these deliberations.

The process of producing this book has been very instructive to us, and we feel that the contributions have explored an exciting new domain at the interface between ecology and the search for management strategies to ensure the sustained utilization of natural resources. Perspectives of the relationship between humans and natural ecosystems are clearly different for each region. These range from an evolved association measured in millennia in the Mediterranean Basin, to the rapid and utilitarian transformation of Western Australian kwongan, mallee and woodland for wheat production. As South Africans, we had yet another perspective. The compilation of this book coincided with a dramatic restructuring of the political order in our country. This process certainly carries the threat of instability, but it also provides exciting opportunities for establishing new social paradigms. Implications for sustainable management of the human environment are obvious. It is well known that South African mediterranean-type ecosystems are a "hotspot" of biodiversity (Cowling et al. 1992). They are also important water catchments, and offer a key to foreign investment through ecotourism. A new social order could hasten the demise of this repository of biodiversity and source of human sustenance. Alternatively, it could initiate a new awareness for nurturing the fragile basis for human survival. The key to conserving these ecosystems lies in promoting its wise use as a resource. Reconciling conservation with exploitation is critical, and the issues addressed in this Volume are totally relevant to that issue. It is against that backdrop that we assembled this Volume, hoping that the future will see a greater interaction between humanity and science, without compromising the foundation of either.

We trust that the hard work and commitment of the chapter authors is as much appreciated by the readership as it is by the editors. Also much appreciated is the input by the chapter reviewers and general advisers, whose sole interest has been the promotion of ecology. We thank William Bond (Botany Department, University of Cape Town), Richard Groves (Division of Plant Industry, CSIRO, Canberra), Peter Grubb (Botany School, University of Cambridge), Ian Macdonald (Southern Africa Nature Foundation, Stellenbosch), June Juritz (Department of Statistical Sciences, University of Cape Town), John Lawton (Imperial College, London), Eugene Moll (Department of Management Studies, Gatten College, University of Queensland), Ian Noble (Research School of Biological Sciences, Australian National University, Canberra), Marcel Rejmanek (Botany Department, University of California, Davis), Jacques Roy (CNRS-CEPE, Centre Louis Emberger, Montpellier), Mike Rutherford (National Botanical Institute, Cape Town), Bob Scholes (Division of Forest Science and Technology, CSIR, Pretoria), E-Detlef Schulze (University of Bayreuth), Dave Wardle (Department of Classics, University of Cape Town), Brian van Wilgen (Division of Forest Science and Technology, CSIR, Stellenbosch), Ed Witkowski (Botany Department, University of the Witwatersrand) and Paul Zedler (Biology Department, San Diego State University). We also thank Shirley Cowling, Ann Green and Debbie Hunter for their help during the production of the book. For covering travel expenses of the foreign delegates, and providing logistic and secretarial support at the Cape Town symposium and Agulhas Plain workshop, we thank the South African Foundation for Research Development in Pretoria. For support of the whole MTE synthesis project we are indebted to the National Botanical Institute and its Director, Brian Huntley. The Institute for Plant Conservation at University of Cape Town is also acknowledged for its direct and indirect support.

We want to thank, amongst others, the following for permission to reproduce original or modified material: Annual Reviews Incorporated (Fig. 5.2); Cambridge University Press (Table 1.1); the Ecological Society of America (Fig. 4.8); the Foundation of Research Development (South Africa) (Fig. 1.5 and Table 1.2); Kluwer Academic Publishers (Figs. 2.8, 4.5 and 4,6); Oxford University Press (Figs. 1.6, 6.1 and 6.3); Springer (Figs. 1.4, 3.16, 3.18, 3.19, 4.2, 5.3 and 5.4); and the Staatliche Museum, Berlin (cover illustration).

References

Cowling RM, Holmes PM, Rebelo AG (1992) Plant diversity and endemism. In: Cowling RM (ed) The ecology of fynbos: Nutrients, fire and diversity, Oxford University Press, Cape Town, pp 62-112

Davis GW, Keeley JE, Lamont BB (1993) Biodiversity and ecosystem function in mediterranean-type ecosystems. Bull S Afr Inst Ecol 11(3): 3-5

Hobbs RJ (ed) (1992) Biodiversity in mediterranean ecosystems in Australia. Surrey Beatty. Chipping Norton, NSW

Richardson DM, Cowling RM (1993) Biodiversity and ecosystem processes: opportunities in mediterranean-type ecosystems. Trends Ecol Evol 8: 79-81

George Davis and Dave Richardson
Cape Town, Autumn 1994

Contents

1	**Mediterranean-Type Ecosystems: Opportunities and Constraints for Studying the Function of Biodiversity** R.J. Hobbs, D.M. Richardson and G.W. Davis (8 Figures)	1
1.1	Focus on Mediterranean-Type Ecosystems	1
1.2	Definitions	3
1.3	Why Study Biodiversity and Ecosystem Function?	5
1.4	Linking Diversity and Ecosystem Function	7
1.5	Mediterranean-Type Ecosystems: Similarities and Differences Between Regions	11
1.5.1	Climate	11
1.5.2	Soils	12
1.5.3	Vegetation	15
1.5.4	Human Transformation	19
1.6	History of Research on Mediterranean-Type Ecosystems	23
1.7	Opportunities for Studying Links Between Biodiversity and Ecosystem Function in MTEs	28
References		32
2	**Biodiversity and Ecosystem Function in the Mediterranean Basin: Human and Non-Human Determinants** J. Blondel and J. Aronson (9 Figures)	43
2.1	Introduction	43
2.2	Geography, Climate and Habitat Diversity	46
2.2.1	Geography	46
2.2.2	Climate	47
2.2.3	Habitat Diversity	49
2.3	Patterns of Mediterranean Diversities in Relation to Space and Time	49
2.3.1	Flora and Vegetation	49
2.3.2	Vertebrates	53

2.4 The Historical Load and Mediterranean Adaptations 57
2.4.1 Life History Traits of Mediterranean Plants: Historical Effects or Adaptations? 57
2.4.2 Intraspecific Variation 59
2.4.3 Tests of Convergence Between Bird Communities of Mediterranean Regions of France, Chile and California . . . 61
2.5 Humans and the Mediterranean Biota 62
2.5.1 Fire 63
2.5.2 Deforestation 64
2.5.3 Land-Use Patterns 65
2.5.4 The Case of the Disappearing Black Truffles 67
2.5.5 Differences in Human Pressures, North and South 70
2.6 Biodiversity and Ecosystem Function at the Community Level 71
2.6.1 Disturbances as Key Events in Community Dynamics: a Case Study of Post-Fire Succession 71
2.6.2 Regeneration Traits 75
2.6.3 Changing Tree Cover: Oaks, Pines and Soil Fertility 75
2.6.4 Aromatic Compounds and Insect Pollinators 77
2.6.5 Extinctions and Invasions 77
2.7 Functional Groups and Redundancy 81
2.7.1 Legumes, Grasses and Possible Interactions among Them . . . 81
2.7.2 Avian Frugivores and Bird-Dispersed Plants in Mediterranean Shrublands 87
2.7.3 Species Equivalence and Density Compensation 91
2.7.4 Earthworms: Keystone Species at Risk 92
2.7.5 The Spatial Dynamics of Predatory Ants 94
2.8 Ecological and Evolutionary Consequences of Patchiness . . . 96
2.9 The Future 99
2.9.1 Regional and Global Changes 99
2.9.2 Restoration, Rehabilitation and Reallocation 100
2.9.3 Positive Feedback Switches in Mediterranean Ecosystems . . 103
References 105

3 Biodiversity and Ecosystem Functioning in Mediterranean-Climate California
J.E. Keeley and C.C. Swift (20 Figures) 121

3.1 Introduction 121
3.2 Patterns and Determinants of Diversity 122
3.2.1 Historical Determinants of Floristic Diversity 122

3.2.2 Historical Determinants of Vertebrate Diversity 125
3.2.3 Present Patterns of Plant Diversity 127
3.2.4 Present Patterns of Insect Diversity 128
3.2.5 Present Patterns of Vertebrate Diversity 131
3.2.6 Diversity Patterns of Fungi 133
3.2.7 Patterns of Genetic Diversity 134
3.2.8 Effect of Wildfire Frequency on Landscape Diversity 136
3.2.9 Effect of Alien Species' Invasions on Native Flora and Fauna 137
3.2.10 Effect of Habitat Fragmentation on Patterns of Diversity . . . 140
3.3 Biodiversity and Community Structure and Function 144
3.3.1 Plant Linkages with Other Trophic Levels: Oaks as a Case History 144
3.3.2 Vertebrate Communities 146
3.3.3 Invertebrate Communities 150
3.4 Biodiversity and Ecosystem Function 153
3.4.1 The Annual Grassland Ecosystem 153
3.4.2 The Chaparral Ecosystem 161
3.5 Conclusions 171
3.6 Future Research Needs 173
References 173

4 Functional Approaches to Biodiversity in the Mediterranean-Type Ecosystems of Central Chile
E.R. Fuentes, G. Montenegro, P.W. Rundel, M.T.K. Arroyo, R. Ginocchio and F.M. Jaksic (13 Figures) 185

4.1 Introduction 185
4.2 Biodiversity in Chile 186
4.2.1 Vascular Plants 186
4.2.2 Mammals 188
4.2.3 Birds 189
4.2.4 Reptils and Amphibians 190
4.2.5 Invertebrates 190
4.3 Mediterranean-Type Ecosystems in Central Chile 191
4.4 The Role of Diversity in Ecosystem Function 192
4.5 Plant Species Diversity and Growth-Form Diversity 194
4.6 Environmental Resources as a Factor in Diversity 197
4.6.1 Soil Moisture Resources 197
4.6.2 Nutrient Availability 201
4.6.3 Fire Frequency and Biodiversity 203

4.7 Storage of Species Diversity 206
4.8 Biodiversity in Vertebrate Predator/Prey Systems 209
4.9 Production of Honey at the Community Level 216
4.9.1 Species Representation in Pollen Load Samples 217
4.10 Impact of Changes in Climate and Land-Use on Biodiversity 223
4.11 Conclusions 224
References 226

5 Function of Biodiversity in the Mediterranean-Type Ecosystems of Southwestern Australia
R.J. Hobbs, R.H. Groves, S.D. Hopper, R.J. Lambeck, B.B. Lamont, S. Lavorel, A.R. Main, J.D. Majer, and D.A. Saunders (7 Figures) 233

5.1 Introduction 233
5.2 Mediterranean-Type Ecosystems in Australia 233
5.2.1 Physical Characteristics 233
5.2.2 Vegetation Patterns 236
5.2.3 Floristic Patterns 241
5.2.4 Faunal Richness 243
5.2.5 Landscape Patterns 244
5.2.6 Response to Disturbance 246
5.2.7 Response to Environmental Constraints 248
5.2.8 Human Transformation 251
5.3 Ecosystem Function of Biodiversity 253
5.3.1 Environmental Change and the Loss of Megafauna 255
5.3.2 A Comparison of Natural and Agricultural Systems 258
5.3.3 Loss of Multiple Ecosystem Components: Mistletoes in Eucalypt Woodlands 260
5.3.4 Complex Interactions: Cockatoos, Banksias and Grubs, and Marsupials and Mycorrhizae 262
5.3.5 Ecosystem Processes 264
5.3.6 Functional Groups: Useful or Not? 266
5.3.7 Keystone Species: *Banksia prionotes* as a Keystone Plant . . 266
5.4 Research Opportunities 267
5.4.1 Removal of a Functional Group: Jarrah Forest Fire Regimes 269
5.4.2 Removal of a Functional Group: Effects of *Phytophthora cinnamomi* 269
5.4.3 Additions of Species / Functional Groups in Mine Site Rehabilitation 270

5.4.4 Addition and Removal of Species / Functional Groups: Fragmentation and Revegetation in Agricultural Areas 271
5.4.5 Comparisons Between Species-Rich and Species-Poor Native Plant Communities 271
5.5 Conclusions 272
References 275

6 Links between Biodiversity and Ecosystem Function in the Cape Floristic Region
D.M. Richardson, R.M. Cowling, W.J. Bond, W.D. Stock and G.W. Davis (5 Figures) 285

6.1 Introduction 285
6.1.1 The Link Between Biodiversity and Ecosystem Function .. 285
6.1.2 Defining Biodiversity, Ecosystem Function and Scales of Measurement 286
6.1.3 Opportunities at the Cape 287
6.2 Geography of the Region 288
6.3 Patterns of Biodiversity in the Cape Floristic Region 290
6.3.1 Broad-Scale Patterns of Biodiversity 290
6.3.2 Patterns and Determinants of Diversity: Plant Species and Functional Groups 291
6.3.3 Patterns and Determinants of Diversity: Animal Species and Functional Groups 297
6.4 Biodiversity and Ecosystem Function: Community-Level Perspectives 299
6.4.1 Plant Growth Forms as Functional Groups 299
6.4.2 Regeneration Traits of Plants 300
6.4.3 An Ecophysiological Approach to Functional Grouping in Plants 301
6.4.4 Empty Functional Groups 307
6.4.5 Animal Communities 307
6.4.6 Keystone Taxa 310
6.5 Biodiversity and Ecosystem Function: Landscape-Level Perspectives 315
6.5.1 Within Formations 315
6.5.2 Between Formations 316
6.6 The Ecology of Formation Shifts 319
6.6.1 Forest-Fynbos 321
6.6.2 Fynbos-Renosterveld 321
6.6.3 Fynbos-Grassland 321

6.6.4 Fynbos-Alien Trees 321
6.6.5 Formation Shifts and Ecosystem Function 322
6.7 Conclusions 322
References 324

7 Ecosystem Function of Biodiversity: Can We Learn From the Collective Experience of MTE Research?
G.W. Davis and M.C. Rutherford (4 Figures) 335

7.1 Introduction 335
7.2 Mediterranean-Type Ecosystems as a Focus 338
7.3 Diversity and System Function: a Distillation of MTE Thinking 340
7.3.1 Similarities and Differences in Approach 341
7.4 The Way Ahead 345
7.4.1 Continuing the Search for a New Paradigm 346
References 348

Subject Index 351

Taxonomic Index 359

Contributors

Aronson, J., Centre Nationale de la Recherche Scientifique, Centre d'Ecologie Fonctionelle et Evolutive, B.P. 5051, 34033, Montpellier cedex, France

Arroyo, M.T.K., Departamento de Biología, Facultad de Ciencias, Universidad de Chile, Casilla 653 Santiago, Chile

Blondel, J., Centre Nationale de la Recherche Scientifique, Centre d'Ecologie Fonctionelle et Evolutive, B.P. 5051, 34033, Montpellier cedex, France

Bond, W.J., Botany Department, University of Cape Town, Rondebosch 7700, South Africa

Cowling, R.M., Institute for Plant Conservation, Botany Department, University of Cape Town, Rondebosch 7700, South Africa

Davis, G.W., Stress Ecology Research Programme, National Botanical Institute, Private Bag X 7, Claremont 7735, South Africa

Fuentes, E.R., Departamento de Ecología, Pontificia Universidad Católica de Chile, Casilla 114-D Santiago, Chile

Ginocchio, R., Departamento de Ecología, Pontificia Universidad Católica de Chile, Casilla 114-D Santiago, Chile

Groves, R.H., CSIRO Division of Plant Industry, GPO Box 1600, Canberra, ACT 2601, Australia

Hobbs, R.J., CSIRO Division of Wildlife and Ecology, Locked Bag No 4, P O Midland WA 6056, Australia

Hopper, S.D., Department of Conservation and Land Management, western Australian Wildlife Research Centre, P.O. Box 51, Wanneroo, WA 6065, Australia

Jaksic, F.M., Departamento de Ecología, Pontificia Universidad Católica de Chile, Casilla 114-D Santiago, Chile

Keeley, J.E., Department of Biology, Occidental College, Los Angeles, CA 90041, USA

Lambeck, R.J., School of Environmental Biology, Curtin University of Technology, P.O. Box U1987, Perth 6000, Australia

Lamont, B.B., School of Environmental Biology, Curtin University of Technology, P.O. Box U1987, Perth 6000, Australia

Lavorel, S., Ecosystem Dynamics Group, Research School for Biological Sciences, Australian National University, GPO Box 475, Canberra ACT 2601, Australia

Main, A.R., Zoology Department, University of Western Australia, Nedlands, WA 6009, Australia

Majer, J.D., School of Environmental Biology, Curtin University of Technology, P.O. Box U1987, Perth 6000, Australia

Montenegro, G., Departamento de Ecología, Pontificia Universidad Católica de Chile, Casilla 114-D Santiago, Chile

Mooney, H.A., Department of Biological Sciences, Stanford University, Stanford, California 94305, USA

Richardson, D.M., Institute for Plant Conservation, Botany Department, University of Cape Town, Rondebosch 7700, South Africa

Rundel, P.W., Department of Biology and Laboratory of Biomedical and Environmental Sciences, University of California, Los Angeles 90024-1786, USA

Rutherford, M.C., Stress Ecology Research Programme, National Botanical Institute, Private Bag X 7, Claremont 7735, South Africa

Saunders, D.A., CSIRO Division of Wildlife and Ecology, LMB 4, PO Midland, WA 6056, Australia

Stock, W.D., Botany Department, University of Cape Town, Rondebosch 7700, South Africa

Swift, C., Biology Department, Whittier College, Whittier, CA 90602, USA

1 Mediterranean-Type Ecosystems: Opportunities and Constraints for Studying the Function of Biodiversity

R.J. Hobbs, D.M. Richardson and G.W. Davis

1.1 Focus on Mediterranean-Type Ecosystems

The Mediterranean Basin has played a prominent role in humanity's biological and cultural evolution. From archaeological sites scattered throughout the region, and from the rich written records of historical times, we can trace our development from hunters and gatherers to a species able to reap the benefits of pastoralism, agriculture, and finally sophisticated technology – from harvesters of the wild to managers of our environment. People of the Mediterranean Region have, through the ages, been pioneers in the search for a balance between exploitation and conservation of natural resources. Their efforts have resulted in the domestication of livestock, the development of high-yield cereal grains, and the establishment of important cultivation techniques. The relatively benign winter-rainfall climate of post-Pleistocene times, and the rich diversity of terrestrial and marine biota, were both important factors in human social evolution.

There are four other areas of the world with climates similar to that of the Mediterranean Basin, all centred between 30 and 40° north or south of the equator (Fig. 1.1), and all influenced by similar oceanic and atmospheric circulation patterns. The climates and biota of these regions provided aboriginal people with resource bases that were later to be radically altered by European colonisers. These later settlers arrived in waves of migration appropriately initiated in the Mediterranean Region by the sea faring nations of Spain and Portugal in the 15th and 16th centuries. Since then science-driven technology has improved humanity's ability to extract resources from natural systems, separating dominant land-use practices even further from those of the first peoples. These intensified human activities may now, however, be impairing the ability of natural systems to provide the raw materials needed to maintain and regulate the natural processes essential for sustaining human life. We proceed into the 21st century knowing that: (1) natural systems can sustain human life, even when transformed to a great extent ; (2) the global set of natural systems contains a large amount of diversity (Wilson 1992); and (3) current human activities are drastically reducing that diversity (Wilson 1988). What we do

G.W. Davis and D.M. Richardson (Eds.)
Ecological Studies, Vol.109

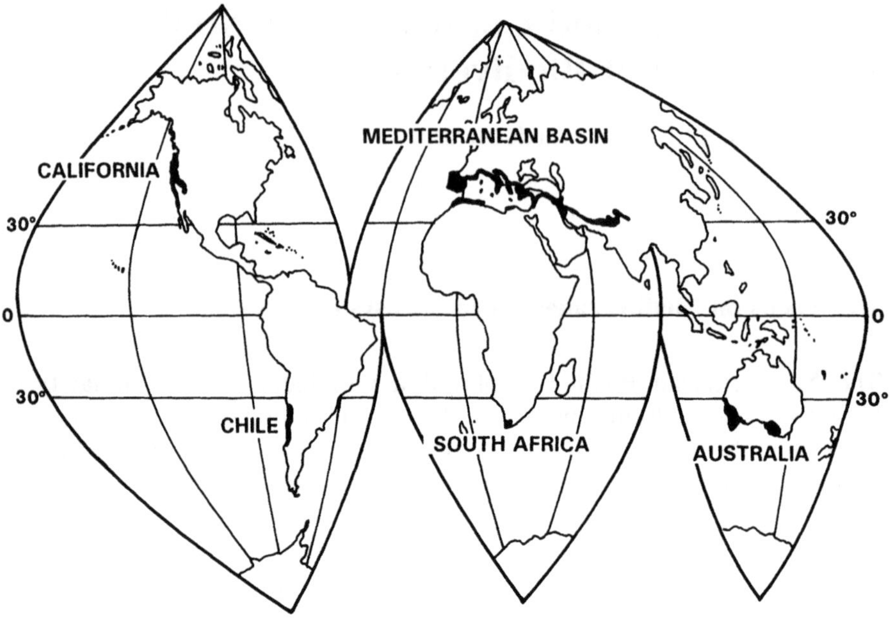

Fig. 1.1. The five mediterranean-climate regions of the world, based on the Cs climate zones of Köppen (1931) (after Müller 1982)

not know, however, is: (1) whether there is some minimum area of the Earth's surface which we need to leave undisturbed; and (2) whether there is a direct link between biological diversity and the maintenance of ecological processes that are necessary for the continued survival of humankind. The questions are of international importance (Younés 1992), and the process of scientific assessment is already well under way (Solbrig 1991, 1992a; Schulze and Mooney 1993).

A major question now being asked is: "how does biodiversity influence the functioning of ecosystems?". The complete answer is probably not simple, but dedication to the search, and uncompromising application of scientific methodology cannot fail to throw some light on this important property of the human environment. This book adds to the debate on the role of biodiversity in ecosystem function by establishing the set of mediterranean-type regions as a focus for comparative review. This chapter, in turn, places that exercise in context by providing a conceptual background to the study of biodiversity and ecosystem function, and an overview of the characteristics of mediterranean-type ecosystems and past research on them.

1.2 Definitions

Biodiversity encompasses the heterogeneity that occurs at all levels in a region's biota. It includes not only the variety of species of plants, animals and other organisms, but also the genes they contain and the communities and ecosystems of which they form part. Patterns of biodiversity form a nested hierarchy, with genotypic variability forming the basis upon which diversity at the population, species, assemblage and ecosystem levels is built. Biodiversity can be considered to encompass not only the compositional aspects at each hierarchical level, but also structural and functional aspects (Noss 1990).

The term *ecosystem functioning* refers to the capacity of an ecosystem to carry out the primary ecosystem processes of capturing, storing and transferring energy, carbon dioxide, nutrients and water (Solbrig 1991; Woodward 1993). Embedded in these overall ecosystem processes are many more population and community level processes that ensure the continued functioning of ecosystems (May 1989; O'Neill 1989). Woodward (1993) refers to these as "endogenous properties accruing from functions". Ecosystem functioning may, therefore, also be considered in an hierarchical manner (Fig. 1.2). Some of the processes or functions considered central to ecosystem functioning are listed in Table 1.1. Thus, for example, nutrient cycling within an ecosystem may depend on the response of the plant community to disturbance, grazing by herbivores, nitrogen fixing symbioses, plant/pollinator interactions and so on. A complete description of a particular function requires stating what is transferred, the genotypes and species involved, the transfer pathway between them or with their abiotic environment, and the extent of transfer under prevailing conditions.

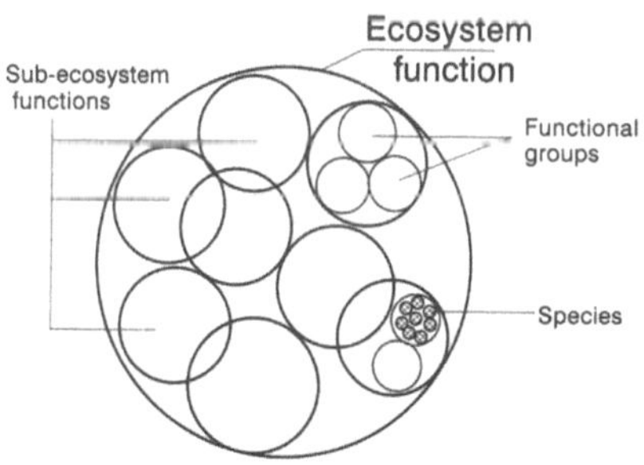

Fig. 1.2. A conceptual model illustrating a hierarchical approach to ecosystem functioning, illustrating how a primary function (nutrient, water or energy transfer) is made up of a number of different sub-ecosystem functions (e.g. decomposition, N fixation, species interactions etc.), each of which is carried out by one or several functional groups consisting of functionally equivalent species (Hobbs 1992)

Table 1.1. Primary ecosystem functions, and examples of secondary functions (or processes) by which primary functions are completed. (Hobbs 1993)

Ecosystem functions:
Energy transfer
Nutrient transfer
Water transfer
CO_2 transfer
Processes involved:
Soil formation/transformation
Nutrient uptake/release
Decomposition
Water uptake/evapotranspiration
Photosynthesis
Herbivory
Pollination
Propagule dispersal/storage
Predation/parasitism/pathogen attack
Other species interactions
Response to disturbance

Transfers are often mediated by organisms other than the species of prime interest, e.g. pollen transport by animals, or nutrient uptake via symbiotic microbes (Lamont 1992). Ecosystem resilience and response to disturbance, and the components thereof (Westman 1986), are also important aspects of ecosystem functioning. Species that perform similar roles can be regarded as belonging to the same *functional group*, but membership of such groups may vary from one system process to the next. These concepts are expanded in the chapters that follow.

Mediterranean-type ecosystems (MTEs) occur in areas characterised by winter rainfall and summer drought. Five regions of the world are recognised as possessing MTEs, namely the Mediterranean Basin, California, central Chile, the southern and southwestern Cape Province of South Africa, and southwestern and parts of southern Australia (Fig. 1.1). Throughout this volume, "Mediterranean" refers to the geographical area around the Mediterranean Sea, while "mediterranean" is adopted as an adjective often used to describe features of the five regions listed above.

1.3 Why Study Biodiversity and Ecosystem Function?

Conservation of biodiversity can be promoted from several perspectives, ranging from the ethical and aesthetic arguments that rest on human values, through the role that genetic diversity plays in evolutionary processes, to the practical consideration of potential uses for previously unexploited species. A question that remains unanswered, however, is whether biological diversity plays a significant role in the functioning of natural ecosystems, and to what extent that diversity can be reduced without causing malfunction of the ecosystem. Science may in time supply an accurate understanding, but unfortunately human pressures are altering many natural systems at a rate faster than ecological investigation can supply definitive answers. To maximise the insight that ecology can provide, both for academic understanding and for sustaining the human environment, there is an urgent need to clarify existing concepts of biodiversity and ecosystem function, and to erect tractable hypotheses for guiding empirical research.

We still do not know to the nearest order of magnitude how many species of organisms there are on earth (May 1992; Wilson 1992). The advent of powerful computing techniques has, however, allowed ecologists to develop quantitative methods, and to extend the power of descriptors. The basic measure of species density in an area of uniform habitat (i.e. species richness) is termed the alpha diversity. This has been expanded in a number of ways to include the turnover of species with habitat change (beta diversity), and the turnover of species between similar, but geographically separated habitats (gamma diversity) (Bond 1989). The relative importance of a species within a community is also a useful qualifier to the measure of diversity, and is often introduced in terms of the contributed biomass. It is often incorporated as a factor in various descriptive indices, of which the Shannon-Wiener Index (Whittaker 1975) is the most frequently used. Such values tell us how evenly importance is spread between species comprising the diversity. Refinements are continually being sought and implemented. These include the quantitative measures of pattern and mosaic diversity derived from affinity analysis (Scheiner 1992), and the sophistication of statistical techniques that improve the predictive value of quantitative models based on empirical data.

Systems ecology is a less well defined field than the study of diversity, although probably much older. Theophrastus and Aristotle studied plant-animal interactions in the 4th and 3rd centuries B.C., both as a contribution to natural science, and as applied research for agriculture (Thanos 1992a). Indeed, human understanding of the functional aspects of biological systems has grown empirically, driven by the need to exploit the system's productive capacities. In societies dependent on intensive agriculture, management of primary production occurs through manipulation of water and nutrient cycles, artificial dispersal of propagules, and physical disturbance of substrates by tillage. Over time this has created a particular perspective and information base for the human understanding of nature, including ecosystem function. The study of ecosystem function therefore has to its advantage a wealth of empirical information. However, much of this was derived from practical work with a limited diversity of domesticated organisms, of which 25 species provide 90% of all human sustenance and international commerce in foodstuffs (Solbrig 1992a). We now need to widen our understanding. Global changes in our mineral resource base, in the composition of the atmosphere, and in our patterns of land use, have reduced the possibilities of acquiring a working knowledge of system function on a trial-and-error basis. Now, as we approach the 21st century, the ability of natural systems to absorb human effluent, provide arable replacements for degraded farmland, and offer the raw materials of human industries is limited. It is not yet known to what extent declining biodiversity is a critical factor.

Due mainly to the complexity of natural systems, it is proving difficult for ecologists to understand the exact roles that species play in the functioning of ecosystems, and to what extent different species may overlap in their contribution to a particular ecosystem function. Working within the hypothetico-deductive framework of conventional science, it is reasonable to postulate two alternative approaches for investigating the suspected link between biodiversity and ecosystem function. In the first instance we can adopt the precautionary stance advocated by Ehrlich and Ehrlich (1992), and propose that the maintenance of important ecosystem functions is dependent on biotic diversity. Validation of this hypothesis would involve demonstrating that functional attributes are negatively affected by the loss of member species. The alternate approach, put forward as a foil by Solbrig (1992b) in the context of human dependence on ecosystem function, proposes that simple systems are just as capable of delivering the necessary products as are species-rich ones. The relevant null hypothesis is that most species in a system, over and above a few key members, are redundant. The experimental approaches that stem from these hypotheses need careful consideration before they can be implemented. A Type II error (acceptance of a false null hypothesis) is very likely to occur in ecology, because of the difficulty of experimental replication. In the case of the first approach (Ehrlich and Ehrlich 1992), this may place unnecessary constraints on environmental development. It would, however, not sanction the system degradation that is possible with the latter approach (Solbrig 1992b). Correct formulation of a scientific investigation of the problem needs to take all of these factors into account.

Biodiversity is not a resource per se, but a property of biological systems. One way to express this property is through the structure that the components provide. As an analogy, the raw materials and tools of the building trade dictate an overall pattern of architectural design. Logs and saws give us log cabins, savannas and pastoralism give us grass-roofed huts, mineral resources and the technology to exploit them give us the concrete, steel and glass skyline of the urban environment. It is an open question whether structural similarity between different natural systems necessarily implies a degree of functional similarity. We may thus ask whether the observed similarities in structure of vegetation in the shrublands of the Mediterranean Basin, California, Chile, South Africa and southern Australia (Mooney et al. 1970) have any functional parallels. A second question is whether ecosystems depend directly on diversity for their continued function. The latter question is the one with which contemporary ecologists, conservationists and land managers are currently grappling, and the one which the chapters of this book scrutinise in the context of MTEs.

1.4 Linking Diversity and Ecosystem Function

Processes that define the function of an ecosystem can usually be described in terms of the flow of matter or energy along flowpaths which connect system patches (Fig. 1.3). An ecosystem can thus be regarded as "an interconnected system of parts" (Woodmansee 1988). Nutrients, water, photosynthates and other products of energy capture all have their unique flowpaths through any particular system, and each flowpath describes a different aspect of the system's function. Biotic diversity and complex biotic interactions may be important in terms of the maintenance of community structure. However, the link between these features and the functions of water, energy and nutrient transfers is not immediately apparent. Indeed, it could be postulated that ecosystem functions can be carried out by a relatively small number of species. For example, a mono-specific pine plantation functions in a similar way to a diverse eucalypt forest, in that energy, nutrient and water transfers occur in each (although rates may differ). In other words, one hypothesis is that biodiversity is not important in ecosystem functioning. In fact, it could be argued that no clear link between diversity and function has yet been demonstrated (e.g. Vitousek and Hooper 1993), despite the fact that most biologists (including ourselves) would intuitively assume that the two are interlinked.

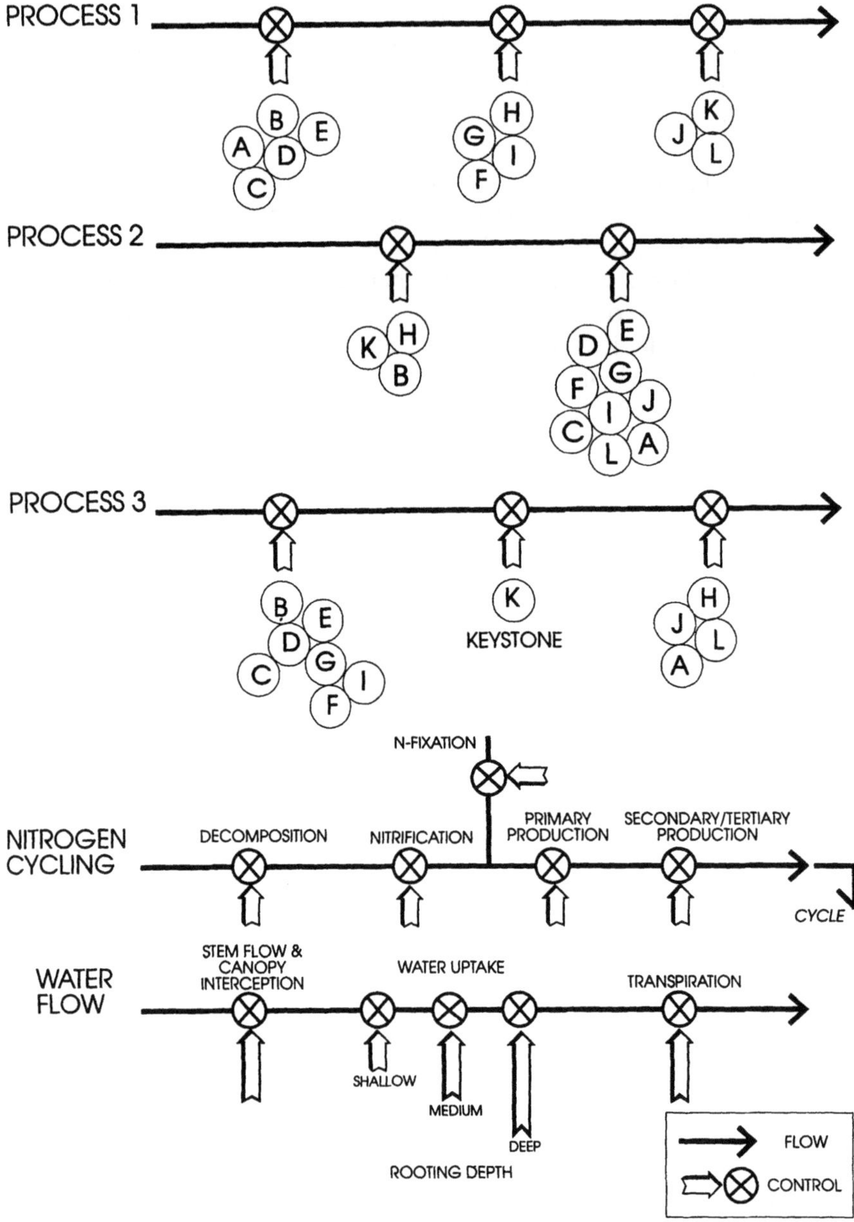

Fig. 1.3. Ecosystem function as the flow of energy and matter through a system, facilitated by species arranged in functional groups. The top three flowpaths depict different hypothetical processes, with the different groupings of species (A - L) relative to each particular process. In the third process, species K would be regarded as a 'keystone' species whose absence would disrupt the entire flowpath. The bottom two flowpaths illustrate how this model might be applied to real world system processes, without any attempt to portray the species in each functional group.

The failure to establish a link between diversity and function can be related almost directly to the historical development of ecology into separate fields of endeavour. O'Neill et al. (1986) suggested that there was a basic distinction between the "population-community" or "biotic" approach to ecology and the "process-functional" approach. The former considers an ecosystem in terms of its constituent species, or populations, organised into guilds and communities, whereas the latter considers it in terms of the functions discussed above. The two streams of ecology have developed with little cross-fertilisation, and we are now faced with the problem of reconciling the two. This will be a difficult task. O'Neill et al. (1986) state that "... the problem is that the relationship between organisms in a system and the functions of the system is not always clear". Functional phenomena do not always correspond in any simple manner to biotic composition (MacMahon et al. 1978, 1981). Functional components need not be a single species, or even an aggregate of species. Similarly, different species can perform the same or similar functions, the same species can perform several different functions at the same time, or the same species can perform different functions at different times or places (O'Neill et al. 1986). Ecologists interested in functions have seldom considered the relevance of biotic diversity or interactions, and similarly, community and population ecologists rarely think about how their data and theoretical constructs can feed into a consideration of ecosystem functions. The problem then becomes one of fitting organism-oriented concepts such as biodiversity into the process-functional ecosystem framework.

One approach is the breakdown of the overall ecosystem functions of energy, nutrient and water transfers into their constituents (Hobbs 1992). These broad functions require a larger number of functions or processes that usually operate at the sub-ecosystem level. These functions or processes are predominantly biotic, and operate at the population or community level. It is thus at this level that biodiversity must be considered. For each function at this sub-ecosystem level, there will be a number of functional groups, each comprising a number of species. Questions of resilience are also more easily tackled at the community level, and Grubb and Hopkins (1986) have provided a preliminary theoretical appraisal of the effect of species richness on resilience. From this approach it becomes clearer that, whereas biodiversity may appear relatively unimportant at the broad level of ecosystem function, it may have a role by contributing to the various subfunctions and through affecting the various components of ecosystem resilience. This provides a framework into which data collected for different purposes could be fitted in order to assess the importance of diversity for ecosystem function.

The above approach assumes that species can be grouped according to their ecological function. In so far as biological species facilitate a particular flow, there may be groups of species whose members all perform similar functions; these are termed *functional groups* (Walker 1992). Solbrig (1991) subdivides species in an ecosystem into different functional types or groups which contain functionally similar taxa or functional analogues. The question then becomes on what basis to classify functional types. In this context, a group of functionally related species may participate in the same pathway but not necessarily to the same extent. For a different process, the same species could conceivably be grouped differently. It has yet to be seen whether sensible functional groupings can be created to incorporate some or all of the various aspects of ecosystem function and/or adaptive responses to environmental variables (Smith et al. 1994).

The grouping of species into functional types raises the issue of functional redundancy or substitutability (i.e. the extent to which species within functional groups may be interchangeable), and whether species diversity is higher than strictly required for ecosystem functioning (Lawton and Brown 1993). Walker (1992) considers that a knowledge of ecological redundancy is essential for the proper evaluation of a decline in biodiversity. He argues that we need to ascertain which aspects of diversity and which kinds of species are most important to ecosystem function. The identification of functional groups containing little redundancy may aid in directing conservation priorities. If a functional group had a single member (only one taxon that was able to facilitate a particular flow) and if the removal of this member would radically alter processes within the system, then this species would obviously be important for ecosystem functioning. Species with no functional analogues have been called "keystone" or "driver" species (Lamont 1992, Bond 1993), and it is important to identify and characterise these species.

However, the definition of a species as a keystone in terms of ecosystem function may not match the more usual definition of keystones in terms of biotic interactions. Lamont (1992) argues that the relative importance of a species depends on the number of species with which it interacts directly or indirectly. This is a continuum, but analyses could be helped by grouping species into those essential for the continued presence of just one other species in the ecosystem, essential for two or more (n-1) species, and essential for all (n) species, i.e. maintenance of the ecosystem. The last category has been described as a *keystone species* (Paine 1969). Its multiplier effect is so great that in its absence the ecosystem would collapse. This could be by interacting with most species, directly or indirectly or through using most resources (the dominant species). It could simply be a keystone for another species essential for the functioning of the ecosystem.

The relationship between biodiversity and function becomes increasingly complex at higher levels of the biodiversity hierarchy. It is much easier to identify the functional significance of the individual components of biodiversity (for example the role of a particular species), than it is to identify the significance of biodiversity per se. The functional significance of biodiversity is likely to be greater than the sum of the contributions of the individual components that make up that diversity. For example, the removal or addition of any individual organism could result in some small change in the rate of a given process. However, the cumulative impact of a succession of such changes is not simply the sum of the impacts associated with the removal or addition of each individual. The impact of losses or additions at any level of the biodiversity hierarchy will also be a consequence of how other species respond to such changes in abundance. Analysis of the role of biodiversity in ecosystem functioning must therefore consider more than the role of the individual components of that diversity. It must take into account the interactions between the various taxa, and how the different components of the region's biodiversity respond to changes in the abundance of other members of the community. The importance of diversity at larger scales is only just beginning to be considered. For example, landscape patchiness and diversity may be an important component in ecosystem functioning, but has received little attention.

1.5 Mediterranean-Type Ecosystems: Similarities and Differences Between Regions

1.5.1 Climate

What is a mediterranean-type climate? A bi-seasonality in temperature and precipitation – cool wet winters with low solar irradiance and hot dry summers with high solar irradiance – forms the basis of Köppen's (1931) definition of a mediterranean-type climate that has been adopted by most geographers. Winter rainfall is the key characteristic of mediterranean regions but the amount and distribution of rainfall create very different conditions, particularly with regard to the degree of summer dryness (Orshan 1983). Milder mediterranean-type climates, with less severe summer drought, are much more favourable to plant growth than the more typical mediterranean climate with a pronounced summer drought. Regions with a mediterranean-type climate differ most markedly in terms of total rainfall. For example, parts of central Chile and southern California have an annual rainfall of between 250 and 350 mm a^{-1} (Miller 1983), whereas parts of "mediterranean-climate" South Africa receive over 2000 mm a^{-1} and a summer rainfall similar to annual totals for California and Chile (Versfeld et al. 1992).

In classifying climatic diversity, di Castri (1981) adopted the bioclimatic classification of Emberger (1955). Six types of mediterranean climate were defined on the basis of aridity and the severity of the summer drought as expressed by rainfall and temperature. Regions vary from very arid (e.g. Baghdad in Iraq) to super-humid (e.g. Pemberton in Western Australia; see Nahal 1981). A further separation is based on the mean minimum temperature of the coldest month, ranging from extremely cold (< -7 °C) to hot (> 7 °C). A useful index for classifying climates on the basis of aridity is Emberger's (1955) pluviometric quotient (Q): $Q = (2000 \times R)/(M^2 - m^2)$, where R = annual rainfall (mm); M = mean maximum temperature for the hottest month (°C); m = mean minimum temperature for the coldest month (°C). Pluviometric quotients for representative study sites in the five MTEs are shown in Fig. 1.4. The diversity of climates must be taken into account when comparing results from different mediterranean-type ecosystems.

1.5.2 Soils

The five mediterranean climate regions have different geologies and physiographic histories, and therefore provide different substrates for plant growth. In spite of some large differences between substrates of the different regions, physiognomy of vegetation in the different regions has been regarded as evolutionarily convergent (Mooney and Dunn 1970). In this section we review very briefly the origins and character of soils in the five regions. South Africa and southern Australia generally comprise much older landscapes, each consisting of an inland mass of geologically older material, surrounded by a discontinuous strip of Tertiary and Quaternary marine deposits. Landscapes of the Mediterranean Basin, California and Chile are much younger, and reflect mountain-building events that occurred as recently as the Tertiary and Quaternary. These different geological histories have given rise to very different sets of soils in the five regions. What they have in common is that they provide a growth substrate which is seasonally droughted.

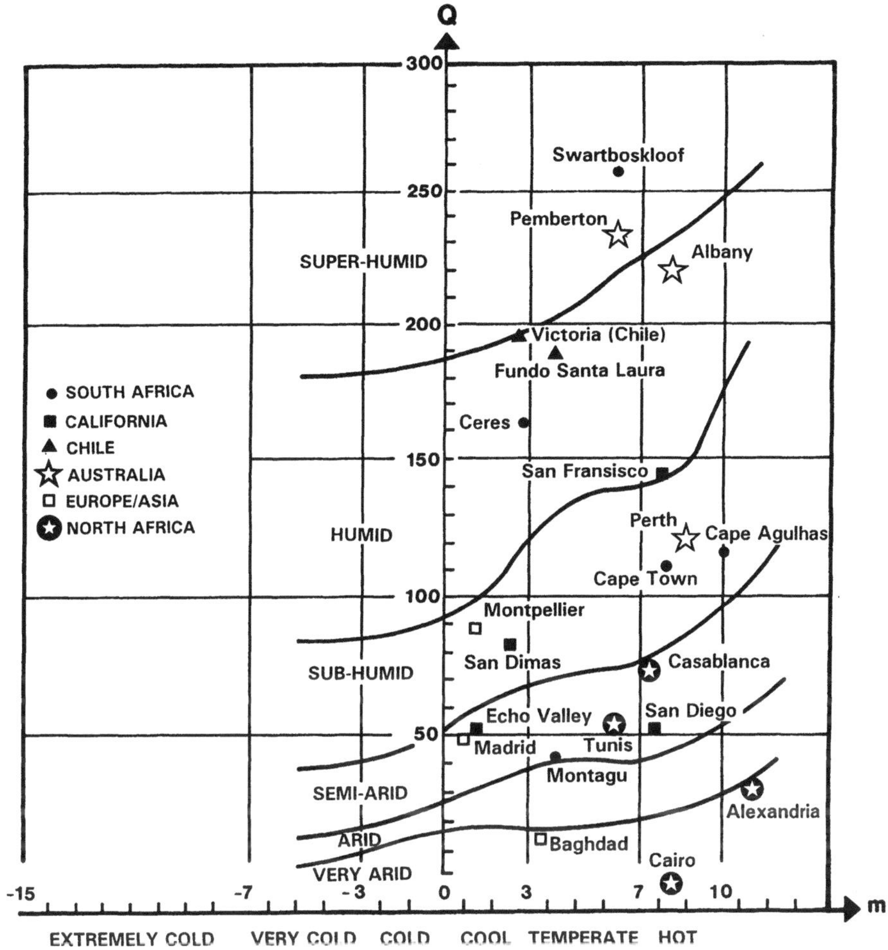

Fig. 1.4. A climate diagram showing Emberger's (1955) pluviometric quotient (**Q**; see text), versus mean minimum temperature for the coldest month (**m**) for stations with mediterranean-type climates. Data for South African sites are from the Weather Bureau, and others extracted from Miller et al. (1977) and Nahal (1981) (After Versfeld et al. 1992)

1.5.2.1 South Africa

Formation of the Cape folded mountains occurred during the Palaeozoic, when strata of sandstones and shales were buckled anticlinally to create ranges with elevations of up to 2300 m. Soils of the uplands are generally highly leached lithosols with little structure. Soils on the lowland shales form the best substrates for agriculture. Calcareous soils of limestone origin occur to a limited extent in coastal regions, while unconsolidated sands of marine origin are more widespread and support a unique coastal dune vegetation.

1.5.2.2 Australia

The geology of southern Australia is similar to that of South Africa in that the older formations generally occur inland, and the coastal zones are of more recent origin. No orogenic movement has occurred on the continent since the end of the Jurassic Period, although warping took place in Western Australia during the Tertiary, and uplifting into the Quaternary. However, much of the landscape has been exposed to weathering since Palaeozoic, and even Precambrian times. The mediterranean-climate regions of Australia (Aschmann 1973a) include parts of the Western and the Central-Southern soil landscape regions (CSIRO 1983). They do not have the same topographical relief as South African mediterranean-climate landscapes, reaching only 730 m at Mount Lofty near Adelaide. Soils on older substrata are characterised by laterites and the process of podzolization, while coastal deposits of the Tertiary and Quaternary give rise to soils ranging from calcareous sands through to decalcified humus podzols and bleached sands, depending on their age. Some terra rossa soils occur on calcareous parent material in the Central-Southern region (Northcote 1983).

1.5.2.3 The Mediterranean Basin

Physiography of the Mediterranean Basin reflects much more recent geological activity than either Australia or South Africa. Major orogenic activity took place during the Tertiary, but subsidence and uplifting extended into the Quaternary. Continental glaciation during the Pleistocene also contributed much erosional material to the Mediterranean Basin. As a result, the landscapes of the Mediterranean Region are very varied, with both great topographical relief (maximum elevation of 3000 m), and a greatly dissected coastline. Pedogenic processes, acting on predominantly limestone rocks, have given rise to a mosaic of soil types. A common soil of the lowland areas is the terra rossa, which is clay-rich and relatively fertile, although it contains much rock debris. Leached podzols occur in the highland areas where precipitation is high enough, and brown forest soils occur; these soils probably formed when forest was more widespread (Thrower and Bradbury 1973).

1.5.2.4 California and Chile

Geological events in the Cordilleran mountain chain down the west coasts of North and South America are associated with violent tectonic activity during the late Tertiary and early Quaternary. This has given rise to landscapes even more rugged than those in the Mediterranean Basin, with maximum elevations of 4500 and 7000 m for California and Chile respectively. Soils of the upland areas are generally coarse textured lithosols which closely reflect their parent materials. Alluvial deposits in the major inland valleys (the Great Valley of California and the Central Valley of Chile) give rise to the more fertile soils of these regions. In Chile, the soils have been markedly influenced by the north-south precipitation regime, with podzolisation being marked in soils of the more humid south.

1.5.2.5 General Soil Characteristics in Mediterranean-Type Ecosystems

The predominant pattern of winter rainfall in all mediterranean-climate regions persisted throughout most of the Cenozoic era (Deacon 1983). Pedogenic processes dependent on water-driven erosion and leaching have therefore produced many similarities between soils of the different regions, in spite of the sometimes large differences in substratum geology between some regions (Table 1.2). Much of the natural vegetation of the five regions occurs on calcareous or moderately to strongly leached soils, with low availability of several nutrients, but especially phosphorous (Fig. 1.5). Nitrogen, which is also limiting for plant growth in many MTEs, is greatly affected by fire. Losses can occur by volatilisation or through physical removal in the post-burn phase. Balance of this nutrient is therefore probably maintained in the long term by either symbiotic or rhizospheric nitrogen fixation (Rundel 1983; see also van Wyk et al. 1992).

1.5.3 Vegetation

The vegetation of MTEs has frequently been discussed in terms of convergent evolution, especially with respect to the basic form and function of the sclerophyllous shrubs that dominate many of the plant communities (Specht 1969a,b; Mooney 1977; Cody and Mooney 1978; Cowling and Campbell 1980; Milewski 1983). More recently, however, the extent to which vegetation convergence occurs has been questioned, and Barbour and Minnich (1990) suggest that significant differences can be found in physiognomy, leaf and life form spectra, species richness, morphological

Table 1.2. The relative frequencies[a] of the main lithological substrates from which soils of the mediterranean-climate regions have formed. (After Groves et al. 1983)

Substratum	W Austr	S Afr	Med	Calif	Chile
Siliceous (eg sandstones, quartzites) and acid igneous rocks	♦♦♦	♦♦♦	♦	♦	nr
Argillaceous and mafic igneous rocks	T	♦♦	T	♦♦♦	♦♦♦
Calcareous rocks	T	♦	♦♦♦	T	nr
Ultramafic rocks (e.g. serpentines)	nr	nr	T	♦	nr

[a] nr = not recorded.
T = Trace.
♦ = occasional.
♦♦ = frequent.
♦♦♦ = common.

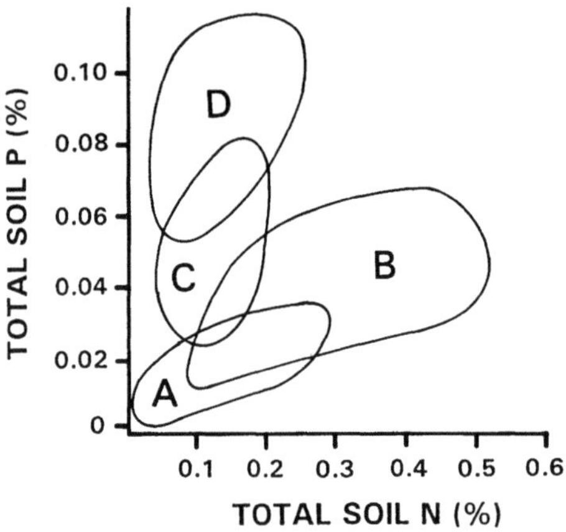

Fig. 1.5. Nutrient levels in surface soils of mediterranean-climate regions, based on total soil nitrogen and total soil phosphorus. **A** Australia and South Africa; **B** Mediterranean Basin (France); **C** California; **D** Chile. (After Groves et al. 1983)

adaptations to fire, phenology, biomass and productivity. They also point out that characteristic "mediterranean" vegetation types actually also occur in non-mediterranean environments, e.g. chaparral also occurs in Arizona (Mooney and Miller 1985), and fynbos-like vegetation occurs in the Afromontane Region of Africa (Killick 1979).

A potential reason for this apparent dichotomy of opinion perhaps lies in the choice of vegetation types used in comparisons. The vegetation types considered as "typically mediterranean" are usually chaparral in California, maquis and garrigue in the Mediterranean Basin, mattoral in Chile, kwongan in Australia, and fynbos in South Africa. As Barbour and Minnich (1990) correctly point out, there are many differences between these vegetation types. Many communities in all regions conform with Specht's (1979) definition of heathlands - evergreen sclerophyllous communities on nutrient-poor soils with, but not necessarily dominated by, heaths of the order Ericales. The heathland concept has been criticised because it demands the simplistic categorisation of shrubland types in terms of soil nutrient status (Cowling and Holmes 1992). Within each region there is a considerable degree of variation in the composition and structure of the shrub vegetation, and a range of other vegetation types are also present (Table 1.3). For example, Tomaselli (1981a) lists 15 variants of matorral vegetation in the Mediterranean Basin, and Hanes (1981) gives 9 variants of chaparral in California. The basic fynbos and kwongan types can be split into innumerable different floristic and geographic variants, and there are a range of other shrub types present in both areas (George et al. 1979; Beard 1984; Cowling and Holmes 1992).

Depending on the variant of sclerophyllous shrubland chosen for comparison in each region, it is possible to find strong similarities or dissimilarities. For instance, southwestern Australian kwongan in some cases strongly resembles some types of Cape fynbos, but is quite dissimilar to most types of chaparral. However, shrubland dominated by *Allocasuarina* spp. in southwestern Australia is very different in form and function to kwongan but quite similar to chaparral. Simple comparisons of individual vegetation types from each region are thus unlikely to yield a true picture of the extent to which the vegetation in MTEs converges or otherwise.

While much emphasis has been placed on between-region comparisons of the sclerophyllous shrublands, considerably less attention has been given to other important vegetation types that are present in all the regions. Shrublands generally occur in a mosaic of other types, including grassland, woodland and forest. Grasslands have received detailed examination in California (Huenneke and Mooney 1989; Heady et al. 1992), as have *Eucalyptus* forests in southwestern Australia (Dell et al. 1989), and *Quercus ilex* woodland in the Mediterranean Basin (Barbero et al. 1992, and other papers in that Vol.). In South Africa, vegetation other than fynbos (e.g. renosterveld) is receiving increasing attention (Cowling et al. 1986).

There has been considerable discussion of the floristic diversity of shrublands, particularly in Australia and South Africa (see Hobbs et al., Richardson et al., Chaps. 5 and 6, this Vol.). What it is frequently overlooked is that the heavily grazed woodlands and shrublands of the Mediterranean Basin have perhaps the greatest alpha diversity of any temperate plant community (Naveh and Whittaker 1979; see also Blondel and Aronson, Chap. 2, this Vol.). The diversity in these communities

Table 1.3. Major vegetation types described for each mediterranean-type region, following local classifications. Levels of detail vary between regions. Major structural and/or floristic types of shrub vegetation are given for each region

Mediterranean basin (Tomaselli 1981a, b)
1. Broadleaved-evergreen forest
2. Coniferous forest
3. Oak woodlands
4. Savanna
5. Matorral
   High matorral (= maquis, espinal, macchia alta)
   Middle matorral (= garrigue, jaral)
   Low matorral (= phrygana, batha)
7. Steppe/grassland
8. Semi-desert

California (Barbour and Major 1977; Hanes 1981)
1. Coniferous forest
2. Broadleaved-evergreen forest
3. Oak woodland/savanna
4. Riparian woodland
5. Pinyon-juniper woodland
6. Chaparral
   Chamise (*Adenostoma fasciculatum*)
   Ceanothus (*Ceanothus* spp.)
   Manzanita (*Arctostaphylos* spp.)
   Scrub-oak (*Quercus dumosa*)
   Serpentine chaparral
   Red shanks (*Adenostoma sparsifolium*)
   Woodland chaparral
   Desert chaparral
   Island chaparral
7. Coastal sage scrub
8. Grassland

Chile (Rundel 1981)
1. Matorral
   typical
   coastal
   montane
2. Espinal (savanna)
3. Hygrophilous forest

Cape Province (Campbell 1985; Cowling and Holmes 1992)
1. Forest and thicket:
   Afromontane forest
   Eastern forest and thicket
   Western thicket
   Dune thicket
   Succulent thicket
   Tall succulent shrubland
2. Karroid and renoster shrubland:
   Tall succulent shrubland
   Succulent shrubland
   Clanwilliam karroid shrubland
   Renoster veld
3. Grassland and grassy shrubland:
   Grassy shrubland
   Grassland
4. Cape fynbos shrubland:
   Grassy fynbos
   Asteraceous fynbos
   Restioid fynbos
   Ericaceous fynbos
   Proteoid fynbos
   Closed scrub fynbos

Southwestern Australia (Beard 1984; Beard and Sprenger 1984)
1. Tall forest and woodland (*Eucalyptus diversicolor*)
2. Forest (*Eucalyptus marginata*)
3. Woodlands
   E. loxophleba, E. wandoo,
   E. salmonophloia, E. salmonophloia – E. salubris, E. calophylla – E. wandoo
4. Low forest
5. Low woodland (*Banksia* spp.)
6. Low woodland (others)
7. Thicket
   Acacia-Allocasuarina-Melaleuca
   Other thickets and scrub
8. Mallee
9. Heath
   Scrub-heath with low trees
   Mallee-heath
   Scrub-heath
   Heaths
   Coastal wind-pruned heaths
   Other heaths on coastal limestone
   Heaths on laterite
   Heaths on winter-wet sites
10. Reed swamps
11. Halophytes

South Australia (Specht 1972)
1. Open forest (with heath understorey)
2. Low open forest to woodland (with heath understorey)
3. Open scrub (with heath understorey) (= mallee)
4. Woodland to open forest (with herbaceous understorey)
5. Open scrub (= mallee)
6. Tall open shrubland
7. Low woodland
8. Low shrubland (= shrub steppe)
9. Tussock grassland
10. Heath

derives from the large numbers of annuals capable of surviving multiple stresses of drought, fire, grazing and cutting.

Climatic and edaphic control of vegetation distribution at regional and local scales has been examined in Western Australia (Beard 1983), California (Bakker 1971) and South Africa (Cowling and Holmes 1992). The spatial and temporal interrelationships between different mosaic elements have received considerable attention in the Mediterranean Basin, where different elements are thought to form successional and regressional series in response to disturbance and human use (Braun-Blanquet 1964; Debussche et al. 1980; Tomaselli 1981b; Escarré et al. 1983; Naveh and Lieberman 1984). These ideas are now being applied in Californian and South African systems (George et al. 1992; Huntsinger and Bartolome 1992; Manders and Richardson 1992; Manders et al. 1992), and to a limited extent in Australia (Hopkins and Robinson 1981; Grubb and Hopkins 1986). Since the variety of vegetation types present in each region contributes to the overall biodiversity of that region, more detailed work on the total vegetation mosaic both within each region and comparatively across regions seems appropriate.

1.5.4 Human Transformation

The MTEs of the world have very different histories of human occupation and transformation of natural ecosystems. Indigenous agriculture and animal husbandry was practised in the Mediterranean Basin from 10 000 years ago, and the vegetation is now predominantly human-modified (Naveh and Kutiel 1990). The other regions were occupied by indigenous hunter-gatherers and pastoralists until relatively recently. Chile was colonised by the Spanish in 1550, South Africa by the Dutch in 1652, California by the Spanish in 1769, and southwestern Australia by the British in 1829. European grains and livestock were introduced to all regions soon after colonisation; in Chile this was nearly 450 years ago, in South Africa 330 years ago, and in California 215 years ago. Australia has a much shorter history of European settlement than the other regions - only 150 years, although the impact has been profound and rapid (Aschmann 1973b; Fox 1982). Cultural differences between the colonising nations, inherent differences between regions (e.g. environment, resource availability and relationships between colonisers and indigenous peoples), and differences in the time since settlement, have interacted to produce marked contrasts in current patterns of transformation and the status of the

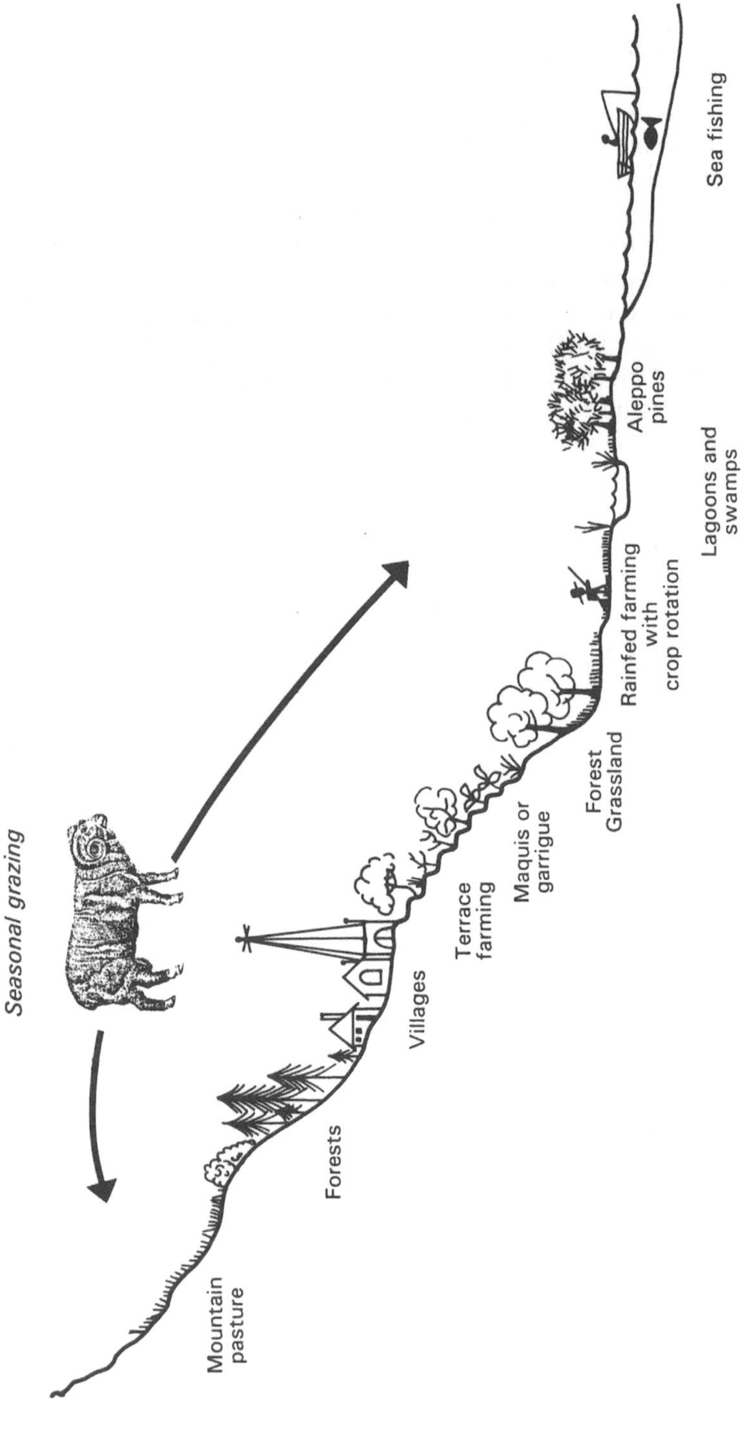
Seasonal grazing
Mountain pasture
Forests
Villages
Terrace farming
Maquis or garrigue
Forest Grassland
Rainfed farming with crop rotation
Lagoons and swamps
Aleppo pines
Sea fishing

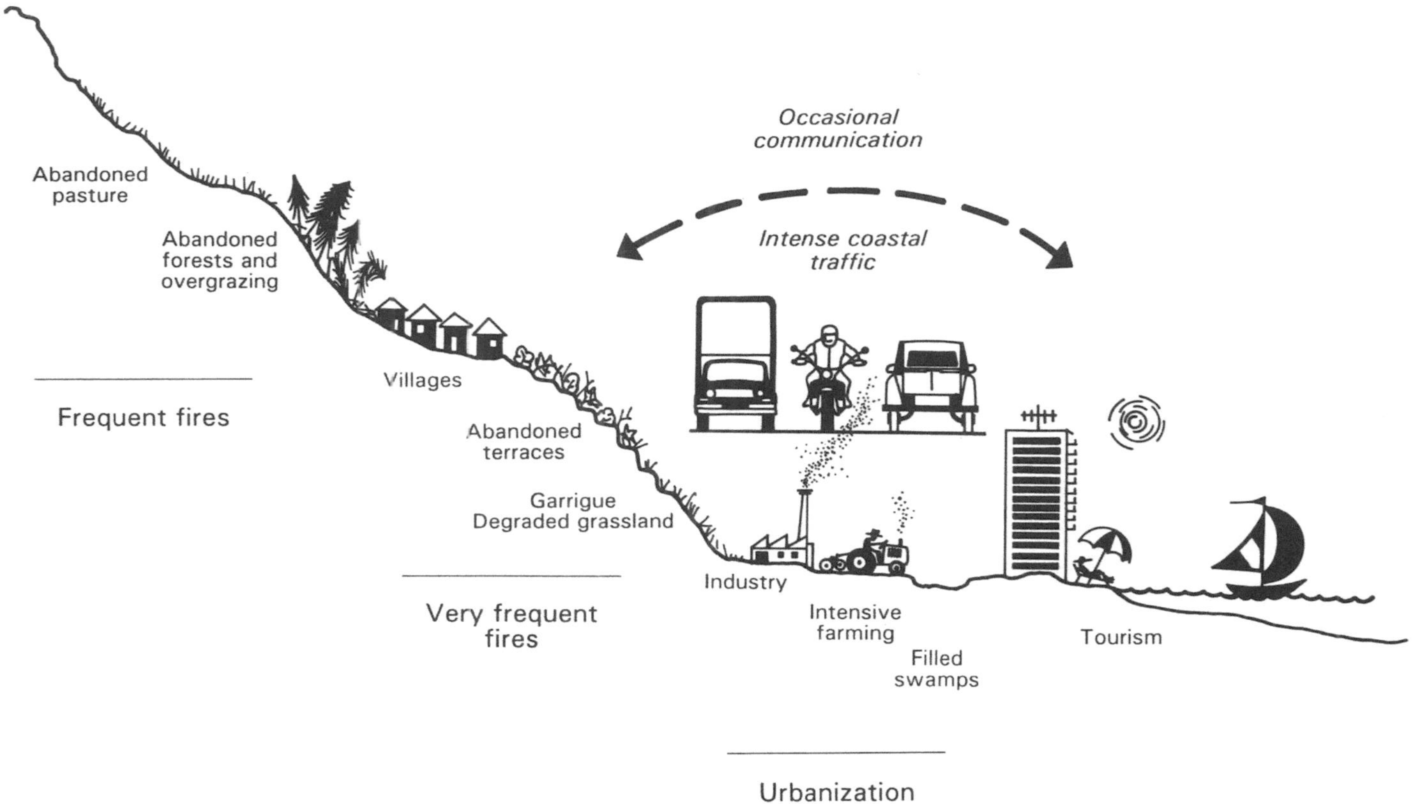

Fig. 1.6. Changes in land-use patterns result in landscape fragmentation. The diagram depicts the situation in the Mediterranean Basin, where increasing urbanization and a reduced economic reliance on agrarian activities results in degeneration of the land-use mosaic. (Based on Grenon and Batisse 1989)

natural environment. Examples are: changes to the fire regime (increased fire frequency overall but with periods of fire suppression, and changes in seasonal distribution of fires in all regions), grazing regime (increased grazing pressure in all regions), elimination of large native herbivores (notably in Australia and the Cape), rapid clearing of indigenous forests (especially in Australia and California), and planting of alien trees (notably at the Cape) (Kruger et al. 1989).

Part of the process of fragmentation in the Mediterranean Basin is depicted in the two "snapshot" diagrams in Fig 1.6. It should be borne in mind that the earlier impression of land use is only a particular stage of landscape transformation between hunter-gatherer times and the present. Massive landscape shifts had already occurred before the pastoral/agrarian landscape depicted here, the most notable probably caused by deforestation during the rise of the sea-faring empires (Attenborough 1987; see also Blondel and Aronson, Chap. 2, Sect. 2.5.3, this Vol.). Unlike the other mediterranean-climate regions, however, the same net shift occurred over millennia instead of over centuries, a fact noted for Australia by Hobbs and Hopkins (1990). The result is that although Mediterranean Basin landscapes have been greatly altered from their pristine states, the intervening economic and land use structures have acted as a buffer to slow the impact of transformation, and have in many cases promoted the retention of diversity at the landscape level.

Most natural ecosystems in all the regions persist as fragments in an agricultural/urban matrix (e.g., see Hobbs et al., Chap. 5, Fig. 5.6, this Vol.). The effects of this fragmentation on ecosystem processes have been best studied in Australian MTEs. Extensive removal of native vegetation has led to dramatic changes in many ecosystem components and processes, particularly in terms of the hydrological regime, vegetation dynamics, invasions by non-native species, and survival of faunal populations (Saunders and Hobbs 1992; Hobbs 1993; Hobbs et al. 1993; McFarlane et al. 1993; Saunders 1993; see also Hobbs et al., Chap. 5, this Vol.). The montane areas of the Cape are probably the closest to pristine of all MTEs (Aschmann 1973b), although these are rapidly being degraded by invasive alien plants. The history of introduction of alien organisms to the five regions has been well documented (Kruger et al. 1989; Groves and di Castri 1991). Different taxonomic groups are prominent in the invasive alien biotas of the five regions. For example, trees are over-represented in the invasive alien flora of the Cape, whereas grasses and herbs are over-represented in Australia, California and Chile. Several alien mammals have been very successful in MTEs, notably those in Australia (Myers 1986), but none has invaded natural ecosystems at the Cape (Richardson et al. 1992). Differences between MTEs in the magnitude of invasion by alien organisms, and the impact of the invaders on ecosystem functioning, can be ascribed to differences in the inter-region transfer of biota due to geographical distance, cultural ties, resource requirements, and to poorly understood differences in the susceptibility of different MTEs to invasion by different groups (e.g. Richardson and Cowling 1992).

1.6 History of Research on Mediterranean-Type Ecosystems

The mediterranean-climate regions of the world boast a long history of ecological research. In addition to the research carried out within individual regions, there has been a relatively long history of cooperation in research between the five regions (Table 1.4). During the 1950s the different regions joined together under the auspices of programmes such as the UNESCO Arid Zone Project, FAO/UNESCO projects, MAB, IBP and SCOPE to conduct comparative and synthetic studies of various aspects of the ecology of MTEs (di Castri et al. 1988a). Strong links between pairs of regions, particularly Chile - California and Australia - South Africa have also fostered inter-regional exchange and comparison. From 1971 a series of international conferences on mediterranean-type ecosystems (MEDECOS) has drawn together workers from all regions. These and a range of other meetings have produced a wealth of detailed information (di Castri and Mooney 1973; Mooney and Conrad 1977; Conrad and Oechel 1982; Kruger et al. 1983; Dell 1984; Dell et al. 1986; di Castri et al. 1988b; Thanos 1992b; Arianoutsou and Groves 1993; Arroyo et al. 1994). The most neglected fields in between-MTE comparisons are vegetation structure (with the exception of some intensive South Africa–Australia, California–Mediterranean and California–Chile comparisons), plant-animal interactions (with the exception of the South Africa-Australian comparative work), phenology, management, animal species composition, and plant-water relations (Fig. 1.7). The aspects of MTEs that are most evenly compared across the five regions are plant species composition and biological invasions (Table 1.5).

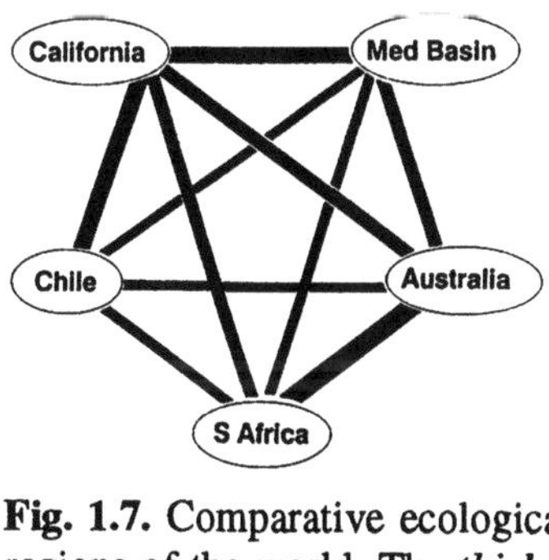

Fig. 1.7. Comparative ecological studies between the five mediterranean climate regions of the world. The *thickness of the lines* joining any two regions indicates the relative frequency of comparisons between the pairs of regions. (Based on data in Table 1.4)

Table 1.4. Comparative studies between different regions with a mediterranean-type climate. Only studies providing data from more than one region are presented. The "general" category includes edited volumes which contain comparative material, but not those which are colections of studies from individual regions.

Feature studied	Med	Aust	S. Afr	Calif	Chile
General and multiple topics:					
Emberger (1959)	♦	♦			
di Castri and Mooney (1973)	♦	♦	♦	♦	♦
Mooney (1977)				♦	♦
Thrower and Bradbury (1977)				♦	♦
Cody and Mooney (1978)	♦		♦	♦	♦
di Castri (1981)	♦	♦	♦	♦	♦
Miller (1981)				♦	♦
Specht and Moll (1983)	♦	♦	♦	♦	♦
Milewski (1983)		♦	♦		
Kruger et al. (1983)	♦	♦	♦	♦	♦
Milewski and Cowling (1985)		♦	♦		
Mooney (1988)	♦	♦	♦	♦	♦
Specht (1988)	♦	♦	♦	♦	♦
Barbour and Minnich (1990)	♦	♦	♦	♦	♦
Keeley (1992)			♦	♦	
Arroya et al. (1993)		♦		♦	♦
Cowling et al. (1994)		♦	♦		
Climate:					
Emberger (1958)	♦	♦			
Aschmann (1973a)	♦	♦	♦	♦	♦
Russell and Moore (1976)		♦	♦		
Milewski (1979)		♦	♦		
Versfeld et al. (1992)	♦	♦	♦	♦	♦
Soils and geology:					
Thrower and Bradbury (1973)	♦	♦	♦	♦	♦
Zinke (1973)	♦			♦	
Day (1983)	♦	♦	♦	♦	♦
Plant species composition and diversity:					
Naveh (1967)	♦			♦	
Gulmon (1977)				♦	♦

Table 1.4. *Continued*

Feature studied	Med	Aust	S. Afr	Calif	Chile
Keeley and Johnson (1977)				♦	♦
Naveh and Whittaker (1979)	♦	♦	♦	♦	♦
Shmida (1981)	♦			♦	
Cowling et al. (1992)	♦	♦	♦	♦	♦
Huntsinger and Bartolome (1992)	♦			♦	
Le Maitre and Midgley (1992)	♦	♦	♦	♦	♦
Vegetation structure:					
Specht (1969a)	♦	♦		♦	
Mooney et al. (1970,1972)				♦	♦
Parsons and Moldenke 1975)				♦	♦
Parsons (1976)				♦	♦
Milewski (1979, 1983)		♦	♦		
Cowling and Campbell (1980)			♦	♦	♦
Barbour et al. (1981)	♦			♦	
Shmida and Barbour (1982)	♦			♦	
Lamont et al. (1985)		♦	♦		
Hoffman et al. (1989)				♦	♦
Quézel and Barbero (1989a,b)	♦			♦	
Richardson and Cowling (1992)		♦	♦		
Cowling et al. (in press)		♦	♦		
Carbon uptake/allocation:					
Specht (1969b)	♦	♦		♦	
Mooney and Dunn (1970)				♦	♦
Oechel et al. (1981)				♦	♦
Mooney (1983)	♦	♦	♦	♦	♦
Phenology:					
Kummerow (1983)	♦	♦	♦	♦	♦
Orshan et al. (1988)	♦		♦		♦
Nutrient relations:					
Specht (1969b)	♦	♦		♦	
Gray and Schlesinger (1981)	♦	♦	♦	♦	
Marion (1982)	♦	♦	♦	♦	♦
Rundel (1982,1983)	♦	♦	♦	♦	♦
Lamont (1982,1983)		♦	♦		
Groves (1983)		♦	♦		

Table 1.4. *Continued*

Feature studied	Med	Aust	S. Afr	Calif	Chile
Shaver (1983)				♦	♦
Day (1983)	♦	♦	♦	♦	♦
Read and Mitchell (1983)	♦	♦	♦	♦	♦
Lamont (1994)		♦		♦	♦
Kruger (1987)		♦	♦		
Water relations:					
Poole et al. (1981)				♦	♦
Stock (1988)		♦	♦		
Smith et al. (1992)		♦	♦	♦	
Animal species composition:					
Fuentes (1976)				♦	♦
Glanz (1977)				♦	♦
Milewski (1981)		♦	♦		
Blondel et al. (1984)	♦			♦	♦
Fox et al. (1985)		♦	♦	♦	
Invasions:					
Groves (1986,1991)	♦	♦	♦	♦	♦
Macdonald et al. (1988)	♦	♦	♦	♦	♦
Kruger et al. (1989)		♦	♦	♦	♦
Fox (1990)	♦	♦	♦	♦	♦
Response to fire and other disturbance:					
Mooney and Conrad (1977)					
Gill and Groves (1981)	♦	♦	♦	♦	
Specht (1981)	♦	♦		♦	
Kruger (1983)	♦	♦	♦	♦	♦
Fox and Fox (1986)	♦	♦			
Keeley (1992)			♦	♦	
Scott and van Wyk (1992)			♦	♦	
Plant-animal interactions:					
Milewski and Bond (1982)		♦	♦		
Collins and Rebelo (1987)		♦	♦		
Grubb (1992)		♦	♦		

Table 1.4. *Continued*

Feature studied	Med	Aust	S. Afr	Calif	Chile
Stock (1992)		♦	♦		
Cox et al. (1993)	♦	♦	♦	♦	♦
Management and human impacts:					
Aschmann (1973b)	♦	♦	♦	♦	♦
Conrad and Oechel (1982)	♦	♦	♦	♦	♦

Table 1.5. Fields of comparative study most commonly undertaken between regions, as revealed by the pairwise comparisons of data in Table 1.4, and using only comparisons comprising more than two component studies. The three top ranking comparisons are listed (or four where the third and fourth ranks are the same). Third ranked comparisons in the last four entries were too weak for inclusion.

Regions compared[a]	Fields of comparative study
AU-SA	Plant-animal interactions; nutrients; vegetation structure/plant species composition
CA-CH	Plant species composition; vegetation structure; invasion
MB-CA	Plant species composition; vegetation structure; invasion; disturbance
AU-CA	Invasion; plant species composition; disturbance; nutrients
MB-AU	Disturbance; invasion; plant species composition; climate
SA-CA	Invasion; disturbance; plant species composition
MB-SA	Invasion; plant species composition
SA-CH	Invasion; plant species composition
AU-CH	Invasion; plant species composition
MB-CH	Invasion; plant species composition

[a] AU = Australia
CA = California
CH = Chile
MB = Mediterranean Basin
SA = South Africa

While the list of comparative work is impressive, it should be noted that there are some weaknesses in the comparative approach to date. Relatively few studies (27%) were actually conducted over all five regions, with much research focussing specifically on comparison between two regions. Although many of the studies chose well paired sites for comparison, this was not always the case, and mismatches in climate, substrate or vegetation reduce the value of some comparisons. Note, for example, the marked climatic differences between intensive study sites in California (Echo Valley, San Dimas, San Diego), Chile (Fundo Santa Lauro), the Mediterranean Basin (Montpellier), and South Africa (Swartboskloof) (Fig. 1.4). Comparative work has tended to focus on particular topics and particular vegetation types, to the exclusion of others. In particular, shrub-dominated vegetation has been considerably more studied than grassland, woodland and forest types, all of which occur in several of the regions. In spite of these limitations, the body of work represents a considerable resource from which to develop further comparative and coordinated research directions. Grubb (1992) recently emphasised the value of inter-continental comparisons for throwing light on otherwise inexplicable phenomena, a feature which may prove very useful in the study of the ecological function of biodiversity.

1.7 Opportunities for Studying Links between Biodiversity and Ecosystem Function in MTEs

Mediterranean-type ecosystems offer good opportunities for testing hypotheses relating biodiversity to ecosystem function at all scales; they provide the opportunity to explore the effects of various diversities and functions simultaneously (Richardson and Cowling 1993). MTEs have a rich history of comparative ecological and evolutionary research (Table 1.4). They also draw together the experience and perspectives of researchers from both industrialised and developing countries. MTEs embrace a spectrum of systems ranging from pristine through fragmented to totally transformed landscapes, with a wide array of other gradients (e.g. topography, nutrients, moisture availability, species richness, management intensity). They are convergent in many respects and different in others, and are therefore valuable for research into system processes. MTEs have many species-rich plant genera (e.g. *Banksia*, *Eucalyptus* and *Hakea* in Australia, *Cistus* in the Mediterranean Basin, *Erica* and *Aspalathus* in South Africa, *Arctostaphylos* and *Ceanothus* in California), and some have a high ratio of species:growth form. These provide opportunities to explore the concept of functional redundancy or substitutability, i.e. whether taxa within genera, growth forms or variously defined functional groups, may be interchangeable. MTEs also have a range of natural and artificial vegetation formations comprising diverse mixtures of growth forms and taxa. This facilitates evaluation of the role of biodiversity at different levels (e.g. species diversity, growth form diversity) on ecosystem functioning. There are many replicated treatments for landscapes at similar levels of patch diversity within and between MTEs. Examples of such "treatments" are landscapes variously transformed by plantations of *Pinus radiata* and *Eucalyptus* spp, pastures, crops, vineyards and orchards of citrus and deciduous fruit (Fig. 1.8). At the highest scale, MTEs replicate each other.

Unfortunately, resources for doing research are unevenly distributed among the five regions. The United States and some of the Mediterranean Basin countries have highly sophisticated research infrastructures and a well established research culture. Chile, South Africa and other Mediterranean Basin countries, on the other hand, are embedded in the developing world, with fewer material resources and human skills to channel into ecological studies. Not only are the resource bases of the regions different, but the political expectations, and the economic circumstances of people living there vary from one region to another (Table 1.6). This makes a difference in the ability of research communities to motivate for public funds, and to advance basic ecological knowledge in their regions. These differences are compounded by the proportion of land within national boundaries that experience a mediterranean-type climate, and by the economic importance of those areas. The Californian MTEs, for example, occupy less than 0.5% of the land area of the United States, and are in many cases marginal in terms of the national economy. In North Africa, on the other hand, MTEs are of high national importance as the only land with sufficient precipitation for arable agriculture.

It is also worth considering the trajectories of development expected in the different regions, as dictated by economic and sociological conditions. In terms of environmental changes, the Mediterranean Basin is probably more stable than most other areas for the immediate future, simply because most land has already been influenced by human activity (and enjoys some form of protection). As technology and management techniques change, however, this may cease to be true. Highlighting this concern about land degradation in the Mediterranean Basin is the existence of a European Community (EC) project that is investigating Mediterranean land use and desertification (MEDALUS). This multi-disciplinary project, involving 17 institutes from the member countries of the EC, is investigating the processes that lead to land degradation. It aims to contribute guidelines to ensure sustainable agriculture.

Resolution of the problems that arise when we attempt to balance human needs and natural renewable resources is an increasingly important area for ecological research - human needs are escalating, resources are in many cases dwindling, and, with hard work, our understanding of ecosystems is

Fig. 1.8a-b. Convergence in biodiversity at the landscape scale in mediterranean-type ecosystems of **a** Tuscany, Italy (above) and **b** the southwestern Cape Province of South Africa (below) (Photos D.M. Richardson)

Table 1.6. Sociological and economic factors influencing the development of comparative research programms involving the global set of MTE's

	Total area[a] (km²)	MTE area[b] (%)	MTE area[c] (%)	Popul. density[a] (per km²)	Birth rate[a] (per 1000 p.a.)	Death rate[a] (per 1000 p.a.)	GNP[a] (US$/capital)
Mediterranean Basin							
Albania	27,398	72	45	113.1	24.7	5.7	930
Algeria	2,381,741	9	6	10.5	35.5	8.3	2,170
Cyprus	9,231	100	100	75.3	18.3	8.6	7,050
Egypt	99,738	0	0	53.3	38.8	9.5	630
France	543,965	12	6	104.1	13.5	9.4	17,830
Greece	131,957	92	92	77.8	10.1	9.3	5,340
Israel	21,501	<5	54	219.7	22.2	6.2	9,750
Italy	301,277	58	75	191.7	9.8	9.3	15,150
Lebanon	10,452	>95	>95	258.4	31.7	8.7	-
Libya	1,775,500	<1	<5	2.1	46.0	7.0	5,310
Morocco	710,850	33	38	35.5	35.5	9.8	950
Portugal	91,949	>95	>95	111.9	11.5	9.3	4,890
Spain	504,782	66	71	77.9	10.7	8.2	10,920
Syria	184,050	32	35	67.6	30.8	3.2	990
Tunisia	154,530	35	36	48.3	25.2	6	1,420
Turkey	779,452	60	25	72.5	29.4	8.4	1,630
Australia	7,682,300	4	8	2.2	14.9	7.4	14,440
Chile	756,626	19	17	17.7	23.3	6.0	1,770
S. Africa	1,221,037	5	<5	27.4	32.1	9.9	2,520
USA	9,372,614	4	3	27	16.7	8.6	21,700
California	411,049	30	59	72.4	-	-	-

[a] Data from Europa World Yearbook (1992), Europa Publications Ltd, London.
[b] Derived from Müller (1982) according to system of Köppen and Geiger.
[c] Derived from Müller (1982) according to Troll and Paffen.
[-] No information available

improving. A prerequisite for full understanding of ecosystem function, however, is an ability to link all levels of scale in the hierarchy of the natural world, from the molecule to the landscape. MTEs, with their wide range of biotic diversity and distinctive ecosystem processes involving fluxes of water, nutrients and energy, are ideal sites for testing hypotheses relating biodiversity to ecosystem function at all scales (Richardson and Cowling 1993). Linking of north and south, and east and west, across the five regions offers an important global perspective to this important exercise. The chapters that follow set the scene for much important work to follow.

Acknowledgements. We thank the following people for assistance, information and critical comment on drafts of the manuscript: Richard Cowling, Byron Lamont, Bob Scholes and Paul Zedler. Institutes supporting the work were the CSIRO in Australia, the Institute for Plant Conservation, University of Cape Town, and the National Botanical Institute of South Africa.

References

Arianoutsou M, Groves RH (eds) (1993) Plant-animal interactions in mediterranean-type ecosystems. Kluwer, Dordrecht

Arroya MTK, Zedler PH, Fox MD (eds) (1994) Ecology and biogeography of mediterranean ecosystems in Chile, California and Australia. Springer, Berlin Heidelberg New York (in press)

Aschmann H (1973a) Distribution and peculiarity of mediterranean ecosystems. In: di Castri F, Mooney HA (eds) Mediterranean-type ecosystems. Origin and structure. Ecological Studies 7. Springer, Berlin Heidelberg New York, pp 11-19

Aschmann H (1973b) Man's impact on several regions with mediterranean climates. In: di Castri F, Mooney HA (eds) Mediterranean-type ecosystems. Origin and structure. Ecological studies 7. Springer, Berlin Heidelberg New York, pp 363-371

Attenborough D (1987) The first Eden. The Mediterranean world and man. Collins, London

Bakker E (1971) An island called California. University of California Press, Berkeley

Barbero M, Loisel P, Quézel P (1992) Biogeography, ecology and history of Mediterranean *Quercus ilex* ecosystems. Vegetatio 99/100: 19-34

Barbour M, Major J (eds) (1977) Terrestrial vegetation of California. Wiley, New York, pp 417-469

Barbour MG, Minnich RA (1990) The myth of chaparral convergence. Isr J Bot 39: 453-463

Barbour MG, Shmida A, Johnson AF, Holton B (1981) Comparison of coastal dune scrub in Israel and California: physiognomy, association patterns, species richness, phytogeography. Isr J Bot 30: 181-198

Beard JS (1983) Ecological control of the vegetation of Southwestern Australia: moisture versus nutrients. In: Kruger FJ, Mitchell DT, Jarvis JUM (eds) Mediterranean-type ecosystems. The role of nutrients. Ecological studies 43. Springer, Berlin Heidelberg New York, pp 66-73

Beard JS (1984) Biogeography of the kwongan. In: Pate JS, Beard JS (eds) Kwongan. Plant life of the sandplain. University of Western Australia Press, Nedlands, pp 1-26

Beard JS (1990) Plant life of Western Australia. Kangaroo Press, Kenthurst, NSW

Beard JS, Sprenger BS (1984) Geographical data from the vegetation survey of Western Australia. Vegetation Survey of Western Australia. Occasional Paper No 2. Vegmap Publ, Perth

Bettenay E (1983) Australian soil and landscape regions: western region. In: Soils - an Australian viewpoint. CSIRO, Melbourne & Academic Press, London, pp 179-187

Blondel J, Vuillermeir F, Marcus LF, Terouanno E (1984) Is there ecomorphological convergence among Mediterranean bird communities of Chile, California and France? Evol Biol 18: 141-213

Bond WJ (1989) Describing and conserving biotic diversity. In: Huntley BJ (ed) Biotic diversity in southern Africa: concepts and conservation. Oxford University Press, Cape Town, pp 2-18

Bond WJ (1993) Keystone species. In: Schulze E-D, Mooney HA (eds) Ecosystem function of biodiversity. Ecological studies 99. Springer, Berlin Heidelberg New York, pp 237-253

Braun-Blanquet J (1964) Pflanzensoziologie. Springer, Vienna, New York

Campbell BM (1985) A classification of the mountain vegetation of the fynbos biome. Mem Bot Surv S Africa 50: 1-115

Cody ML (1986). Diversity, rarity and conservation in mediterranean-climate regions. In: Soulé ME (ed) Conservation biology. The science of scarcity and diversity. Sinauer, MA, pp 122-152

Cody ML, Mooney HA (1978) Convergence versus nonconvergence in mediterranean-climate ecosystems. Annu Rev Ecol Syst 9: 265-321

Collins BG, Rebelo AG (1987) Pollination biology in the Proteaceae of Australia and southern Africa. Aust J Ecol 12: 95-129

Conrad CE, Oechel WC (eds) (1982) Dynamics and management of mediterranean-type ecosystems. Pacific Southwest Forest and Range Experiment Station, Berkeley

Cowling RM (ed) (1992) The ecology of fynbos: nutrients, fire and diversity. Oxford University Press, Cape Town

Cowling RM, Campbell BM (1980) Convergence in vegetation structure in the mediterranean communities of California, Chile and South Africa. Vegetatio 43: 191-197

Cowling RM, Holmes PM (1992) Flora and vegetation. In: Cowling RM (ed) The ecology of fynbos: Nutrients, fire and diversity. Oxford University Press, Cape Town, pp 23-61

Cowling RM, Pierce SM, Moll EJ (1986) Conservation and utilization of South Coast renosterveld, an endangered South African vegetation type. Biol Conserv 12: 363-377

Cowling RM, Holmes PM, Rebelo AB (1992) Plant diversity and endemism. In: Cowling RM (ed) The ecology of fynbos: nutrients, fire and diversity. Oxford University Press, Cape Town, pp 62-112

Cowling RM, Witkowski ETF, Milewski AV, Newbey KR (1994) Taxonomic, edaphic and biological aspects of plant endemism on matched sites in mediterranean Australia and South Africa. J Biogeogr (in press)

Cox GW, Contreras LC, Milewski AV (1994) Role of fossorial animals in community structure and energetics of mediterranean systems. In: Arroya MTK, Zedler PH, Fox MD (eds) Ecology and biogeography of mediterranean ecosystems in Chile, California, and Australia. Springer, Berlin Heidelberg New York (in press)

CSIRO Division of Soils (1983) Soils: an Australian viewpoint. CSIRO, Melbourne/Academic Press, London

Day JA (ed) (1983) Mineral nutrients in mediterranean ecosystems. S Afr Nat Sci Prog Rep 71. CSIR, Pretoria

Deacon HJ (1983) The comparative evolution of mediterranean-type ecosystems: a southern perspective. In: Kruger FJ, Mitchell DT, Jarvis JUM (eds) Mediterranean-type ecosystems. The role of nutrients. Ecological studies 43. Springer, Berlin Heidelberg New York, pp 3-40

Debussche M, Escarré J, Lepart J (1980) Changes in mediterranean shrub communities with *Cytisus purgans* and *Genista scorpius*. Vegetatio 43: 73-82

Dell B (ed) (1984) Medecos IV. Proceedings of the 4th International Conference on Mediterranean Ecosystems. University of Western Australia, Nedlands

Dell B, Hopkins AJM, Lamont BB (eds) (1986) Resilience in mediterranean-type ecosystems. Junk, Dordrecht

Dell B, Havel JJ, Malajczuk N (1989) The jarrah forest. A complex mediterranean ecosystem. Kluwer, Dordrecht

di Castri F (1981) Mediterranean-type shrublands of the world. In: di Castri F, Goodall DW, Specht RL (eds) Mediterranean-type shrublands. Elsevier, Amsterdam, pp 1-52

di Castri F, Mooney HA (eds) (1973) Mediterranean-type ecosystems. Origin and structure. Ecological studies 7. Springer, Berlin Heidelberg New York

di Castri F, Younés T (1992) Ecosystem function of biological diversity. Biol Int Spec Issue 22

di Castri F, Floret C, Rambal S, Roy J (1988a) Preface. In: di Castri F, Floret C, Rambal S, Roy J (eds) Time scales and water stress. IUBS, Paris, pp v-viii

di Castri F, Floret C, Rambal S, Roy J (eds) (1988b) Time scales and water stress. IUBS, Paris

Ehrlich PR, Ehrlich AH (1992) The value of biodiversity. Ambio 21: 219-226

Emberger L (1955) Project d'une classification biogeographique des climats. In: Les divisions ecologiques du monde. Centre de la Recherche Scientifique, Paris, pp 5-11

Emberger L (1958) Afrique du Nord et Australie méditerranéene - un parallèle bioclimatologique. Colloque du Comité des zones arides de l'UNESCO, Canberra 1956, pp 141-146

Emberger L (1959) La place de l'Australie méditerranéene dans l'ensemble des pays méditerranéens du vieux monde. Monogr Biol 8: 259-273

Escarré J, Houssard C, Debussche M, Lepart J (1983) Evolution de la végétation et du sol aprés abandon cultural en region méditerranéenne: étude de la succession dans les garrigues du Montpéllierais (France). Acta Oecol Oecol Plant 4: 221-239

Europa World Yearbook (1992) Europa Publ, London

Fox BJ, Fox MD (1986) Resilience of animal and plant communities to human disturbance. In: Dell B, Hopkins AJM, Lamont BB (eds) Resilience in mediterranean-type ecosystems. Junk, Dordrecht, pp 39-64

Fox BJ, Quinn RD, Breytenbach GJ (1985) A comparison of small-mammal succession following fire in shrublands of Australia, California and South Africa. Proc Ecol Soc Aust 14: 179-198

Fox MD (1982) Vegetation changes in Mediterranean Australia since European settlement. In: Conrad C, Oechel WC (eds) Dynamics and management of mediterranean-type ecosystems. USDA For Serv Gen Tech Rep PSW-58. Pacific Southwest Forest and Range Experiment Station, Berkeley, California, pp 112-117

Fox MD (1990) Mediterranean weeds: exchanges of invasive plants between the five Mediterranean regions of the world. In: di Castri F, Hansen AJ, Debussche M (eds) Biological invasions in Europe and the Mediterranean Basin, Kluwer, Dordrecht, pp 179-200

Fuentes ER (1976) Ecological convergence of lizard communities in Chile and California. Ecology 57: 1-17

George AS, Hopkins AJM, Marchant NG (1979) The heathlands of Western Australia. In: Specht RL (ed) Heathlands and related shrublands. Descriptive studies. Elsevier, Amsterdam, pp 211-230

George MR, Brown JR, Clawson WJ (1992) Application of non-equilibrium ecology to management of Mediterranean grasslands. J Range Manage 45: 436-440

Gill AM, Groves RH (1981) Fire regimes in heathlands and their plant-ecological effects. In: Specht RL (ed) Ecosystems of the world 9B. Heathlands and related shrublands. Analytical studies. Elsevier, Amsterdam, pp 61-84

Glanz W (1977) Comparative ecology of rodent communities in California and Chile. PhD thesis, University of California, Berkeley

Gray JT, Schlesinger WH (1981) Nutrient cycling in mediterranean-type ecosystems. In: Miller PC (ed) Resource use by chaparral and matorral. A comparison of vegetation function in two mediterranean-type ecosystems. Springer, Berlin Heidelberg New York, pp 259-285

Grenon M, Batisse M (eds) (1989) Futures for the Mediterranean Basin: the Blue Plan. Oxford University Press

Groves RH (1983) Nutrient cycling in Australian heath and South African fynbos. In: Kruger FJ, Mitchell DT, Jarvis JUM (eds) Mediterranean-type ecosystems. The role of nutrients. Ecological studies 43. Springer, Berlin Heidelberg New York, pp 179-191

Groves RH (1986) Invasion of Mediterranean ecosystems by weeds. In: Dell B, Hopkins AJM, Lamont BB (eds) Resilience in mediterranean-type ecosystems. Junk, Dordrecht, pp 129-145

Groves RH (1991) The biogeography of mediterranean plant invasions In: Groves RH, di Castri F (eds) Biogeography of mediterranean invasions. Cambridge University Press, Cambridge, pp 427-438

Groves RH, di Castri F (eds) (1991) Biogeography of Mediterranean invasions. Cambridge University Press, Cambridge

Groves RH, Beard JS, Deacon HJ, Lambrechts JJN, Rabinovitch-Vin A, Specht RL, Stock WD (1983) Introduction: the origins and characteristics of mediterranean ecosystems. In: Day JA (ed) Mineral nutrients in mediterranean ecosystems. S Afr Nat Sci Prog Rep 71, CSIR, Pretoria, pp 1-17

Grubb PJ, Hopkins AJM (1986) Resilience at the level of the plant community. In: Dell B, Hopkins AJM, Lamont BB (eds) Resilience in mediterranean-type ecosystems. Junk, The Hague, pp 21-38

Grubb PJ (1992) A positive distrust in simplicity - lessons from plant defences and from competition among plants and among animals. J Ecol 80: 585-610

Gulmon SL (1977) A comparative study of the grassland of California and Chile. Flora 166: 261-278

Hanes TL (1981) California chaparral. In: di Castri F, Goodall DW, Specht RL (eds) Mediterranean-type shrublands. Elsevier, Amsterdam, pp 139-174

Heady HF, Bartolome JW, Pitt MD, Stroud MC, Savelle GD (1992) California prairie. In: Coupland RT (ed) Natural grasslands. Introduction and Western Hemisphere. Elsevier, Amsterdam, pp 313-335

Hobbs RJ (1992) Is biodiversity important for ecosystem function? Implications for research and management. In: Hobbs RJ (ed) Biodiversity of mediterranean ecosystems in Australia. Surrey Beatty, Chipping Norton, NSW, pp 211-229

Hobbs RJ (1993) Effects of landscape fragmentation on ecosystem processes in the Western Australian wheatbelt. Biol Conserv 64: 193-201

Hobbs RJ (1994) Can we use plant functional types to describe and predict responses to environmental change? In: Smith TM, Shugart HH, Woodward FI (eds) Plant functional types. Cambridge University Press, Cambridge, (in press)

Hobbs RJ, Hopkins AJM (1990) From frontier to fragments: European impact on Australia's vegetation. Proc Ecol Soc Aust 16: 93-114

Hobbs RJ, Saunders DA, Lobry de Bruyn LA, Main AR (1993) Changes in biota. In: Hobbs RJ, Saunders DA (eds) Reintegrating fragmented landscapes. Towards sustainable production and nature conservation. Springer, Berlin Heidelberg New York, pp 65-106

Hoffman AJ, Teillier S, Fuentes ER (1989) Fruit and seed characteristics of woody species in mediterranean-type regions of Chile and California. Rev Chil Hist Nat 62: 43-60

Hopkins AJM, Robinson CJ (1981) Fire-induced structural change in a Western Australian woodland. Aust J Ecol 6: 177-188

Huenneke LF, Mooney HA (eds)(1989) Grassland structure and function: California annual grassland. Kluwer, Dordrecht

Huntsinger L, Bartolome JW (1992) Ecological dynamics of *Quercus*-dominated woodlands in California and Spain: a state-transition model. Vegetatio 99/100: 299-305

Keeley JE (1992) A Californian's view of fynbos. In: Cowling RM (ed) The ecology of fynbos: nutrients, fire and diversity. Oxford University Press, Cape Town, pp 372-388

Keeley SC, Johnson AW (1977) A comparison of the pattern of herb and shrub growth in comparable sites in Chile and California. Am Midl Nat 97: 120-132

Killick DJB (1979) African mountain heathlands. In: Specht RL (ed) Heathlands and related shrublands of the world. Elsevier, Amsterdam, pp 97-116

Köppen W (1931) Die Klimate der Erde, Grundriss der Klimakunde 2nd edn. Berlin

Kruger FJ (1983) Plant community diversity and dynamics in relation to fire. In: Kruger FJ, Mitchell DT, Jarvis JUM (eds) Mediterranean-type ecosystems. The role of nutrients. Ecological studies 43. Springer, Berlin Heidelberg New York, pp 446-472

Kruger FJ (1987) Responses of plants to nutrient supply in mediterranean-type ecosystems. In: Tenhunen JD, Catarino FM, Lange OL, Oechel WC (eds) Plant responses to stress: functional analysis in mediterranean ecosystems. Springer, Berlin Heidelberg New York, pp 415-428

Kruger FJ, Mitchell DT, Jarvis JUM (eds) (1983) Mediterranean-type ecosystems. The role of nutrients. Ecological studies 43. Springer, Berlin Heidelberg New York

Kruger FJ, Breytenbach GJ, Macdonald IAW, Richardson DM (1989) The characteristics of invaded Mediterranean climate regions. In: Drake JA, Mooney HA, di Castri F, Groves RH, Kruger FJ, Rejmánek, Williamson M (eds) Biological invasions. A global perspective. Wiley, New York, pp 181-213

Kummerow J (1983) Comparative phenology of mediterranean-type plant communities. In: Kruger FJ, Mitchell DT, Jarvis JUM (eds) Mediterranean-type ecosystems. The role of nutrients. Ecological studies 43. Springer, Berlin Heidelberg New York, pp 300-317

Lamont BB (1982) Mechanisms for enhancing nutrient uptake in plants, with particular reference to mediterranean South Africa and Western Australia. Bot Rev 48: 597-689

Lamont BB (1983) Strategies for maximizing nutrient uptake in two Mediterranean ecosystems of low nutrient status. In: Kruger FJ, Mitchell DT, Jarvis JUM (eds) Mediterranean-type ecosystems. The role of nutrients. Ecological studies 43. Springer, Berlin Heidelberg New York, pp 246-273

Lamont BB (1992) Functional interactions within plants - the contribution of keystone and other species to biological diversity. In: Hobbs RJ (ed) Biodiversity of mediterranean ecosystems in Australia. Surrey Beatty, Chipping Norton NSW, pp 95-127

Lamont BB (1994) Mineral nutrient relations in mediterranean regions of California, Chile and Australia. In: Arroyo MTK, Zedler PH, Fox MD (eds) Ecology and biogeography

of mediterranean ecosystems in Chile, California and Australia. Springer, Berlin Heidelberg New York (in press)
Lamont BB, Collins BG, Cowling RM (1985) Reproductive biology of the Proteaceae in Australia and South Africa. Proc Ecol Soc Aust 14: 213-224
Lawton JH, Brown VK (1993) Redundancy in ecosystems. In: Schulze E-D, Mooney HA (eds) Ecosystem function of biodiversity. Springer, Berlin Heidelberg New York, pp 277-270
Le Maitre DC, Midgley JJ (1992) Plant reproductive ecology In: Cowling RM (ed) The ecology of fynbos: nutrients, fire and diversity, Oxford University Press, Cape Town, pp 135-174
Macdonald IAW, Graber DM, DeBenedettii S, Groves RH, Fuentes E (1988) Introduced species in nature reserves in mediterranean-type climatic regions of the world. Biol Conserv 44: 37-66
MacMahon JA, Phillips DL, Robinson JV, Schimpf DJ (1978) Levels of biological organisation: an organism-centered approach. BioScience 28: 700-704
MacMahon JA, Schimpf DJ, Andessen DC, Smith KG, Bayn RL (1981) An organism-centered approach to some community and ecosystem concepts. J Theor Biol 88: 287-307
Manders PT, Richardson DM (1992) Colonization of Cape fynbos communities by forest species. For Ecol Manage 48: 277-293
Manders PT, Richardson DM, Masson PH (1992) Is fynbos a stage in succession to forest? Analysis of the perceived ecological distinction between two communities. In: van Wilgen BW, Richardson DM, Kruger FJ, van Hensbergen HJ (eds) Fire in South African mountain fynbos. Springer, Berlin Heidelberg New York
Marion GM (1982) Nutrient mineralization processes in mediterranean-type ecosystems. In: Conrad CE, Oechel WC (eds) Dynamics and management of mediterranean-type ecosystems. Pacific Southwest Forest and Range Experiment Station, Berkeley, pp 313-320
May RM (1989) Levels of organisation in ecology. In: Cherrett JM, Bradshaw AD, Goldsmith FB, Grubb PG, Krebs JR (eds) Ecological concepts. The contribution of ecology to an understanding of the natural world. Blackwell, Oxford, pp 339-363
May RM (1992) How many species inhabit the earth? Sci Am 267: 42-48
McFarlane DJ, George RJ, Farrington P (1993) Changes in the hydrological balance. In: Hobbs RJ, Saunders DA (eds) Reintegrating fragmented landscapes. Towards sustainable production and nature conservation. Springer, Berlin Heidelberg New York, pp 146-186
Milewski AV (1979) A climatic basis for the study of convergence of vegetation structure in mediterranean Australia and southern Africa. J Biogeogr 6: 293-299
Milewski AV (1981) A comparison of reptile communities in relation to soil fertility in the mediterranean and adjacent parts of Australia and Southern Africa. J Biogeogr 8: 493-503
Milewski AV (1983) A comparison of ecosystems in mediterranean Australia and southern Africa. Annu Rev Ecol Syst 14: 57-76
Milewski AV, Bond WJ (1982) Convergence of myrmecochory in mediterranean Australia and South Africa. In: Buckley RC (ed) Ant-plant interactions in Australia. Junk, The Hague, pp 89-98
Milewski AV, Cowling RM (1985) Anomalies in the plant and animal communities in similar environments at the Barrens, Western Australia, and the Caledon Coast, South Africa. Proc Ecol Soc Aust 14: 199-212

Miller PC (ed) (1981) Resource use by chaparral and matorral. A comparison of vegetation function in two mediterranean-type ecosystems. Ecological studies 39. Springer, Berlin Heidelberg New York

Miller PC (1983) Canopy structure of mediterranean-type shrubs in relation to heat and moisture. In: Kruger FJ, Mitchell DT, Jarvis JUM (eds) Mediterranean-type ecosystems. The role of nutrients. Ecological studies 43. Springer, Berlin Heidelberg New York, pp 133-166

Miller PC, Bradbury DE, Hajek E, LaMarche V, Thrower NJW (1977) Past and present environment. In: Mooney HA (ed) Convergent evolution in Chile and California. Dowden, Hutchinson & Ross, Stroudsburg, pp 27-72

Mooney HA (ed) (1977) Convergent evolution in Chile and California: mediterranean climate ecosystems. Dowden, Hutchinson and Ross, Stroudsburg, PA

Mooney HA (1983) Carbon-gaining capacity and allocation patterns of mediterranean-climate plants. In: Kruger FJ, Mitchell DT, Jarvis JUM (eds) Mediterranean-type ecosystems. The role of nutrients. Ecological studies 43. Springer, Berlin Heidelberg New York, pp 103-119

Mooney HA (1988) Lessons from mediterranean-climate regions. In: Wilson EO, Peter FM (eds) Biodiversity. National Academy Press, Washington DC

Mooney HA, Conrad CE (eds) (1977) Environmental consequences of fire and fuel management in mediterranean ecosystems. USDA For Serv Gen Tech Rep WO-3

Mooney HA, Dunn EL (1970) Photosynthetic systems of mediterranean climate shrubs and trees of California and Chile. Am Nat 104: 447-453

Mooney HA, Miller PC (1985) Chaparral. In: Chabot BF, Mooney HA (eds) Physiological ecology of North American plant communities. Chapman and Hall, New York, pp 213-231

Mooney HA, Dunn, EL, Shropshire F, Song L (1970) Vegetation comparisons between the mediterranean climate areas of California and Chile. Flora 159: 480-496

Mooney HA, Dunn, EL, Shropshire F, Song L (1972) Land-use history of California and Chile as related to the structure of the sclerophyll scrub vegetation. Madroño 21: 305-310

Müller MJ (1982) Selected climate data for a global set of standard stations for vegetation science. Junk, The Hague

Myers K (1986) Introduced vertebrates in Australia, with emphasis on the mammals. In: Groves RH, Burdon JJ (eds) Ecology of biological invasions. Cambridge University Press, Cambridge, pp 120-136

Nahal I (1981) The mediterranean climate from a biological viewpoint. In: di Castri F, Goodall, DW, Specht RL (eds) Ecosystems of the world 11: mediterranean-type shrublands. Elsevier, Amsterdam, pp 63-86

Naveh Z (1967) Mediterranean ecosystems and vegetation types in California and Israel. Ecology 48: 345-359

Naveh Z, Kutiel P (1990) Changes in the Mediterranean vegetation of Israel in response to human habitation and land use. In: Woodwell GM (ed) The earth in transition. Patterns and processes of biotic impoverishment. Cambridge University Press, Cambridge, pp 259-299

Naveh Z, Lieberman S (1984) Landscape ecology. Theory and application. Springer, Berlin Heidelberg New York

Naveh Z, Whittaker RH (1979) Structural and floristic diversity of shrublands and woodlands in northern Israel and other mediterranean areas. Vegetatio 41: 171-190

Northcote KH (1983) Australian soil and landscape regions: central southern region. In: Soils – an Australian viewpoint. CSIRO, Melbourne & Academic Press, London, pp 211-217

Noss RF (1990) Indicators for monitoring biodiversity: a hierarchical approach. Conserv Biol 4: 355-364

Oechel WC, Lawrence W, Mustafa J, Martínez J (1981) Energy and carbon acquisition. In: Miller PC (ed) Resource use by chaparral and matorral. A comparison of vegetation function in two mediterranean-type ecosystems. Springer, Berlin Heidelberg New York, pp 151-183

O'Neill RV (1989) Perspectives in hierarchy and scale. In: Roughgarden J, May RM, Levin SA (eds) Perspectives in ecological theory. Princeton University Press, Princeton, pp 140-156

O'Neill RV, DeAngelis DL, Waide JB, Allen TFH (1986) A hierarchical concept of ecosystems. Princeton University Press, Princeton

Orshan G (1983) Approaches to the definition of Mediterranean growth forms. In: Kruger FJ, Mitchell DT, Jarvis JUM (eds) Mediterranean-type ecosystems. The role of nutrients. Ecological studies 43. Springer, Berlin Heidelberg New York, pp 86-100

Orshan G, Le Floc'h E, Le Roux A, Montenegro G (1988) Plant phenomorphology as related to summer drought in mediterranean-type ecosystems. In: di Castri F, Floret C, Rambal S, Roy J (eds) Time scales and water stress. IUBS, Paris, pp 111-124

Paine RT (1969) A note on trophic complexity and community stability. Am Nat 103: 91-93

Parsons DJ (1976) Vegetation structure in the mediterranean scrub communities of California and Chile. J Ecol 64: 435-447

Parsons DJ, Moldenke AR (1975) Convergence in vegetation structure along analagous climatic gradients in California and Chile. Ecology 56: 950-957

Poole DK, Roberts SW, Miller PC (1981) Water utilization. In: Miller PC (ed) Resource use by chaparral and matorral. A comparison of vegetation function in two mediterranean-type ecosystems. Springer, Berlin Heidelberg New York, pp 123-149

Quézel P, Barbero M (1989a) Altitudinal zoning of forest structures in California and around the Mediterranean: a comparative study. Berlin Beilfelder Oekol Beitr 4: 25-44

Quézel P, Barbero M (1989b) Zonation altitudinale des structures forestières de végétation en Californie méditerranéenne: leur interpretation en fonction des méthodes utilisées pour le pourtour méditerranéen. Ann Sci Forest 46: 233-250

Read DJ, Mitchell DT (1983) Decomposition and mineralization processes in mediterranean-type ecosystems and in heathlands of similar structure. In: Kruger FJ, Mitchell DT, Jarvis JUM (eds) Mediterranean-type ecosystems. The role of nutrients. Ecological studies 43. Springer, Berlin Heidelberg New York, pp 208-232

Richardson DM, Cowling RM (1992) Why is mountain fynbos invasible and which species invade? In: van Wilgen BW, Richardson DM, Kruger FJ, van Hensbergen HJ (eds) Fire in South African mountain fynbos. Springer, Berlin Heidelberg New York, pp 161-181

Richardson DM, Cowling RM (1993) Biodiversity and ecosystem processes: opportunities in mediterranean-type ecosystems. Trends Ecol Evol 8: 79-81

Richardson DM, Macdonald IAW, Holmes PM, Cowling RM (1992) Plant and animal invasions. In: Cowling RM (ed) The ecology of fynbos: nutrients, fire and diversity. Oxford University Press, Cape Town, pp 271-308

Rundel PW (1981) The matorral zone of central Chile. In: di Castri F, Goodall DW, Specht RL (eds) Mediterranean-type shrublands. Elsevier, Amsterdam, pp 175-201

Rundel PW (1982) Nitrogen utilization efficiencies in mediterranean-climate shrublands of California and Chile. Oecologia 55: 409-413

Rundel PW (1983) Impact of fire on nutrient cycles in mediterranean-type ecosystems with reference to chaparral. In: Kruger FJ, Mitchell DT, Jarvis JUM (eds) Mediterranean-type ecosystems. The role of nutrients. Ecological studies 43. Springer, Berlin Heidelberg New York, pp 192-207

Russel JS, Moore AW (1976) Classification of climate by pattern analysis with Australasian and Southern African data as examples. Agric Meteorol 16: 45-70

Saunders DA (1993) A community-based observer scheme to assess avian responses to habitat reduction and fragmentation in south-western Australia. Biol Conserv 64: 203-218

Saunders DA, Hobbs RJ (1992) Impact on biodiversity of changes in land-use and climate. In: Hobbs RJ (ed) Biodiversity of mediterranean ecosystems in Australia. Surrey Beatty, Chipping Norton, NSW, pp 61-75

Scheiner SM (1992) Measuring pattern diversity. Ecology 73: 1860-1867

Schulze E-D, Mooney HA (eds) (1993) Biodiversity and ecosystem function. Springer, Berlin Heidelberg New York

Scott DF, van Wyk DB (1992) The effects of fire on soil water repellency, catchment sediment yields and streamflow. In: van Wilgen BW, Richardson DM, Kruger FJ, van Hensbergen HJ (eds) Fire in South African mountain fynbos. Springer, Berlin Heidelberg New York, pp 216-233

Shaver GR (1983) Mineral nutrient and nonstructural carbon pools in shrubs from Mediterranean-type ecosystems of California and Chile. In: Kruger FJ, Mitchell DT, Jarvis JUM (eds) Mediterranean-type ecosystems. The role of nutrients. Ecological studies 43. Springer, Berlin Heidelberg New York, pp 286-299

Shmida A (1981) Mediterranean vegetation in California and Israel: similarities and differences. Isr J Bot 30: 105-123

Shmida A, Barbour MG (1982) A comparison of two types of mediterranean scrub in Israel and California. In: Conrad CE, Oechel WC (eds) Dynamics and management of Mediterranean-type ecosystems. Pacific Southwest Forest and Range Experiment Station, Berkeley, pp 100-106

Smith RE, van Wilgen BW, Forsyth GG, Richardson DM (1992) Coexistence of seeders and resprouters in a fire-prone environment. The role of ecophysiology and soil moisture. In: van Wilgen BW, Richardson DM, Kruger FJ, van Hensbergen HJ (eds) Fire in South African mountain fynbos. Springer, Berlin Heidelberg New York, pp 108-122

Smith TM, Shugart HH, Woodward FI (eds) (1994) Plant functional types. Cambridge University Press, Cambridge

Solbrig OT (1991) Biodiversity: scientific issues and collaborative research proposals. MAB Digest 9, UNESCO, Paris

Solbrig OT (1992a) Biodiversity: an introduction. In: Solbrig OT, van Emden HM, van Oordt PGWJ (eds) Biodiversity and global change. Monogr 8, International Union of Biological Sciences, Paris, pp 13-20

Solbrig OT (1992b) Biodiversity, global change and scientific integrity. J Biogeogr 19: 1-2

Specht RL (1969a) A comparison of the sclerophyllous vegetation characteristic of mediterranean-type climates in France, California and southern Australia. I. Structure, morphology and succession. Aust J Bot 17: 277-292

Specht RL (1969b) A comparison of the sclerophyllous vegetation characteristic of mediterranean-type climates in France, California and southern Australia. II. Dry matter, energy and nutrient accumulation. Aust J Bot 17: 293-308

Specht RL (1972) The vegetation of South Australia (2nd edn). Government Printer, Adelaide

Specht RL (1979) Heathlands and related shrublands of the world. In: Specht RL (ed) Ecosystems of the world, vol 9A. Heathlands and related shrublands of the world: descriptive studies. Elsevier, Amsterdam, pp 1-18

Specht RL (1981) Primary production in mediterranean-climate ecosystems regenerating after fire. In: di Castri F, Goodall DW, Specht RL (eds) Mediterranean-type shrublands. Elsevier, Amsterdam, pp 257-267

Specht RL (ed) (1988) Mediterranean-type ecosystems: a data source book. Kluwer, Dordrecht

Specht Rl, Moll EJ (1983) Mediterranean-type heathlands and sclerophyllous shrublands of the world: an overview. In: Kruger FJ, Mitchell DT, Jarvis JUM (eds) Mediterranean-type ecosystems. The role of nutrients. Ecological studies 43. Springer, Berlin Heidelberg New York, pp 41-65

Stock WD (1988) Nutrients and water stress as selective forces in the evolution of the fynbos and kwongan floras. In: di Castri F, Floret C, Rambal S, Roy J (eds) Time scales and water stress. IUBS, Paris, pp 623-629

Stock WD (1992) Mechanical defence costs in South African and Australian members of the Proteaceae. In: Thanos CA (ed) Plant-animal interactions in mediterranean-type ecosystems. University of Athens, Greece, pp 210-215

Thanos CA (1992a) Theophrastus on plant-animal interactions. In: Thanos CA (ed) Plant-animal interactions in mediterranean-type ecosystems: Proc 6th Int Conf Mediterranean-Climate Ecosystems, Crete 1991. University of Athens, Athens pp 1-5

Thanos CA (ed) (1992b) Plant-animal interactions in Mediterranean-type ecosystems: Proceedings of MEDECOS VI, the 6th International Conference on Mediterranean Climate Ecosystems, Crete 1991. University of Athens

Thrower NJW, Bradbury DE (1973) The physiography of the mediterranean lands with special emphasis in California and Chile. In: di Castri F, Mooney HA (eds) Mediterranean-type ecosystems. Origin and structure. Springer, Berlin Heidelberg New York, pp 37-52

Thrower NJW, Bradbury DE (eds) (1977) Chile-California mediterranean scrub atlas. Dowden, Hutchinson and Ross, Stroudsburg, PA

Tomaselli R (1981a) Main physiognomic types and geographic distribution of shrub systems related to mediterranean climates. In: di Castri F, Goodall DW, Specht RL (eds) Mediterranean-type shrublands. Elsevier, Amsterdam, pp 95-106

Tomaselli R (1981b) Relations with other ecosystems: temperate evergreen forests, savannahs, steppes and desert shrublands. In: di Castri F, Goodall DW, Specht RL (eds) Mediterranean-type shrublands. Elsevier, Amsterdam, pp 123-130

Van Wyk DB, Lesch W, Stock WD (1992) Fire and catchment chemical budgets. In: van Wilgen BW, Richardson DM, Kruger FJ, van Hensbergen HJ (eds) Fire in South African mountain fynbos. Springer, Berlin Heidelberg New York, pp 240-257

Versfeld DB, Richardson DM, van Wilgen BW, Chapman RA, Forsyth GG (1992) The climate of Swartboskloof. In: van Wilgen BW, Richardson DM, Kruger FJ, van Hensbergen HJ (eds) Fire in South African mountain fynbos. Springer, Berlin Heidelberg New York, pp 21-36

Vitousek PM, Hooper DU (1993) Biological diversity and terrestrial ecosystem biogeochemistry. In: Schulze E-D, Mooney HA (eds) Ecosystem function of biodiversity. Springer, Heidelberg, pp 3-14

Walker BH (1992) Biodiversity and ecological redundancy. Conserv Biol 6: 18-23

Westman WE (1986) Resilience: concepts and measures. In: Dell B, Hopkins AJM, Lamont BB (eds) Resilience in mediterranean-type ecosystems. Junk, The Hague, pp 5-19

Whittaker RH (1975) Communities and ecosystems. 2 edn. Macmillan, New York

Wilson EO (1988) The current state of biological diversity. In: Wilson EO (ed) Biodiversity. National Academy Press, Washington, pp 3-18

Wilson EO (1992) The diversity of life. Allen Lane, Penguin, London

Woodmansee RG (1988) Ecosystem processes and global change. In: Rosswall T, Woodmansee RG, Risser PG (eds) Scales and global change. Wiley, London, pp 11-27

Woodward FI (1993) How many species are required for a functional ecosystem? In: Schulze E-D, Mooney HA (eds) Ecosystem function of biodiversity. Ecological studies 99. Springer, Berlin Heidelberg New York

Younés T (1992) Ecosystem function of biodiversity: a progress report on the IUBS-SCOPE-UNESCO Programme. Biol Int 24: 16-21

Zinke PJ (1973) Analogies between soil and vegetation types of Italy, Greece and California. In: di Castri F, Mooney HA (eds) Mediterranean type ecosystems. Origin and structure. Springer, Berlin Heidelberg New York, pp 61-82

2 Biodiversity and Ecosystem Function in the Mediterranean Basin: Human and Non-Human Determinants

J. Blondel and J. Aronson

2.1 Introduction

The rapid loss of genetic diversity is of great concern to most modern societies. This depletion is occurring at several scales, ranging from gene assemblages in local populations to complex communities, ecosystems and landscapes. The stock of information in DNA molecules, which is the basis of biological diversity and the raw material for evolution and adaptation, is being lost at a rate probably without precedent (with the possible exception of the five major extinction events during the Phanerozoic era). The processes and mechanisms of the current dramatic loss of diversity have been reviewed in many recent books and papers (e.g. Wilson 1988, 1992), and the predicted consequences of such a precipitous decline of diversity for the planet, and for humans, is much debated at present (Schulze and Mooney 1993).

Unfortunately, very few data exist that connect losses of biodiversity to major ecosystem processes or community functioning. In the absence of such data, some analogies have been proposed to illustrate the presumed importance of the various parts of an ecological system and what could be the consequences of rapid and massive loss of diversity on ecosystem functioning and services. The best known of these is the *rivet-popper hypothesis* (Ehrlich and Ehrlich 1981), which compares the role of species to that of numerous, virtually identical rivets in the robustness of an aircraft. Another analogy is that of the differential and complementary components of an automobile for its continued functioning and comfort to the passengers (Schulze 1989). Such concerns are of course not new: more than four decades ago, Aldo Leopold (1949), one of the fathers of the modern conservation movement in the United States, wrote: "The first step in intelligent tinkering is to save all the parts".

Other ecologists, including Elton, Holling, MacArthur, Margalef, May, McNaughton, Odum, Paine, Pimm, and Hurd and Wolf, addressed the same general question at the ecosystem level from the 1950s to the mid- 1970s (see McNaughton 1993 and references therein). The concern during this period was centred around testing the long-standing hypothesis that

G.W. Davis and D.M. Richardson (Eds.)
Ecological Studies, Vol.109

increasing complexity (or biodiversity) of an ecosystem should be correlated with increased stability. Only recently did attention turn to investigating the possibility that some ecosystem functions may be simultaneously achieved by different species which are "ecological equivalents" (Bond 1993). This preoccupation has now gained wide attention, no doubt as a result of concern over global loss of diversity and species extinctions. Main (1981) and Lamont (1982) addressed these issues from a novel point of view, i.e. the various mechanisms for enhancing nutrient uptake in plants and their relevance for "plugging leaks" in nutrient losses from ecosystems. These important contributions are only rarely cited (e.g. Pate and Hopper 1993), as is the case for the numerous important papers on diversity and stability that arose from a symposium including ecologists and geneticists at the Brookhaven National Laboratory (Woodwell and Smith 1969).

In this chapter, biodiversity will be defined as the amount of genetic information which occurs in nested sets of living systems, from local populations to whole ecosystems, and sets of ecosystems (landscapes). All or part of this genetic information is inevitably assumed to be necessary for the functioning of any given natural system. This does not mean, however, that all the elements of biodiversity are equally important for ecosystem functioning and stability. As pointed out by Harper (in Schulze and Mooney 1993, p. 504) "... at the heart of any useful concept of biodiversity is the range of genetic information present - a new species added to a community adds to biodiversity by the extent to which it adds new genetic information". To this should be added that the range of new interspecific interactions, or the "interaction strength" of those interactions, made possible by the addition (or removal) of any new species to an existing community, depends on the identity of the species added and those already present in a system (see Odum 1969; Paine 1992). For the sake of simplicity, three components or levels of biodiversity can be identified (Barbault 1991): (1) intra-specific diversity, i.e. genetic variation within species and populations, (2) inter-specific diversity or community diversity, i.e. species richness and community structure, and (3) functional group diversity, which is useful for scaling up from population to ecosystem level analyses (Körner 1993). A fourth level could be added to cover functional diversity among ecosystems within a landscape.

By ecosystem functioning, we mean both biogeochemical cycles (capture, storing and transfer of carbon, water and nutrients), and the population and community processes within the system (Vitousek and Hooper 1993; Woodward 1993). As regards the former, we tentatively agree with Main (1981) and Lamont (1982) that nutrient "leaks" can be an indication of ecosystem malfunctioning. Population or community processes in an ecosystem include those interactions among organisms which constitute the adaptive responses of individual species, co-evolved species, and species assemblages to a given set of environmental conditions. In other words, a fully developed ecosystem has a certain stock of genetic and ecological "information" stored in the constituent biological parts that insures the long-term survival of the ecosystem itself. In theory this means that the maintenance of all three types of an ecosystem's biodiversity (intra- and interspecific, and among functional groups) is inseparable from its ongoing biogeochemical processes in a given trajectory (see May 1973). To address the relationship between biodiversity and ecosystem function, then, we need to : (1) know how energy, nutrients and water flow in the system; and (2) identify the relevant organisms and the number and strength of interactions within populations and among species that assure the uptake, utilisation and throughflow of solar energy and other resources. Since redundancy may well exist, it also makes sense to delineate, if only tentatively, *functional groups* in the ecosystem under study (Körner 1993). Redundancy means that ecosystems may have a built-in buffering capacity to compensate for species loss (Lawton and Brown 1993). It is, however, reasonable to assume, along with May (1977) and Ehrlich and Ehrlich (1981), that there exists a threshold beyond which species loss will lead a biological system to shift to different levels of function and altered feedback systems. We add that what holds true of an ecosystem, or any given "functional group" or "keystone species" at a given time, does not necessarily apply at other stages of the system's development (succession) (see Aronson et al. 1993a).

In this chapter we will not dwell on whether so-called *keystone species* (Paine 1966; Lamont 1992; Bond 1993) exert such a disproportionate role in ecosystems that their loss would destabilize the functioning of the whole system. The available information on Mediterranean ecosystems does not facilitate the identification of kcystone species. Rather, we will discuss the third type of biodiversity, genetic variation within species, in some detail. Based on information for a variety of organisms, we can attempt some inductions, deductions and speculations. As Stebbins (1970, p. 59) pointed out, "most of the variation within species with respect to visible characters is associated with adaptation to climatic and edaphic factors of the environment." Moreover, numerous studies over the past decades have shown that fitness-related traits can change rapidly (over a few generations) if, by so doing, they confer greater fitness to the organism. The maintenance of genetic variation thus hinges, at least partly, on the interactions between genotypes and environments. Under fluctuating environmental conditions, and in heterogeneous landscapes, such as are common in the Mediterranean region, maintaining intraspecific variation enables populations to respond to unpredictable changes in environmental conditions (e.g. plant-animal relationships) and, therefore, ensure their ongoing participation in ecosystems.

Some baseline data on biodiversity are available for many communities of organisms in most mediterranean-type ecosystems (MTEs). For this reason, it may appear that there is a good potential for determining the relationships between diversity and ecosystem functioning in MTEs. However, all MTEs, and the Mediterranean Basin in particular, present an enormous challenge to researchers, not only because of the size of the areas and their physical complexity, but also because of each one's particular geological and anthropological history. In the case of the Mediterranean Basin, the diversity of species has been reasonably well documented for plants and some groups of animals, especially vertebrates. Much less is known, however, about the variation in diversity at different levels (i.e. α-, ß-, γ- and δ-diversity, which roughly correspond to a spatial scaling). Almost nothing is known of the role of diversity in nested biological systems, from populations to communities, on ecosystem function in any given region. When reviewing the huge literature dealing with biodiversity in the Mediterranean Basin, we were surprised by how few studies link biodiversity at any level with ecosystem function. In light of the limited information available, we must be content with outlining some regional and local diversity patterns in selected groups of plants, animals and, to a lesser extent, other organisms with respect to their geographical and historical determinants. We shall then extrapolate what we can about their possible impacts on ecosystem functioning. Omissions in this review are our fault, of course, but if this chapter gives rise to experimental studies designed to tackle some of the issues raised, it will have achieved its goal.

2.2 Geography, Climate and Habitat Diversity

2.2.1 Geography

The first point of overwhelming importance for understanding patterns of diversity in the Mediterranean Basin is its geographic location at the intersection of two major landmasses, Eurasia and Africa. The second is its complex geological history, especially in terms of the changes associated with plate tectonics, volcanic activity and the uplifting of the Alps, and the climatic oscillations that have occurred since the Miocene-Pliocene (the period from which we date the establishment of a mediterranean-type climate in the region) (Suc 1984). Thus, historical events produced an unusual geographical and topographical diversity, with high mountains (more than 4500 m elevation), peninsulas, and one of the largest archipelagos in the world (46 000 km of coastline). Most of the lithological substrata are calcareous with only a few places of siliceous substrata. The Mediterranean Basin includes several hundred islands and islets totalling 103 000 km², about 4% of the Mediterranean region as a whole, and about 18 000 km of coastline, 39% of the total for the region (Kolodny 1974). In this chapter, we follow Quézel (1978), who defined a Mesogean sub-Empire (within the Holarctic Empire) including three units: the Mediterranean, the Irano-Turanian and the Saharo-Arabian regions. The Mediterranean region thus defined extends over about 2.3 x 10^6 km² (Quézel 1985, see Fig. 2.1). It is thus nearly ten times greater in size than any other MTE.

2.2.2 Climate

The critical climatic features of the Mediterranean region are the particular combination of temperatures and rainfall which lead to summer dryness and winter moisture. Bagnouls and Gaussen (1953) and Walter and Lieth (1960) defined a dry month as one in which the mean temperature (in °C) is more than twice the amount of rainfall (in mm). During these months, hydric and thermic stresses both influence mediterranean biota because potential evapotranspiration largely exceeds precipitation. However, the physiographic diversity of the region results in a wide range of local climates, since mean average temperatures range from 5 °C to 18 °C and rainfall from < 100 up to 3 000 mm (Le Houérou 1990a). Nahal (1981) summarized the main features of mediterranean climates from a biological point of view.

Given that mediterranean-type climates progressively intergrade with mesic climates to the north, and hot desert climates to the south, the Mediterranean Basin is not a clear-cut biogeographical unit. Classical methods for delimiting the Mediterranean region include: (1) distributional limits of dominant and abundant plant species, e.g. *Quercus ilex, Olea europaea, Stipa tenacissima* (Braun-Blanquet 1936, 1952; Quézel 1985), (2) vegetation analysis (Flahault 1937); and (3) climatic variation (de Martonne 1927; Emberger 1943; Gaussen 1954; Walter and Lieth 1960; Péguy 1970; Daget 1977). Bioclimatic criteria that take into account both the cool, wet winters and the period of effective physiological drought in summer are utilized by most biologists (Emberger 1930; Bagnouls and Gaussen 1953; Sauvage 1963; Daget 1977, 1980; Quézel 1985). Furthermore, the average minimum temperature of the coldest month (January) can be conveniently used to distinguish between ecological zones within the larger Mediterranean bioclimatic region (Le Houérou 1991). In the Mediterranean Basin, these "thermal belts" vary more or less regularly as a function of latitude

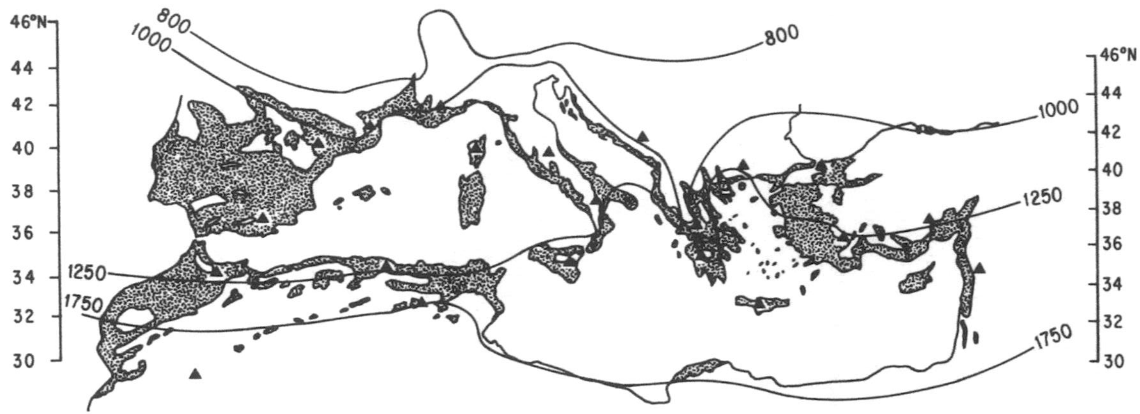

Fig. 2.1. Area map of the Mediterranean Basin. The *shaded area* is the distribution of maquis and garrigues (Tomaselli 1976); *black triangles* indicate some mountain peaks above 2000 m; lines indicate mean annual potential evapotranspiration (UNESCO 1977); thermic gradients consist of ca. 0.60 °C per degree in latitude and 0.55 °C per 100 m elevation, while the corresponding rainfall increases are 40-60 mm and 10% respectively. (Le Houérou 1990a)

and altitude (Fig. 2.1). Precipitation decreases in the same direction. One additional feature to be emphasized is the climatic unpredictability of the region, which, combined with very large inter-annual variations of both temperatures and rainfall, imposes heavy constraints on biota (Nahal 1981; Aronson and Shmida 1992). Autumn thunderstorms may be particularly violent, with local precipitation amounting up to 100 mm h^{-1} (Le Houérou 1990a). Such precipitation falling on hot, dry soils is of no use for most plants and may have catastrophic consequences.

2.2.3 Habitat Diversity

The size and physiographic complexity of the Mediterranean region results in extremely high diversity of vegetation types and communities, depending on a complicated set of factors such as climate, geomorphology, pedology, hydrology and land-use practices. These range from moist fir-beech (*Abies-Fagus*) forests in some mountains of the mainlands and larger islands to arid steppes in North Africa and the Near East. Biogeographical zonation and climatic variation along latitudinal and altitudinal gradients produce a zonation of vegetation belts which has been formalized by Quézel (1985) from the most arid to the most mesic/alpine formations (see Fig. 2.2). Despite the broad regional patterns expressed in Fig. 2.2 and, in more detail, by Zohary (1973) and Le Houérou (1990a), the combination of high topographic diversity and multiple land-use patterns results in a kaleidoscope of habitat patches that vary greatly in size and vegetation structure and composition. Braun-Blanquet (1952) identified 149 major phyto-ecological groupings in southern France alone. Temporal dynamics of the various patches vary as much as the spatial aspects. Like all MTEs, the Mediterranean Basin is highly heterogeneous in both temporal and spatial dynamics, especially where there is a long history of human activities.

2.3 Patterns of Mediterranean Diversities in Relation to Space and Time

2.3.1 Flora and Vegetation

The flora of the Mediterranean Basin includes approximately 25 000 species of flowering plants and ferns (Quézel 1985). This figure is very high when compared with the 6 000 species of non-Mediterranean Europe, an area nearly four times as large (9×10^6 km²; Quézel 1985). However,

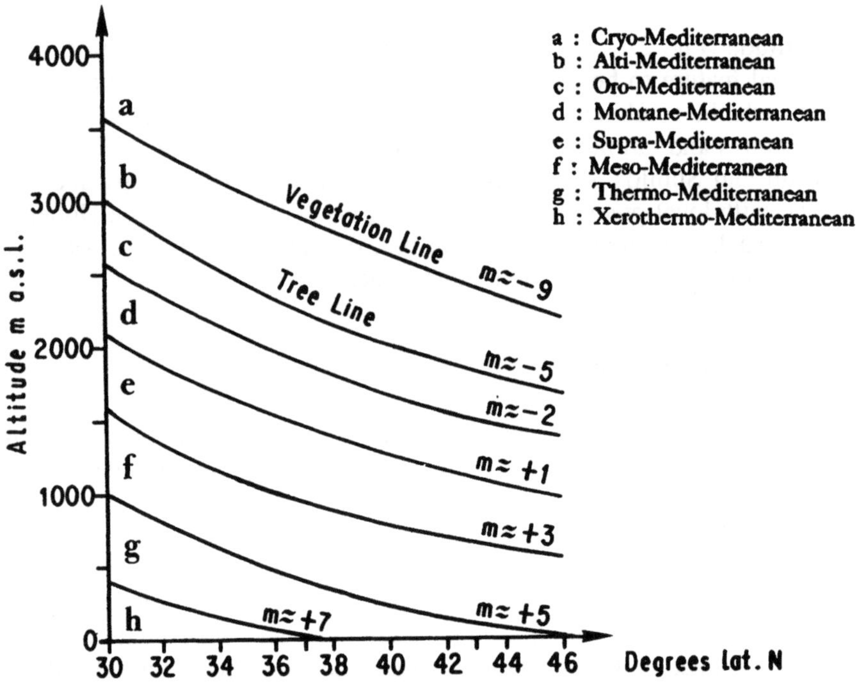

Fig. 2.2. Two altitudinal/latitudinal gradients indicating where the various vegetation belts defined by Quézel (1985) occur in southern France and in North Africa; **m** mean daily temperature of the coldest month. (Le Houérou 1990a)

it is rather low in taxa per km² when compared to other MTEs, especially South Africa. More than half of the plant species are endemic; indeed, four out of five European endemics are Mediterranean species (Gomez-Campo 1985). There are more than 100 common tree species in the Mediterranean Basin (many of them endemic to the region), compared with no more than 30 in Central Europe (Le Houérou 1980). No family is strictly endemic to the Mediterranean region, however, because of the composite origin of the flora and the pluri-regional chorotypes of many taxa. Mediterranean endemics appear to be uncommon or non-existent among mushrooms and other lower plant groups.

The existence of Mediterranean vegetation assemblages first appears in the fossil record in the mid-Pliocene, some 6 Ma B.P. (Suc 1980; Pons and Quézel 1985), although many Mediterranean taxa originated during the Oligocene (Medus and Pons 1980; Pons and Suc 1980; Suc 1980; Herrera 1992). Since that time, Mediterranean taxa have increased substantially in number as a result of climatic changes, especially the decrease in summer rainfall (Suc 1980, 1984).

The extant flora of the Mediterranean Basin is an intricate mixture of taxa with disparate biogeographic origins, ages, and evolutionary histories (Zohary 1973; Raven 1973; Axelrod 1975; Pignatti 1978; Quézel et al. 1980; Pons 1981; Palamarev 1989; Herrera 1992). In essence, however, the contemporary flora includes elements from two broad categories, "Paleo-" and "Neo-Mediterranean". Although many elements which evolved under tropical climatic conditions prevailing during the pre-Pliocene or Pliocene periods were eliminated during the Pleistocene glaciations, the survivors of this group still play a prominent role in the flora and vegetation (Raven 1973; Quézel et al. 1980). Examples are *Arbutus, Calluna, Ceratonia, Chamaerops* and *Laurus.* Many "Neo-Mediterranean" genera absent from Pliocene fossil records represent lineages that appeared in the region through immigration or speciation after the establishment of a mediterranean-type climate. Examples include *Amelanchier, Cistus, Clematis, Halimium,* and *Helianthemum.* Thus the extant flora includes taxa originating from tropical, subtropical, and non-tropical Holarctic regions, resulting in high species diversity and the coexistence of relict taxa (paleoendemics) with more recently evolved neoendemics (see Herrera 1992). Non-tropical woody elements that evolved after the onset of a mediterranean-type climate fall into three main floristic groups:

1. Mediterranean elements with a high percentage of endemics. These differentiated in continental tectonic "microplates", especially the Ibero-Mauritanian, the Balkanic and the Anatolian, and to a lesser degree the Tyrrhenian and the Cyreanican. High mountains promoted the differentiation of an important orophilous flora (Pons and Quézel 1985).
2. Mesogean elements of cold steppe, Irano-Turanian origin, such as *Artemisia, Astragalus* and *Ephedra.* In addition, arid conditions in North Africa and the Near East allowed xerophilous taxa of African origin to join the Mediterranean flora.
3. Holarctic and Eurasian elements such as *Acer, Betula, Cercis, Corylus, Fagus, Fraxinus, Pinus, Quercus, Tilia* and *Ulmus* are prominent components of the present flora. They differentiated either under temperate to warm-temperate or cold climates and are an important part of the post-glacial floras of southern Europe, as they are in North America (Pons 1964; Van Campo et al. 1968; Axelrod 1973; Suc 1978; Quézel 1985).

At the most severe phases of the glacial periods, virtually no arboreal vegetation could have persisted to the north of the Pyrénées, the Alps and the Carpathians (Pons 1981; Huntley and Birks 1983). All forest biota of the western Palearctic must have withdrawn to local refugia, mostly within the limits of the Mediterranean region. This was made possible by the topographical and climatic diversity in the region during glacial and interglacial periods - hence the coexistence on a regional scale of all the European vegetation belts and their associated faunas. Pollen analyses have shown that during the most severe glacial episodes, refugia existed along mountain slopes and in the valleys of large rivers (for instance the Rhône river) which were then uncovered because sea level was 100-150 m lower than today. Local climatic conditions for the survival of Mediterranean taxa existed in such places both in the western (Triat 1978, 1979) and the eastern portions of the Mediterranean Basin (Wijmstra 1969; Niklewski and Van Zeist 1970; Florschultz et al. 1971). However, the penetration of some Mediterranean taxa into the Sahara is indicated by the existence of various remnants in disjunct localities of the Hoggar mountains, more than 1500 km from the Mediterranean-Maghreb zone (Quézel 1965). Conversely, some Afro-tropical elements from the Sudanian and Sahelo-Sudanian regions penetrated the Mediterranean region in North Africa (Quézel 1965) and parts of the Near East ("Near East" and "Middle East" are referred to here in relation to the geographic position of the Mediterranean sea), either as Miocene relicts, or as recent (Holocene) "immigrants" (Shmida and Aronson 1986). Furthermore, numerous taxa with easily transported propagules, especially among those with the annual, biennial or facultative perennial growth forms, have spread and continue to spread from the western or, especially, eastern portions of the Mediterranean Basin, invading the Canary Islands and other MTEs further afield. Pollen analyses carried out along altitudinal gradients in the mountains of the former Yugoslavia reiterate the pervasive importance of forests in the Pleistocene history of the Mediterranean (Beug 1967, 1975). It is, therefore, not surprising that floristic and faunal elements originating from the mesic climates of central Europe are major components of the present-day biota. From the early postglacial period, 12 000 B.P., the progressive development of deciduous oak forests has been a general pattern throughout the whole Mediterranean Basin (Ménendez-Amor and Florschultz 1964; Reille 1975; Van Zeist and Bottema 1977; Van Zeist et al. 1975; Pons and Quézel 1985). These deciduous forests have been partly replaced by evergreen oak (*Quercus ilex, Q. calliprinos*) forests in more recent times (see Sect. 2.6.3 for further discussion).

Reconstructing the history of mediterranean shrublands is more difficult than for forests, since most taxa are weak pollen producers. However, macrofossil remains (Vernet 1973; Pons 1981) and pollen analyses (Van Zeist et al. 1975; Reille et al. 1980) reveal close relationships between percentages of shrubby taxa and human activities in both the western and the eastern portions of the Mediterranean Basin. Most Mediterranean shrublands are the result of human disturbance, including deliberate introductions, especially at mid-altitudes. Nevertheless, there is evidence in some lowlands and coastal regions that shrubby sclerophyllous communities dominated by e.g. *Pistacia* cf. *lentiscus*, *Phillyrea* cf. *angustifolia* and *Juniperus* spp., persisted during the whole Quaternary, long before human influence became important (Reille et al. 1980). Some of these regions (e.g. Cataluña, northeastern Spain) are very rich in endemic species of shrub and chamaephyte growth forms (see Sect. 2.7.1).

2.3.2 Vertebrates

Vertebrate diversities, like plants, are also very high but, mainly for historical and biological reasons and differences in dispersal abilities, endemism rates differ markedly among groups.

2.3.2.1 Mammals

The two main features of the mammalian fauna in the Mediterranean Basin are its multiple biogeographical origin, with elements from Eurasia and Africa, and the large periodic turnovers leading to high ß-diversities over time because of the Pleistocene climatic changes (Cheylan 1991). The present number of land mammal species in the region is about 184, of which 25% are endemic (Cheylan 1991). Being rather weak dispersers, the distribution of non-flying mammals is restricted by geographical barriers, i.e. the Straits of Gibraltar (active as a barrier for the last 5 million years, Jaeger et al. 1987) and the extension of the Sahara. Consequently, the mammalian faunas of the three main subregions (Mediterranean Europe, the Middle East and North Africa) are rather distinct.

In southern Europe, very few subtropical taxa survived after the large extinctions of the first Pleistocene glaciations (examples are *Hystrix, Macaca*) and most extant genera (e.g. *Sus, Cervus, Ursus*) are of boreal origin (Kurten 1968). Present diversities in Mediterranean Europe and the Middle East are mostly a result of immigrations from temperate Asia during the interglacial Riss-Würm (ca. 110 000 to 70 000 B.P.)(Tchernov 1984; Cheylan 1991), whereas species of Afrotropical origin always dominated the fauna in North Africa (Jaeger 1975).

The present mammalian diversity results to a large extent from severe and ancient (Neolithic) human pressures (habitat changes, animal husbandry, hunting and introductions of alien species) that modified natural distributional patterns and led to a decrease in the previously much richer glacial fauna. The sharp decrease in species richness during the late Pleistocene has probably resulted from climatic changes combined with human destruction as was the case for mega-herbivores in North America (Johnson 1977). In western Asia (Tchernov 1984) and North Africa (Jaeger 1975), human impact, combined with desertification during the Holocene and historic times, resulted in the extinction of many species, especially large-hoofed mammals. The mass extinction of large mammals in North America and in the Mediterranean region, including on Mediterranean islands where surprising assemblages of dwarf hippopotamuses and elephants have been eradicated (Diamond 1992, Blondel and Vigne 1993) supports the overkill hypothesis of Martin (1984) whereby prehistoric humans were directly responsible for the extinction of many species of large mammals.

2.3.2.2 Birds

Bird diversity of the Mediterranean region includes about 345 breeding species (Blondel 1988), a number to be compared with the 419 species which breed in the 10 million km^2 of Europe (Voous 1960). Bird diversity in the region may be expressed by the ratio of bird species richness (S) to area (A in km^2), S/A 10^{-5}. This ratio is 11.3 in the Mediterranean region compared with 4.15 for Europe (Blondel 1985). After the extinction in Eurasia of many tropical elements at the end of the Miocene (Brodkorb 1971), the Plio-Pleistocene development of avifaunas involved faunal elements from nine biogeographic units among which two played the most important roles: Eurasia (153 species) and the eremian belts of the south and south-east margins of the region (85 species), where an active radiation took place within the wide belt of semi-arid habitats that encircles the western Palearctic from the Atlantic ocean to the Arabian plate and the steppes of south-central Asia. Thus, in contrast to mammals (Heim de Balsac 1936), the Mediterranean bird fauna has closer affinities with that of Asiatic steppes than with those of tropical Africa.

Although we would expect shrubland bird species to be numerous and dominant in the many types of matorral, a formation which extends over more than half of the region (Tomaselli 1976), this category is poorly represented (12%). By contrast, forest birds of boreal origin are widespread and dominant throughout the region. Studying the dynamics of bird communities along ecological successions, Blondel (1987) has shown that the more structurally complex the vegetation becomes, the lower the number of bird species of Mediterranean origin (Fig. 2.3). A comparison by multivariate analysis of bird communities along several ecological successions in the Mediterranean region and mid-Europe shows a surprising convergence in structure of the communities in old mature forests. Each succession starts with a different set of species as a result of regional speciation processes (Blondel and Farré 1988), but the communities of forest habitats are very similar whatever their location in Europe. This is because of the overriding importance of "core" forest species (Hanski

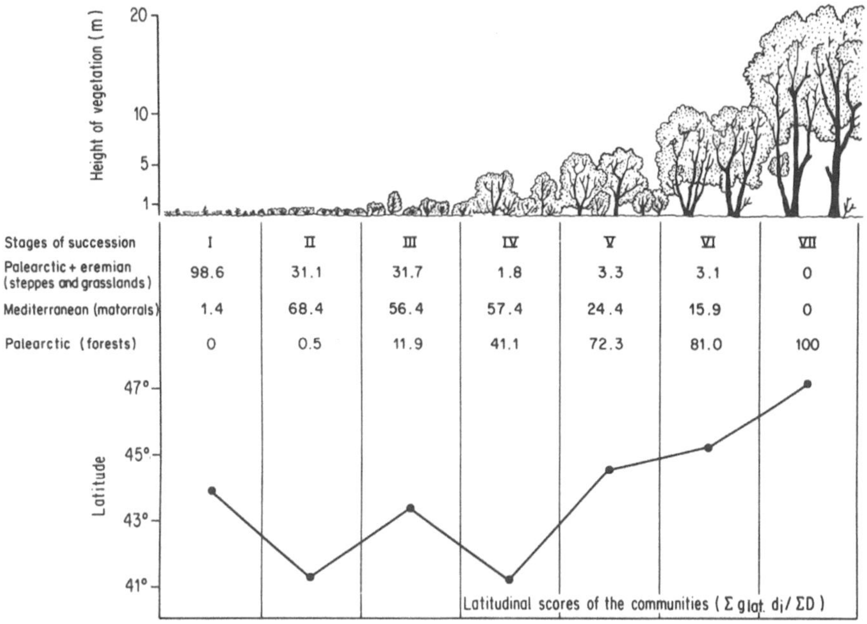

Fig. 2.3. Biogeographic composition (in percentages) of bird species assemblages in seven stages of an ecological succession in southern France. The *solid line* represents the average latitudinal centre of gravity (calculated from the worldwide breeding distribution of the species) of the species assemblages in each seral stage of the succession. On average, those species that occur in early stages have a more southerly worldwide distributional range than those that occur in later stages. (Blondel 1987)

1982), i.e. species that are widespread in the whole Palearctic region and tend to be abundant wherever they occur.

Few speciation events of birds in the Mediterranean region are hypothesized to be directly related to the history of the biota. During the whole Plio-Pleistocene, including glacial episodes, the Mediterranean region offered conditions for survival of all biota of the western Palearctic. Therefore, at a regional level, bird faunas must have been a mixture of many different Eurasian species assemblages, without any clear geographic delimitation between Mediterranean and non-Mediterranean communities. Such a situation did not provide opportunities for geographic isolation such as would have been conducive for allopatric speciation in birds (Blondel 1988; Blondel and Vigne 1993). The telescoping of Eurasian faunas within the Mediterranean during glacial times presumably made α- and ß-diversities very high. However, as explained above, there is historical evidence that more or less isolated patches of shrubland have been part of Mediterranean landscapes during the whole Quaternary (Suc 1978; Pons 1981). Some genera of shrubland birds (e.g. *Sylvia* and *Alectoris*) presumably evolved in such habitats. The hypothesis of successive speciation events in this type of habitat is consistent with paleobotanical findings that demonstrate that the spatial extension of these patches of matorral varied in time according to shifts of temperature and moisture (Pons 1981). As a result of the history of Mediterranean biota during the Pleistocene, only 47 species speciated in the region (i.e. 14% of the avifauna), representing a relatively low level of endemism (Blondel 1988).

2.3.2.3 Reptiles and Amphibians

The reptiles and amphibians of the Mediterranean region include 179 and 62 species respectively (Cheylan and Poitevin 1993) with many more endemic species (62 and 56% respectively) than birds and mammals. Species diversity of reptiles expressed as the ratio of species numbers/log area (km^2) decreases from 6.52 in the Near East to 4.5 in North Africa, 3.9 in the Balkanic peninsula, and 2.5 in the Iberian peninsula. This ratio drops steadily on Mediterranean islands, with 2.19 in Cyprus and 1.21 in Corsica (Cheylan and Poitevin 1993). Amphibian diversity is much lower in Afro-Asiatic Mediterranean areas than in Euro-Mediterranean ones. For both groups, some families are particularly well represented. Examples of amphibians are Discoglossidae (10 species, or 71.4% of the world total), and Salamandridae (19 species; 35.8% of the world total); and of reptiles are Lacertidae (60 species; 30%), Testunididae (4 species; 8%), and Viperidae (14 species; 7.4%) (Cheylan and Poitevin 1993). Many endemic species and genera are archaic lineages which have probably remained unchanged since their differentiation before the Miocene-Pliocene. Given that the metabolic rate of reptiles is, on average, only 5-10% that of endotherms, and their seasonal activity usually does not exceed 4-6 months, their energetic intake probably does not contribute much in ecosystem processes. For instance, the annual prey intake of *Vipera aspis* is 1.15 times its own body mass (Saint-Girons 1983), whereas annual prey intake of the Barn Owl (*Tyto alba*) is no less than 110 times its own body mass (Cheylan and Poitevin 1993). Nevertheless, their role in many ecosystem processes involving food webs is clearly important because most reptiles and amphibians prey on both insects and/or vertebrates and because they also serve as prey for many species of birds and mammals (Valverde 1957, 1967; Cheylan 1979; Cheylan and Poitevin 1993). This is especially true in semi-arid Mediterranean ecosystems where, together with insects, reptiles constitute the most important food supply for many species of vertebrates. Although no quantified studies on energetic pathways involving reptiles and amphibians are available, detailed qualitative studies emphasize their role as prey in many vertebrate communities (Cheylan and Poitevin 1993). Several species are taken as alternative prey when populations of the more usual prey species such as rabbits or micromammals crash (Cheylan 1979).

In summary, the Quaternary histories of different groups of terrestrial vertebrates share many features, but they also differ in important aspects. For amphibians, reptiles, mammals, and birds, most of the sub-regional faunas are not fundamentally different from those of cold- and warm-temperate Europe, which belong mainly to Holarctic faunal types. The only major exception is the mammals of North Africa, which have closer affinities as a group with tropical Africa than with the rest of the Mediterranean Basin. Because non-flying vertebrates are more sensitive to geographical barriers than birds, endemism rates are higher for the former: 25% for mammals, and 55 to 62% for amphibians and reptiles, compared with 14% for birds.

2.4 The Historical Load and Mediterranean Adaptations

Given the particular constraints of the mediterranean climate, major patterns of ecosystem structure and functioning are hypothesized to involve species that share similar sets of adaptations. The literature on alleged convergence in the various MTEs is enormous (see for example Mooney 1977a; Cody and Mooney 1978; Cowling and Campbell 1980; Hobbs et al., Chap. 1, this Vol.). However, those groups of extant Mediterranean species which evolved under non-mediterranean conditions may provide clues about historical and phylogenetic effects on current patterns of diversity. In particular, the persistence of phenotypes and life history traits of plant taxa over long time periods despite large climatic and ecological changes is a demonstration of the importance of considering historical factors when interpreting a variety of traits (Stebbins 1982; Parks and Wendel 1990; Herrera 1992). As we will discuss, these effects may be as important as ecological convergence in determining the composition of species assemblages in a given area.

2.4.1 Life History Traits of Mediterranean Plants: Historical Effects or Adaptations?

Extant Mediterranean plants exhibit a large variety of growth forms and phenological, biochemical and life history traits. It is difficult to determine whether any observed set of these taxon-specific characteristics is attributable to phylogenetic effects or to evolutionary processes driven by the mediterranean-type climate. One example is the series of adaptations to summer drought evident in Mediterranean plants. Margaris and Vokou (1982) distinguished between "maquis" formations consisting primarily of evergreen sclerophylls, and "phrygana" formations consisting primarily of shrubs showing seasonal dimorphism in their aboveground physiognomy (i.e. deciduous shrubs). According to these authors, the maquis usually occurs at the wetter end of the precipitation gradient of the mediterranean-type climate, whereas the phrygana is usually found at the drier end. Although a useful starting point, this distinction seems oversimplified, as shown by a species-specific treatment of the question of MTE plant pheno-morphological diversity (Orshan 1989).

The influence of a mediterranean-type climate on the evolution of taxon-specific characteristics can be determined indirectly by investigating the extent to which ancient lineages (i.e. those that evolved *before* the establishment of a mediterranean-type climate) retain their ancient adaptations as phylogenetic constraints. As described earlier, all extant Mediterranean assemblages of plants include species that originated at different geological times and, for many of them, under more tropical conditions (Herrera 1992). The question thus arises: are these co-occurring taxa homogeneous sets sharing morphologies and life history traits specifically adapted to present-day Mediterranean conditions? If so, the traits of these taxa, and their involvement in local or regional communities, would be the result of sorting processes and/or adaptive changes.

Working with the woody flora of Andalusia in southern Spain, Herrera (1992) found that life history and reproductive traits of the plants of that region covary non-randomly among taxa. By analyzing pairwise associations between general life history traits (summergreenness, spinescence, sclerophylly), flowering biology traits and seed dispersal traits, he showed that historical effects and sorting processes are essential for explaining patterns of co-variation among life history and reproductive features in regional plant communities of southern Spain. For example, ancient Pliocene or pre-Pliocene genera as a group scored differently on a life history reproductive gradient than did "young", post-Pliocene lineages. Herrera, therefore, could find no evidence that present character associations were produced by post-Pliocene selective processes acting on pre-Mediterranean lineages. Instead, patterns of covariation of morphological and life history traits suggest that the set of extant woody genera in Andalusia is a random sample of the entire set of woody genera occurring in southwestern Europe in the late Miocene and Pliocene. This means that character associations which evolved under tropical conditions have survived despite a shift in overall climatic regime from tropical to mediterranean. These ancient "tropics-adapted" taxa have persisted despite the extinction of a substantial fraction of the original communities to which they belonged, and the superimposition of new post-Pliocene taxa that originated after the establishment of a mediterranean climate.

We have no way of knowing how ecosystem functions or individual species' strategies changed as a result of these multiple species losses over a relatively long period. Nevertheless, it would be interesting to investigate the extent to which taxa of older (or paleo-Mediterranean) lineages may contribute to ecosystem processes today as "keystone" species or functional groups. Some of them definitely provide important components of certain plant communities in certain areas. These include *Argania spinosa* in southwestern Morocco, *Ceratonia siliqua* in southern Spain, Greece, Israel, Turkey etc., and *Laurus nobilis* throughout the region.

2.4.2 Intraspecific Variation

For those groups that are fairly well known taxonomically, a large amount of intraspecific variation has been reported in the Mediterranean Basin. However, this variation has only rarely been linked to variation in functional attributes. For instance, bird species that occur as breeders in the Mediterranean Basin are represented by 5.4 subspecies per species at the scale of their worldwide distribution (compiled from Vaurie 1959 - 1965), by 2.3 subspecies per species at the scale of the Palearctic (compiled from Moreau 1966) and by 2.0 subspecies per species in the Mediterranean Basin (Blondel 1985). Weighting these figures by the sizes of areas involved in order to obtain an index of regional subspecific variation, they rise from 0.56 in the whole Palearctic region to 6.73 in the Mediterranean Basin (Blondel 1985).

A large proportion of Mediterranean plant species also shows marked intraspecific variation; there are many subspecies and ecotypes. One example is the *Dactylis glomerata* complex which includes as many as 15 diploid types, 3 tetraploid main types and 1 hexaploid type, the last-mentioned being confined to North Africa (Lumaret 1988). Nine distinct subspecies, including several insular endemics occur in various habitats of the Mediterranean region. Several tetraploid forms (e.g. *Dactylis g. glomerata, D. g. hispanica* and *D. g. marina*) have wide distributional ranges and exhibit high morphological and physiological variation. According to Stebbins and Zohary (1959), the differentiation of closely related tetraploid forms of *D. g. glomerata* resulted from autopolyploidy of diploids from both temperate and Mediterranean groups, thus providing the plants with ecological attributes adapted to both climatic regions. The main point here is that the high environmental heterogeneity of the Mediterranean Basin resulted in a higher ecotypic variation of this species there than in non-Mediterranean parts of its range (Lumaret 1988). As an evolutionary response to the wide range of ecological conditions within the Mediterranean region, this intraspecific variation may thus enlarge the adaptive potential and, possibly, functional attributes of species. It remains to be seen whether such intraspecific variations are more or less common in the Mediterranean Basin than in other MTEs. Adaptations to mediterranean conditions in *Dactylis glomerata* also include morphological traits allowing water-saving mechanisms and seed retention throughout the summer drought. For example, the seeds of *Dactylis g. hispanica* are not shed until autumn, when conditions become suitable for germination (Lumaret 1988). Interestingly, tissue water relations and other physiological characteristics of the grass vary along local transects and broad geographic clines of water stress, the variations being more or less correlated to trends in four different enzyme systems (Roy and Lumaret 1987). Most forms of *D. glomerata* in the Mediterranean group are to some extent adapted to drought conditions and representatives of the complex range from sea level to the alti-Mediterranean vegetation belt with a predominance of tetraploids in more severe ecological conditions. Thus, at least in some cases, tetraploidy may widen the habitat range of this complex at both ends of ecological gradients. Hexapolyploidy in hyper-arid climates of Libya may be a further step of adaptation to extreme ecological conditions. Biochemical characterization of the complex suggests that most of the diploid subspecies belong to ancient, probably Miocene, lineages, whereas the differentiation of tetraploids seems to be much more recent, and occurred after a mediterranean-type climate became established, or even more recently during the Holocene. As new habitats were created by human activities, tetraploids have apparently been more flexible and successful in colonizing new habitats than diploids. Ecotypic variation has also been documented for several other perennial grasses within the Mediterranean Basin, including *Phalaris tuberosa*, and *Brachypodium* spp. (see Roy 1981 for review). Although these variations in physiological responses to differing climatic (and edaphic) conditions around the Basin can help explain the widespread distribution of some species, they do not shed much light on the direct role of these plants in ecosystem functions. All we can say is that intraspecific variation provides the raw material for evolution, which in turn assures the persistence of ecosystems.

Another important feature of many Mediterranean plants is the presence of volatile essential oils in their tissues. Well-known examples are thyme (*Thymus*), mint, basil, parsley, fennel, sage, rosemary, lavender, coriander, oregano, rue, bay leaves (*Laurus nobilis*) and wormwood (*Artemisia* spp.), fenugreek, sesame, saffron, licorice, onions, shallots, chives and garlic. The majority of these are grouped in the Lamiaceae, Apiaceae or Asteraceae. The ecological role of volatile aromatic compounds in leaves (and, to a much lesser extent, in seeds, roots etc.) of Mediterranean plants is complex and not fully understood. Briefly, these oils are highly flammable, which has led to suggestions that they are involved in fire-return feedback dynamics. They may also be implicated in defence against herbivores, bacteria and fungi, but may also inhibit germination of competitor seedlings (allelopathy), mimic insect pheromones as a means of attracting pollinators, and reduce water stress by providing antitranspirant action (Margaris and Vokou 1982).

The key point here is that, at least for the best-studied aromatic species groups (e.g. *Thymus* spp.), there is both significant variation between species and genetically controlled variation within species in the oil content. Building on a large body of previous work, Gouyon et al. (1986) concluded that the distribution of intraspecific variability in oil content (chemotypes) in *Thymus vulgaris* appears to be very strictly determined by environment. Moreover, changes in population composition with regard to chemical polymorphism occur over very short distances, and persist in time through a low migration rate of genes in this gynodioecious species resulting from the predominance of short-distance pollination by bees (J. Thompson, pers. comm.). This is a particularly striking result given the mosaic or patchwork structure of virtually all Mediterranean landscapes.

2.4.3 Tests of Convergence Between Bird Communities of Mediterranean Regions of France, Chile and California

Convergence assumes that communities of phylogenetically unrelated organisms in different regions should converge in attributes such as richness, structure and (eventually) functioning in similar environments as a response to similar selection pressures (Mooney and Dunn 1970; Cody and Mooney 1978). Convergence in form and function has been demonstrated at the levels of taxa and communities in different MTEs of the world, especially for plants (Naveh 1967; Cody and Mooney 1978; Cowling and Campbell 1980; Shmida 1981, but see Barbour and Minnich 1990 for a challenge). This suggests that comparable functional groups may be found in two or more of these different regions. If communities are assemblages of interacting species, the number and population densities of which are mostly determined by the structure of habitats and the amount of available resources, then similar types of habitats should promote convergence in their organization and partitioning into guilds or functional groups. However, from guild structure and ecomorphological configurations (shape of the birds, especially covariation between legs, wings, tarsus and bill) in relation to foraging techniques of the species among mediterranean bird communities of three matched habitat gradients in France, California and Chile, which were then all compared to a non-Mediterranean gradient in France, Blondel et al. (1984) found no evidence of convergence. Using principal components analysis and discriminant function analysis, these authors found no more similarities in patterns of ecomorphological partitioning among mediterranean bird communities than between any of them and the control non-Mediterranean set. This lack of convergence is attributable to differences in the phylogenetic origin and the biogeographic history of the different sets of species. As is the case for woody plants in Spain (Herrera 1992), morphological, physiological and behavioural constraints on different lineages had more influence on species assemblages and species-specific habitat requirements than their sharing a similar type of environment. However, conclusions from tests of convergence depend on the level of similarity and the choice of variables used (Mooney 1977a).

Convergence clearly exists for some community attributes (e.g. numbers of species; Schluter 1986) but not for others, especially those that require large evolutionary changes in species and genera. Although convergence in morphology, structure and, presumably, ecological function is most likely for organisms that depend strongly on climatic variation and seasonal patterns of nutrient cycling (e.g. plants, invertebrates, lizards), it is less likely for homeotherm vertebrates that rely mostly on structural attributes of ecosystems for habitat selection. Thus, convergence in such attributes as species richness and community structure may be an epiphenomenon of similar patterns of resource-sharing in relation to the structure of habitats. As for plants, historical and phylogenetic constraints may limit adaptation to the extent that some animals may not always evolve life history traits tightly adapted to the particular environment in which they now occur.

2.5 Humans and the Mediterranean Biota

There is no other region in the world where the development of ecosystems has been so intimately associated with human social systems for so long. Thus, the most important historical feature of modern Mediterranean biota is surely their anthropogenic structure in all parts of the Mediterranean Basin. However, the impact of humans has differed substantially between the northern and the southern sides of the Mediterranean because of differences in overall aridity, land-use practices, socio-economic constraints and, above all, demographic pressure. The resulting differences in landscapes and terrestrial ecosystems have increased over the last few decades. In this section, we review the principal determinants of biodiversity at the species level or higher that were and are imposed by humans in the Mediterranean region. We then address the possible impact of anthropogenic changes on biodiversity that may have affected ecosystem function.

2.5.1 Fire

No more than 2% of the fires that occur today in the Mediterranean region are of natural origin (Trabaud et al. 1993). People are thought to have commenced using fire some 400 000 years ago (Prodon et al. 1987), although the significant impact of anthropogenic fires on Mediterranean Basin vegetation began during the Neolithic revolution, about 10000 years ago (Thirgood 1981; Trabaud 1984). Humans intentionally used fire intensively for agriculture, to provide livestock with grazing areas, and to protect themselves and their herds from predators. For the last several centuries in particular, most human-induced fires in the Mediterranean region were set for purposes of field preparation for agriculture and temporary improvement of pasture for livestock (Kuhnholtz-Lordat 1938), even though the effects of fire on the quality of vegetation as fodder is controversial (Daubenmire 1968). In this century, the return rate of fire has increased dramatically. In the past decade, about 0.56×10^6 ha of woodland and shrubland have burned every year in the Mediterranean region (ECE/FAO 1990). In France, 30% of all Mediterranean landscapes have burned at least once during the 25 past years (Ramade 1990). The average areas burned annually in Spain, France, Italy and Algeria between 1980 and 1985 were 2460, 406, 1638 and 489 km^2 respectively (Ramade 1990). In some areas, the average fire-return interval may be as little as 5 years (Trabaud et al. 1993).

Fire is, of course, one of the natural disturbance events that initiates cycles of vegetation succession. Hence, fire contributes to the maintenance of many extant MTEs and other ecosystems, provided that its occurrence does not massively exceed the natural fire frequency (Pickett and White 1985). In the Mediterranean region, where natural fires are much less frequent than in other MTEs, especially in California and South Africa (Kruger 1983), a very high rate of man-induced fires completely changed the structure and functioning of ecosystems and contributed to the massive destruction of forest (Naveh and Dan 1973). Among other things, a high frequency of fires in the Mediterranean often results in depauperate communities dominated by species that resprout or germinate quickly after fire, e.g. *Quercus coccifera* (Trabaud 1987), *Cistus* spp. (Roy and Sonié 1992), and *Sacropoterium spinosum* (Naveh 1975). Hence, one characteristic of ecosystems heavily altered by frequent fires is the lack of true successional processes whereby different species assemblages replace one another over time. Vegetation recovery following fire in the Mediterranean region involves endogenous processes of local plant species which progressively return the burned system to a state very similar to that which prevailed before fire, with very few pioneer species colonizing such systems (Trabaud 1987). For instance, the Kermes oak, *Quercus coccifera*, extends over very large areas in southern France. Most of the biomass of this species, which seldom exceeds 1 m in height, is in the form of stumps and roots. While its aboveground vegetation is completely destroyed by fire, vigorous local resprouting from stumps occurs within a few weeks. Thus, many areas are characterized by seemingly "blocked" scrublands dominated by one or a few species. It is not known how long a period without fire would be needed to "release" these stands and make them open to colonization by plant assemblages leading to alternate seral stages. Some stands dominated by *Q. coccifera* that have not burned for about 30 years still have exactly the same overall structure (pers. observ.). In such cases, it may well be that something more than simply halting the fires would be needed to initiate a new trajectory.

2.5.2 Deforestation

The first significant deforestation of Mediterranean forests began as early as 8000 B.P. (Thirgood 1981; Pons and Quézel 1985) and increased dramatically at the end of the Neolithic. Agriculture and animal husbandry began more than 10 000 B.P. in the Eastern Mediterranean (Zohary 1969; Zohary and Spiegel-Roy 1975; Van Zeist 1980), 8 000 B.P. in Greece and Crete (Pons and Quézel 1985), and during the flourishing Cardial civilisation (7500 B.P.) which was an epoch of intense agriculture in the western Mediterranean (Thirgood 1981). Pollen diagrams show that the large-scale Neolithic deforestation in the Alps (Beaulieu 1977) and the Pyrénées (Jalut 1976) coincided with the expansion of cereal culture, first at low altitudes in southern France, at the end of the Atlantic period (7500-4500 B.P.), and then at mid-altitudes towards 5000 B.P. (Triat 1979). Many Mediterranean pollen diagrams (from the Rif in Morocco to the western Taurus through the southern Alps, Corsica and Greece) reveal early and widespread forest clearance, grazing and cultivation, even at high altitudes. As a result, major trends and even some specific events of former Mediterranean civilisations are reflected in pollen diagrams. Examples are the periods of expansion of the Persian, Hellenistic and Egyptian (7000 to 4000 B.P.) civilizations which were all eras of intense clearing and agricultural activity at the expense of forests (Trabaud et al. 1993).

Heavy anthropogenic pressures occurred again as a result of a large population increase during the Calcolithic period, especially the Sub-Boreal (4500 B.P.), and continued up until the decline of the Roman Empire

(Reille et al. 1980). The pollen record shows a reconstitution of dense Mediterranean forest throughout the region that lasted until a resurgence of human activities during the rise of the Venetian Republic and around the time of Arab invasions of North Africa (Thirgood 1981). In Corsica, for example, paleobotanical data show the resurgence of agriculture during the early Middle Ages accompanied by rapid deforestation after the colonization of this island by the Republic of Genova (Pons and Quézel 1985). Finally, the 19th and, especially, the 20th centuries have seen increasingly severe destruction of vegetation throughout most of the Mediterranean region (Naveh 1975; Le Houérou 1980, 1981; Pignatti 1983; Pons and Quézel 1985). Rapidly growing human populations, especially in North Africa and the Near East, mechanization of farming, larger and more frequent fires, excessive use of herbicides, and, more recently, air pollution are among the interrelated causes of this widespread forest destruction. Overgrazing and over-exploitation of the steppe vegetation, drainage of marshes, estuaries and deltas led to the disappearance of many additional ecosystems and agro-ecosystems (e.g. the Pontin Marsh in Italy, and the Rharb marshes and La Calle Lake in North Africa (Quézel 1985). Only in parts of southern Europe, e.g. southern France and parts of northern Spain, is a reverse trend currently underway for reasons discussed in Sect. 2.5.4 and 2.6.2.

2.5.3 Land-Use Patterns

Notwithstanding the predominantly destructive role of humans described above, people have at various times in history also played a role in preserving and maintaining biodiversity in some Mediterranean forests and other ecosystems. This has taken place primarily where and when people lived from subsistence farming, animal husbandry and agroforestry systems organized by more or less autonomous rural communities. Under those conditions, local people exploited a wide variety of forest, pasture and ecotone products, and governed themselves in such a way that the biological and landscape diversity they needed for survival was preserved. Regional variations existed of course: in southern France and Italy, for example, farming, forestry and animal husbandry were generally practised on non-overlapping landscape units. This corresponded to the ancient Roman system *ager-saltus-silva* or "field-pasture-forest", with variants based on altitudinal zones and other factors. In contrast, the now-declining "dehesa" or "montado" system evolved in southern Spain and southern Portugal as an integrated agro-sylvo-pastoral system in a single space (Joffre et al. 1988). Thus, local people in different parts of the Mediterranean have historically chosen how certain ecosystems and landscapes should function. A relatively stable situation existed for biodiversity and landscapes in southern France over much of the Middle Ages and up until the 18th century. Since the mid-18th century, however, land-use patterns there have altered and landscapes have been transformed. Some of the dramatic upheavals that have occurred in the region as a result of changing patterns of land-use and, particularly, utilization of forest products over the past few centuries are shown in Fig. 2.4. In the latter half of the 17th and 18th centuries, the manufacture of glass and glassware was the only industrial activity allowed to the French nobility. Otherwise, they were not permitted to make money except from the products of their lands and forests. For the most part, it was the deciduous oaks (*Quercus pubescens* and *Q. sessilifolius*) that were preferentially cut to make charcoal for the manufacture of glass (and for metallurgy, domestic heating etc.) (Saint-Quirin 1904; Soulier 1993, and references therein), with the result that huge areas formerly dominated by these deciduous trees were clearcut. However, most of the noble glassworks came to a more or less bloody halt with the French Revolution in 1789, and the remainder closed over the following few decades. This resulted in a relaxation of extractive human pressure on the forests.

Firewood was traditionally the top priority among renewable forest resources of villagers in southern France until the use of fossil fuel became generalized just after World War II (WWII). Recently, however, there has been a renewed demand for firewood as decorative fireplaces have become popular. Secondly, until the arrival of imported tannins on the local market in the beginning of the 20th century, the bark of holm oak (*Quercus ilex*) was widely exploited for the fabrication of tannin. A 20-year cycle of clearcutting was found optimum for this purpose (F. Romane, pers. comm.).

Until the middle of the 19th century, cereal culture was the basis of rural economy, especially in low-lying plains, but also on terraces in hilly terrain. Barley and, primarily, wheat were produced as food staples, and as fodder for horses. Between 50 and 80% of the arable land of most communes was devoted to cereal cultivation, at times in rotation with lucerne (*Medicago sativa*). However, when a national rail system was built in the late 1800s, imported wheat became available at cheaper prices than locally produced grain. The export of crops then became feasible, and the production of wine grapes, already well established in the region on a small scale since early Greco-Roman times, expanded greatly (Soulier 1993). Pasture exclosures and livestock management have all but disappeared since WWII, even though they persist in other parts of the region. There is some hope they will be revived as part of a fuel reduction programme and in efforts to reintegrate animal husbandry and forestry (Hubert et al. 1989; Etienne 1991).

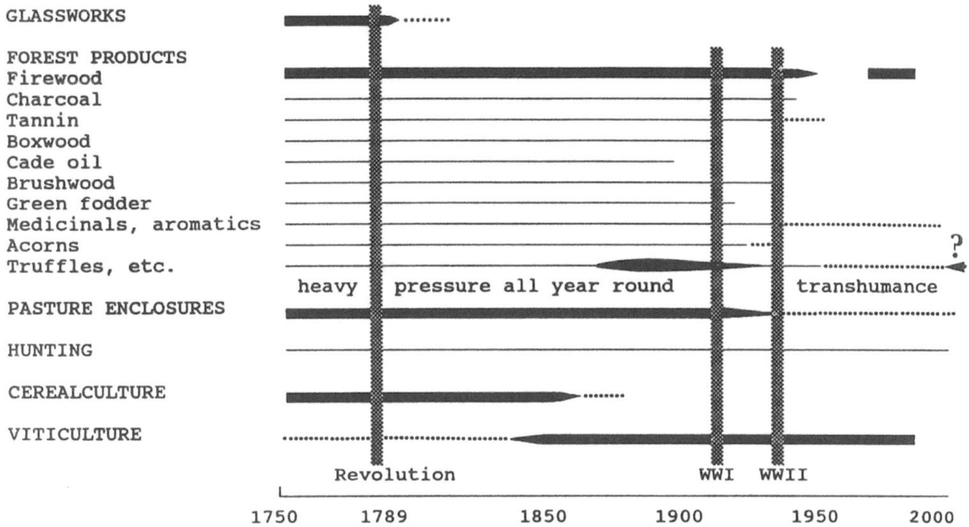

Fig. 2.4. Evolution of the utilization of a typical communal forest in southern France over the past 250 years. Note the large number of forest products utilized prior to the two World Wars. *Dotted lines* indicate a gradual increase or diminution. See text for more details on each product and utilization. (Based partly on de Bonneval 1990)

These changes in land-use patterns have undoubtedly had significant impacts on landscapes, ecosystems and vegetation, and on fauna, fungi and soil biota. Imagine, for example, the powerful but subtle influences imposed on animal and microbial populations by the practice, over several centuries, of a 20-year forest clearcut rotation scheme (see Sect. 2.6.1). Prior to the 18th century, different but equally tumultuous changes in human affairs affected Mediterranean forests. For example, the Roman empire took a huge toll on French, Spanish, North African and other Mediterranean forests for shipbuilding (Thirgood 1981). Much of the North African littoral was transformed to make way for cereal production destined for Roman consumption. This was later replaced by vineyards, just as in the Mediterranean littoral of Europe. The littoral in many Mediterranean countries is currently being transformed under a growing tide of concrete for urban and suburban expansion.

2.5.4 The Case of the Disappearing Black Truffles

To take one example of how fluctuations in the harvesting and management of a single forest product reflect the importance of human impact on both ecosystem structure and functioning in the Mediterranean Basin (and a few adjacent regions), consider the case of the black truffle (*Tuber melanosporum*, and a few close relatives). This highly perfumed, hypogeous fungus is known as the "black diamond" of the French cuisine because of the unique smell which it gives to many dishes. In the second half of the 19th century, there were huge areas (Fig. 2.5) of central and southern France where black truffles grew wild, or were semi-cultivated, especially along the edges of vineyards and oak woodlands. It is our contention that, in many of these areas, humans consciously managed certain ecosystems and landscapes to assure an ongoing supply of truffles. In much earlier days too, truffles occurred abundantly in woodland clearings and other habitats all around the Mediterranean Basin (as recorded by Theophrastus, Pythagorus, Pliny the Elder among others). In this century, however, truffle production has declined dramatically: only 10 tons of black truffles were sold in all of France in 1992, compared with the 1500 to 2000 tons sold annually in international export markets between 1880 and 1910 (Pagnol 1973; Rebière 1982; Delmas 1986). Given the remarkably high prices of fresh black truffles ($400-600/kg in 1992-93), how can this decrease be explained? During the "golden age" of black truffle production (1880-1915), thousands of farmers, villagers and land owners all over France planted fields, field edges and diseased vineyards with acorns from oak trees known to have borne truffles. But, by 1900 resistant *Vitis* rootstocks were imported from California in response to the phylloxera root rot epidemic that attacked vineyards in the early 1870s. Most people went back to growing vines, which are much more reliable, if less lucrative, than truffles. Moreover, and perhaps more importantly, huge areas where natural stands of truffles had been harvested (and managed?) annually, became largely unsuitable for truffles. The key factor involved was certainly the "rural flight" that has, since the beginning of the century, emptied the countryside and left many villages in southern France nearly deserted (see also Sect. 2.5.5). One direct result of this has been the rapid dwindling in the number and size of the vast flocks of sheep that used to keep the understorey of the Mediterranean forests and large surrounding areas clear. In the absence of those flocks - which also fertilized the forest, and browsed the lowest branches of trees - thick, coarse shrubby vegetation quickly filled in, thereby creating an environment inimical to truffles and, incidentally, much more susceptible to forest fires than at any time in the recent past.

Arnolds (1991) and Jaenkie (1991) have recently shown that populations of many ectomycorrhizal fungi, and the average size of individual edible mushrooms harvested, have decreased significantly throughout central Europe over the past four decades. The decline is apparently due to indirect effects of air, soil and water pollution, e.g. massive increases in nitrates, all of which have led to a decrease in forest tree vitality. This explains why

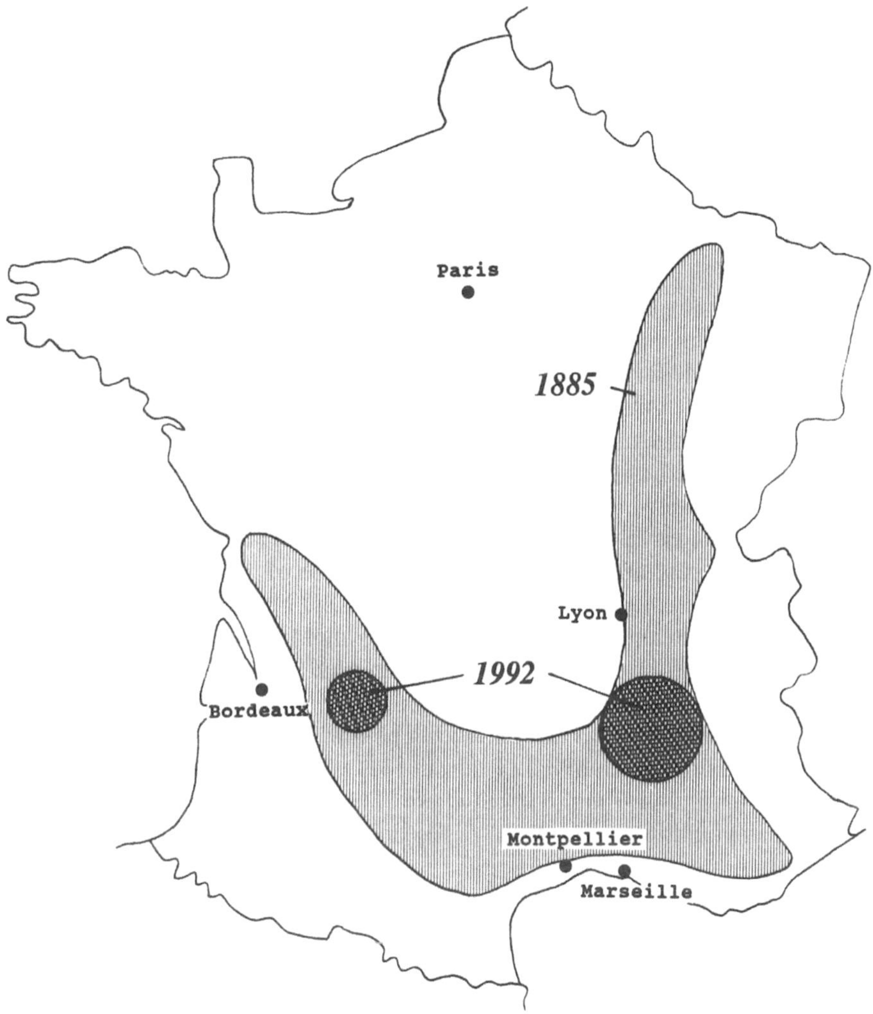

Fig. 2.5. Approximate areas of black truffle production in France in 1885 and in 1992. (Pagnol 1973 and many other sources)

the average size and number of mushrooms have declined. Accordingly, the fungi may be an "early warning signal" and it therefore seems important to pay attention to fluctuating production of large mushrooms in all ecosystems, including Mediterranean ones. The above information strongly suggests that woodland ecosystems in southern France are not functioning as before. A possible role for diffuse pollution of air, soils and water in this decline cannot be entirely ruled out, but we suspect this to be secondary in importance to the changing land-use patterns as described. However, since fungi are associated with trees in complex, presumably symbiotic ways, the dwindling of the hypogeous fungi may indicate changes in profound ecosystem functions affecting the trees as well.

2.5.5 Differences in Human Pressures, North and South

Around the Mediterranean Basin, it is important to note that increasing differences in living standards and development between the Euro-Mediterranean and the Afro-Asiatic-Mediterranean countries have profound effects on land-use practices and the pressure on ecosystems. Modern Euro-Mediterranean countries are increasingly characterized by the development of industry, urbanization, and tourism, and a concurrent decrease of agriculture and pastoralism. In contrast, the main concern in Afro-Asiatic Mediterranean countries at present is the rapidly growing human population, which represents an obstacle to sustainable socio-economic development (Table 2.1; see also Table 1.6 in chapter 1). As a result, forest areas are increasing at a rate of 1 to 1.5% yr^{-1} in the former, and decreasing at roughly 2% yr^{-1} in the latter (Marchand 1990).

In Mediterranean France, for example, most rural areas have been thoroughly transformed by the exodus from rural areas which started at the end of the 19th century and accelerated after WWI and especially after WWII. This migration to the cities can be attributed (if we oversimplify) to the increasing availability of fossil fuel which replaced firewood for industrial, agricultural and domestic use. This trend explains the regrowth of forest vegetation as woodlots, fields and pastures are abandoned. Plant ecologists notice that this "biological regeneration" is accompanied by the replacement of many sclerophyllous Mediterranean species by taxa of more mesic character. For example, deciduous oaks and their associated flora tend to replace the sclerophyllous vegetation (A. Pons, pers. comm.). It is interesting to note that this natural regeneration involves mostly assemblages of species which occurred in the region before the large-scale deforestation by people. Similarly, several species of forest birds recently recolonized the region from the north during the last three decades while several "thermophilous" Mediterranean species withdrew to the south (J. Blondel, pers. observ.).

In sharp contrast, in North Africa, the biggest challenge facing applied ecologists today is the need to restore and rehabilitate degraded ecosystems precisely because human pressure on those systems is greater than ever (see Sect. 2.9). Sub-regional differences aside, one generalization about the entire Mediterranean region is clear. A combination of exponential human population growth in North Africa, urbanization, rural exodus and real estate development (superimposed on an "internationalization" of the world

Table 2.1. Trends of human population sizes (in millions) in some Euro-Mediterranean and Afro-Asiatic Mediterranean countries. (Le Houérou 1991 and INED 1993)

Country	1950	1970	1993	2025[a]
France	41.7	50.8	57.7	59.0
Spain	27.9	33.6	39.1	36.0
Greece	7.6	8.8	10.5	11.0
Algeria	8.7	13.7	27.3	47.0
Morocco	7.5	15.6	28.0	46.0
Iran	16.0	28.7	62.8	162.0

[a] Estimates based on conservative 1993 projections. (INED 1993)

economy), with all the new "historical load" that this imposes on agriculture at the local and regional levels, are creating new and often unprecedented types of land use that are altering all terrestrial and aquatic Mediterranean ecosystems.

2.6 Biodiversity and Ecosystem Function at the Community Level

2.6.1 Disturbances as Key Events in Community Dynamics: a Case Study of Post-Fire Succession

Because most Mediterranean ecosystems are patchy, due to small-scale variation of edaphic and climatic factors and varying land-use practices over several millennia, there is high ß-diversity. Superimposed on this spatial patchiness is great temporal variation produced by disturbance. The importance of this to ecosystem processes and community dynamics is recognized at a landscape level (Sousa 1984; Pickett and White 1985). The diversity of patches within landscapes is relevant to ecosystem processes at various scales. For any given level of γ- (landscape-level) diversity, there is a complex dynamic of α- (within-habitat) and ß- (between-habitat) diversity according to the frequency and magnitude of disturbance events. At a local scale, disturbances are critical events because they cause the local extinction of some species and allow new species to enter the system through new biogenic structures (cf. the "regeneration niche", Grubb 1977).

Post-fire succession provides one example of such community processes. Prodon et al. (1987) investigated the temporal dynamics of communities and habitats under the influence of fire in a catchment of several thousand hectares on metamorphic soils in the Pyrénées. According to the rate of fire return, there is a periodical turnover of four main habitat types: grasslands (e.g. *Brachypodium ramosum* and *Asphodelus* spp.), low maquis (*Cistus* spp., *Lavandula stoechas*), high maquis (*Erica* spp. and *Calycotome* spp.), and oakwoods (*Quercus ilex, Q. suber*). In the spring following a fire, herbaceous vegetation covers 50-70% of the ground. Subsequently, the many seedlings of *Cistus* and *Ulex* spp. grow quickly and eliminate most herbaceous species while taller maquis shrubs (*Erica, Calycotome* etc.) colonize. Finally, oak stems emerge from the maquis and develop into oak woodlands. The response of oak species to fire is quite variable [e.g. resprouting in the Holm oak (*Q. ilex*) vs. recovery of the canopy in the cork oak (*Q. suber*), the trunk of which is protected against fire by its thickened bark]. A common character of the early regenerating stands in the Mediterranean region is their patchiness at a fine scale: bare ground alternates with herbaceous areas and low shrubs or thickets of surviving oaks.

The destruction of the litter deprives the system of both a trophic resource for soil organisms and plants, and a buffer for the soil microclimate. In particular, its combustion results in a dramatic volatilization of carbon and nitrogen, and a disequilibrium in the base exchange capacity. The nitrogen content of the upper soil layers recovers partially in the first few months after fire (mostly because of symbiotic N_2-fixation in the roots of the legumes; see Sect. 2.7.2), decreases later in maquis stages, and increases again slowly up to the stage of reconstituted oak woodlands. A complete recovery of soil biological activity is thus a process requiring more than 20 years (Prodon et al. 1987). This slow recovery rate is associated with the disappearance of soil decomposers, the sensitivity of which depends on several parameters including trophic level, seasonal activity and vertical distribution.

The dynamics of the most important decomposers, the Uropodid mites and Myriapoda, has been studied in detail by Athias-Binche (in Prodon et al. 1987). Some decomposer groups are very sensitive to fire while others are less so. After a typical woodland fire in the eastern Pyrénées, 89% of Glomerids and Iulids were destroyed, 82% of Uropodina, and 52% of Penicillata and Pauropoda (Athias-Binche in Prodon et al. 1987). The less sensitive groups include the humicolous decomposers (*Symphyla*) and such active burrowers as the predacious Geophilids. For example, the densities of *Symphyla*, which may be either decomposers or primary consumers, may recover rapidly some years after fire and become even more abundant than before. Presumably, this is a result of release from interspecific competition and the appearance of new ecological niches associated with post-fire growth of vigorous herbaceous vegetation. As mentioned, for groups such as the Glomerids, Uropodina, and Lithobids, a complete return to pre-fire conditions depends on the rate of litter build-up and may require more than 20 years (Athias-Binche in Prodon et al. 1987). For sites where the return rate of fires is less than the average 20 years, soil communities will never reach an equilibrium. Thus, besides impoverishing soil communities, fires produce important changes in the relative contribution of functional groups of microorganisms and other soil biota. Some true decomposers and detritivores, feeding more or less directly on dead plant material or fungi, recover more rapidly than the litter-dwelling biolytic feeders after fires. Hence, in the long run, a shortened fire-return interval will result in important changes in community structure and dynamics, just as the formerly described 20-year clearcutting practices must have done.

The diversity of small mammals is very low in contemporary Mediterranean systems, with no more than three common species (*Apodemus sylvaticus, Mus spretus* and *Crocidura russula*) in Mediterranean shrublands of southern France (Fons *in* Prodon et al. 1987). Unlike the shrew (*C. russula*), whose populations are completely displaced or destroyed by fire, a few individuals of the two other species are still present just after fire. Both species increase in densities some years after fire and then decline again. Fons (in Prodon et al. 1987) reported that the post-fire population of the Wood Mouse (*A. sylvaticus*) outnumbered that in a control site for about 3 years, presumably as a response to the increase in primary production of herbaceous plants just after fire. After 3 years, however, population density dropped again. The densities of the Wood Mouse after fire may become so high that demographic processes depend more on immigration from unburnt areas than from local reproduction rates. Although the endoparasitic load of the Wood Mouse included the same species in burnt and unburnt areas, the infection rate was lower in the latter, especially for the helminths requiring an intermediate host to complete their life cycle. Thus, fire has an impact on the parasitic load of these mammals, presumably through the destruction of most of the eggs and larvae of the intermediate hosts (Fons *in* Prodon et al. 1987). Such fire-induced processes could have important regulatory functions for micro-mammal populations in Mediterranean ecosystems where a mature sclerophyllous vegetation is unsuitable for high densities of these organisms.

Prodon et al. (1987) recorded 51 birds species in the habitat types described above. These ranged from species of open vegetation (partridges, larks, buntings, finches), to forest species (e.g. *Erithacus rubecula, Sylvia borin* and *Fringilla coelebs*), with species adapted to matorral (warblers) in intermediate stages. A strong turnover of species over the habitat gradient paralleled the post-fire dynamics of the vegetation. At the scale of the whole range of habitats within the mosaic, all 51 species found in the post-fire gradient were already present somewhere within the landscape.

Accordingly, the habitat gradient may be viewed as a closed system within which processes of local extinction and recolonization operate. Species of open vegetation colonize habitats immediately after fire and are then replaced by matorral species as the closed habitats recover. Subsequently, matorral species are replaced by the next group until the recovery process is completed. Alpha diversities are rather high in the first stages of the recovery process, but then decrease as the vegetation becomes more uniform, and finally increase again as the arboreal strata becomes higher and more complex.

At the scale of a landscape, and over a long period of time, the survival of all the species in such a mosaic of habitats involves the existence of a disturbance regime that is unpredictable in time and space in the short term, but predictable in the long term. This regime of disturbance guarantees the survival of all the species at the landscape scale on the grassland ---> forest habitat gradient. In forests of the western Palearctic, for example, the number of bird species which characterizes ecological successions is 50-55 species, irrespective of their geographical location (Blondel 1986). All the species at the level of a landscape constitute a *metacommunity*, the components of which are organized in overlapping local communities. Blondel (1986, 1987) coined the term "*metaclimax*" to define both the spatial scale required for maintaining a self-sustaining system and the disturbance regime that guarantees all the habitat patches required for the dispersal patterns of all the species.

Disturbance events that periodically "rejuvenate" ecosystems appear to be important for many plant species. For example, Pons and Quézel (1985) noticed the near absence of Mediterranean endemics within climax forests. Most Mediterranean endemics seem to have evolved in conditions of frequent disturbance so that they are adapted either to seral stages of successions, or to situations with strong limiting factors (rocky environments, nutrient-limited habitats). Disturbance may be a prerequisite for their survival. However, the dramatically increased rate of disturbances caused by people presumably alters the functioning of metacommunities by breaking down links connecting habitat patches. The present configuration of habitat patches within landscapes may not fit the taxon-specific attributes of colonization, which could result in loss of local diversity and alteration in ecosystem function. One theoretical consequence of such a breakdown in the network of interactions within a metaclimax is that the isolated and species-poor communities could be highly susceptible to biological invasions (Pons and Quézel 1985). Indeed disturbed areas seem especially prone to invasions (Elton 1958; Baker and Stebbins 1965). However, in the Mediterranean Basin this has generally not been the case (see Sect. 2.6.5).

2.6.2 Regeneration Traits

Boxwood (*Buxus sempervirens*) is one example of the many Mediterranean forest products exploited in former times. This species was extensively cut for use as bedding for horses (Fig. 2.4), and its wood was favoured for woodcarving and furniture. It is now uncommon to see full-grown, reproductively mature, individuals of this species. Concurrently, "Cade oil" was extracted from the wood of the Cade tree (*Juniperus oxycedrus*) for the treatment of the feet of horses and sheep against cuts and infection. The species was therefore selectively cut over large areas for this purpose under a variety of exploitation systems. However, unlike the deciduous oaks and *B. sempervirens, J. oxycedrus* coppices to some extent after cutting and resumes seed production after only a few years. Thanks to its bird- and livestock-dispersed seeds, it rapidly reinvades disturbed sites. Its distribution and population genetics were very possibly altered by centuries of selective cutting, and very few large specimens remain in southern France; nevertheless its role in ecosystem processes has probably been little affected.

2.6.3 Changing Tree Cover: Oaks, Pines and Soil Fertility

Major changes in the Mediterranean vegetation as a result of human activities include (1) the replacement of forests by shrublands with concomitant changes in nutrient cycles etc.; (2) the spread of evergreen plants, mainly trees, at the expense of deciduous trees; and (3) complex changes in plant and animal communities, with, among other things, increased α- and ß-diversities in moderately perturbed habitats and a dramatic decrease in highly degraded ones. For example, during the last 2-3 millennia, *Quercus ilex* and *Q. coccifera* have replaced *Q. pubescens* over large areas in southern France (Vernet 1972; Beug 1977; Triat 1979), *Q. ilex, Q. rotundifolia* and *Q. suber* have replaced *Q. canariensis* and *Q. pyrenaica* in Morocco (Reille 1977, Reille et al. 1980), and *Q. ilex* and *Pinus nigra* have replaced *Fagus sylvatica, Q. pubescens* and *Q. petraea* on Corsica (Reille 1975). In Dalmatia, the clearing of deciduous forests resulted in an increase of *Carpinus orientalis* and *Ostrya* and, subsequently, of shrublands dominated by species of *Juniperus, Phillyrea* and *Pistacia* (Beug 1975). A further stage in the degradation of the vegetation is the replacement of tall shrublands by dwarf shrubs, xerophytes and herbaceous vegetation which now cover large areas, especially in North Africa and the Middle East (Le Houérou 1968; Naveh and Dan 1973). Given the ability of many shrubs such as *Quercus coccifera* and its close relatives to resprout vigorously and dominate large areas after fire, these shrublands can be considered as a fire disclimax (Prodon et al. 1987) or "alternative steady state" which is exceptionally resilient in the face of repeated and severe perturbation. In the more arid, and more intensively exploited, Afro-Asiatic Mediterranean areas one last step in ecosystem degradation is marked by what is often presumed to be an irreversible desertification of steppe formations, yielding, for example, the so-called alfa steppe of *Stipa tenacissima* in North Africa (Floret and Pontanier 1982). Essentially, anthropogenic factors have so perturbed the environment that traditional phytosociology and life zone studies produce unreliable results - there is too much intermixing among natural communities. Given this obstacle, Romane (1987) and Floret et al. (1989a,b) have studied transformations of life-form spectra along ecological gradients. Even at that level, few definitive trends were revealed apart from the strong discriminating influence of calcareous vs. acidic substrata.

As a result of large-scale transformation of forests and selective cutting of certain useful woody plants, structural attributes of ecosystems, such as the proportion and distribution of species with drought-adapted growth forms and physiology, have changed dramatically in recent centuries. Many important functional ecosystem attributes, especially their capacity to capture, store, and transfer energy, nutrients, and water (Andersen et al. 1991) must have changed as well. This would be particularly so where the range of nutrient uptake and utilization mechanisms is reduced (Lamont 1982). Morphological and functional adaptations associated with sclerophyl-ly are evolutionary responses to prolonged conditions of water stress (Oppenheimer 1960) and nutrient-poor conditions (Schulze 1982). As a water-saving strategy which enables plants to control transpiration (Mooney and Dunn 1970; Mooney 1983), sclerophylly is an advantage to MTE plants, especially during summer drought (Walter 1983). The balance between photosynthetic gain and the maintenance cost of non-synthesizing tissues is achieved in nutrient- and water-poor supplies by reducing the allocation of photosynthates for leaf growth. This provides a selective advantage for the long-term persistence of leaf biomass and reduced photosynthetic rates (Schulze et al. 1980). Therefore, the substitution of growth forms should be associated with an overall lower primary produc-tivity and a greater resistance of ecosystems to environmental stresses because sclerophyllous growth forms can be presumed more resilient than deciduous ones (De Lillis 1991). Mooney (1977b) reported that net primary production in the evergreen *Q. ilex* forest and in mediterranean shrublands of France is roughly half that found in temperate deciduous forests.

Drastic changes in forest cover and even fire cycles have surprisingly little clear impact on understorey vegetation, nutrient cycling or litter quantity (Billes et al. 1975; Rapp and Lossaint 1981; Leonardi and Rapp 1990). This, along with the ambiguous results for life-form spectra reported above, suggests that various alternative steady states have been established in these long-perturbed Mediterranean ecosystems, incorporating the various different disturbances periodically imposed upon them by people. This may be changing at present however, as suggested in Sect. 2.5.2 in connection with the declining fungi populations throughout Europe. Bottner et al. (1994) showed that C% and C:N ratios decline dramatically from evergreen forests to grasslands, steppes and cultivated areas regardless of the prevailing climate. This suggests declining fertility and decreasing rates of decomposition and mineral turnover in soils and humus.

2.6.4 Aromatic Compounds and Insect Pollinators

The role in community organization and functioning of intraspecific genetic variation in various adaptive traits (type and quantity of volatile compounds, sex ratios within populations etc.) (see Sect. 2.4.3) can perhaps best be approached by considering the coevolution of plants and pollinators. The close evolutionary associations between hymenopteran wasps and many Mediterranean species of ground orchids (*Ophrys*) is a case in point. These orchids attract male pollinators by mimicking sex pheromones of female wasps (Bergström 1978). The chemicals involved are volatile oils. As we discussed previously, such volatile compounds are an important component of Mediterranean plant communities (Sect. 2.4.3). The diversity of these volatile oils is undoubtedly an important mechanism involved in the adaptive radiation of these orchids and aromatic plants in relation to the diversity of pollinator fauna (Bergström 1978). Such within-species genetic variation in the plants was presumably a prerequisite to achieving precisely interlocking interactions with the pollinator. Consequently, the various ecosystem processes they affect depend at least in part on maintenance of this variation.

2.6.5 Extinctions and Invasions

2.6.5.1 Plants

The large proportion of endemic plant species, especially on large peninsulas and islands, raises many problems of conservation (Gomez-Campo 1985). In the western Mediterranean region alone, 1529 species are considered to be rare or threatened according to IUCN criteria, while 25 are now extinct (Leon et al. 1985; Ramade 1990). Since many more species will undoubtedly vanish in the decades to come, there is an urgent need to investigate to what extent rare species may be important in ecosystem processes and services, and whether functional redundancy occurs among apparently similar species (Pate and Hopper 1993).

One consequence of the impoverishment of the vegetation is a potentially higher susceptibility to invasion (Gomez-Campo 1985). Large-scale disturbances caused by humans in the various MTEs have generally favoured invasion by weeds and alien species that occupy potential "regeneration niches" and therefore block (or modify) the successional trajectories of native species and communities (Quézel 1985). Such a situation is largely responsible for the expansion of secondary steppe formations, and may contribute to desertification (Le Houérou 1968). However, the total number of such invasive alien plants in our study area is approximately 250, or 1% of the flora, much less than in the other MTEs (Groves and Burdon 1986; Macdonald et al. 1986; Mooney et al. 1986; di Castri et al. 1990). Most plant invaders in the Mediterranean Basin are herbaceous (Quézel 1985; Wacquant 1990). They, and the few common woody ones [e.g. *Fraxinus ornus* (Thébaud and Debussche 1991), *Broussonetia papyifera* and the nitrogen-fixing *Robinia pseudo-acacia* (J. Aronson, pers. observ.)] seem, for the time being, excluded from most natural habitats except riverbanks (Thébaud and Debussche 1991). They occur mainly in ditches, along roadsides and in other disturbed sites. In southern France, only two alien woody species invade certain abandoned fields and state forests: the Tree of Heaven, *Ailanthus elegantissima*, and the wattle, *Acacia dealbata*. In summary, the Mediterranean Basin has "given" far more invasive species to other regions than it has "received".

2.6.5.2 Animals

The combination of habitat changes and direct persecution had many consequences for Mediterranean vertebrates. Several large mammals became extinct (e.g. *Loxodonta africana, Equus asinus, Gazella rufina, Oryx dammah, Alsephalus busephalus* and *Panthera leo*) and others are on the verge of extinction (*Ursus arctos, Panthera pardus, Monachus monachus*) as a direct consequence of hunting (Cheylan 1991). Very few species of birds have become extinct as a result of human activities, but many large species formerly widespread in the Mediterranean Basin are now threatened generally and confined to small, scattered areas. For both groups, human activities have resulted in large changes in local diversity and have caused many changes in species composition, population density and distribution. For birds and reptiles, the main consequence has not been so much a decrease in overall species richness as a tremendous advantage for species adapted to arid and shrubby habitats, which have benefitted from large-scale deforestation and habitat degradation. Several ecosystem functions and services have changed as a result of changes in community structure. For instance, large-hoofed mammals presumably played a key role in maintaining open areas at a fine-grain scale, and large predators contributed to the density, dispersal patterns, and demography of their prey. On the other hand, recreational hunting has become so popular in most Euro-Mediterranean countries that many forested areas currently yield a higher return from game species (e.g. *Sus scrofa, Cervus elaphus, Capreolus capreolus*) than from any forest product including timber. In southern France, as in parts of Italy and Spain, much of the land formerly devoted to sheep-raising is now leased to hunting societies willing to pay for exclusive hunting rights on large tracts of garrigue (Prudhomme 1986). Many large mammals, supplied with supplemental feed in the wild, or raised in large enclosures and then released before hunting, may raise new problems in forest ecosystems. Besides the loss of genetic diversity, the behaviour of these animals towards humans and other animals may change, as may their feeding habits. Some species, especially the wild boar (*Sus scrofa*) may adopt a keystone role in certain forest areas because of its physical and biological impact on ecosystems and economic importance to hunters. Moreover, they undoubtedly exert a heavy pressure on bulbs, earthworms, plants with succulent roots and certain mushrooms.

Terrestrial non-volant mammals of Mediterranean islands provide a striking example of the influence of humans in changing species diversities and the need for a careful analysis of historical events to interpret biogeographical patterns (Blondel and Vigne 1993). Until the late Pleistocene-early Holocene, Mediterranean islands were inhabited by a few, largely endemic species including pygmy elephants, hippopotamus and deer. These assemblages became extinct and were replaced by introduced species. When people first settled large Mediterranean islands more than 8000 ycars ago, thc combination of hunting, habitat changes and competition from intentionally and inadvertently introduced species pushed all endemic taxa to extinction (Vigne 1992). The massive destruction by man of the "dwarf megamammals" on Mediterranean islands (more than 240 000 bones of the pig-sized hippos have been found in one site on Cyprus) is the best support for the overkill theory in the Mediterranean Basin (Diamond 1992; see also Sect. 2.3.2). The present non-volant mammalian fauna of Mediterranean islands includes three major components, all of them introduced: (1) domestic species (sheep, goat, pig, cattle, dog, horse, donkey and cat); (2) game species such as the red deer, the hare, the rabbit and, presumably, the red fox (Vigne 1988); (3) small mammals (*Crocidura, Suncus, Apodemus, Mus, Rattus, Eliomys* and *Glis*) which were presumably introduced unintentionally as stowaways on ships. Such processes resulted in a three- to fivefold increase in local mammal diversity, but a dramatic overall decrease in genetic diversity since all the endemic species, with the exception of two shrews (Reumer and Payne 1986; Vogel et al. 1990), became extinct in the period following human settlement. A similar process occurred for reptiles and amphibians on other Mediterranean islands as well. For example, seven of the 18 species in Corsica and as many as 13 of the 16 species in the Balearics were introduced (Delaugerre and Cheylan 1992). Concomitantly, many species of mammals entered the Mediterranean Basin as a result of intentional introductions. Examples are the genet (*Genetta genetta*), a mongoose (*Herpestes ichneumon*), rabbit (*Oryctolagus cuniculus*) and deer (*Cervus elaphus*) (Cheylan 1991). The mouflon (*Ovis ammon orientalis*) was first domesticated in the Near East about 8500 B.P., and the resulting domestic sheep (*O. aries*) was rapidly introduced throughout the Mediterranean Basin (Poplin 1979). In some places, especially on large islands, sheep escaped human control and gave rise to feral populations. Two commensal Murid species, the mouse (*Mus musculus*) and the black rat (*Rattus rattus*), invaded human settlements from Asia to the Near East, and thereafter the entire Mediterranean Basin at the beginning of the Holocene (Auffray et al. 1990). In addition, several species (*Myocastor coypus, Ondatra zibethicus* and *Sylvilagus floridanus*) were introduced from the Americas (Cheylan 1991).

Unlike mammals (Cheylan 1991) and plants (di Castri et al. 1990), very few bird species invaded the Mediterranean region from other continents, and none of them succeeded in establishing large populations (Blondel 1991). There are fewer bird invaders in the Palearctic Mediterranean region than in the MTEs of California (Mooney and Drake 1986; Vuilleumier 1991) and, especially, Australia (Groves and Burdon 1986) (no alien birds have invaded natural habitats in South African MTEs; Richardson et al. 1992). An example worth mentioning is that of the Mediterranean partridges (*Alectoris* spp.), which were successfully introduced as game species from the mainland onto Mediterranean islands (Blondel 1988). Historical records show that *A. rufa* occurred on Corsica during the 6th century (Vigne and Marinval-Vigne 1983), and there is evidence that at least two or three, or possibly all four, species of the Mediterranean region were intentionally introduced several times to each of the larger islands. However, there is no paleontological record of partridge occurring on any Mediterranean island before human invasion (Mourer-Chauviré 1975; Vilette 1983). Only one species of *Alectoris* remains on each island: *A. rufa* on Corsica, *A. barbara* on Sardinia, *A. graeca* on Sicily and *A. chukar* on Crete and Cyprus (Blondel 1988).

2.7 Functional Groups and Redundancy

Despite many difficulties inherent in such an approach, any group of taxa bearing a certain set of common structural and/or process features in a given ecosystem may usefully be considered as belonging to a "functional group" (Körner 1993). Lumping species that share similar features into groups such as guilds for animals (Root 1967; Cody 1968, 1974; Simberloff and Dayan 1991), or life-forms for plants (Raunkiaer 1934) is a long-standing practice for categorizing organisms within communities in terms of their systematic affinities, morphology and/or presumed ecosystem function. However, contemporary concern with possible ecosystem "accidents" or "breakdowns" due to species losses has led to a greatly heightened interest in testing the practical reality of functional groups. It is to be hoped that experimental studies on this subject will soon be undertaken in a variety of ecosystems. In the Mediterranean Basin ecosystems, two putative functional groups among plants that lend themselves to such study are the nitrogen-fixing legumes and the grasses.

2.7.1 Legumes, Grasses and Possible Interactions among Them

2.7.1.1 Legume Growth-Form Spectra and Relative Floristic Contribution

Among higher plants, nitrogen-fixing legumes are one of the most obvious candidates for designation as a functional group. However, the more than 18000 legume taxa are a highly diverse, cosmopolitan group, encompassing more than 600 genera in three highly contrasted subfamilies, occurring in almost every major plant growth form. Not all legumes are nitrogen fixers. Of the 3400 species examined thus far (about 20% of the Leguminosae), nodulation and nitrogen fixation are very unevenly distributed among higher taxa in the three subfamilies (Allen and Allen 1981; De Faria et al. 1989), with herbaceous Papilionoideae being most prominent among the nitrogen fixers. Nevertheless, in many semi-arid or arid biomes, large (generally phreatophytic) nitrogen-fixing legumes such as *Acacia, Burkea, Colophospermum, Lysiloma* and *Prosopis* spp. are not only prominent in landscapes, but also play central roles in nutrient cycling and community dynamics (e.g. Belsky et al. 1989; Jarrell and Virginia 1990). It is striking, therefore, that around the Mediterranean Basin and indeed in all the MTEs, with the notable exception of the Australian kwongan (Lamont 1982, 1984), there are few native tall shrub or tree legumes, and that those that do occur are usually not nitrogen-fixers. The only two native legume trees in the Mediterranean Basin, the carob tree (*Ceratonia siliqua*) and the Judas tree (*Cercis silicastrum*) are both non-nitrogen-fixers. *Faidherbia albida* (formerly *Acacia albida*) occurs sporadically along the coast in Israel, but is either a paleorelict (Zohary 1973) or a recent arrival there (Shmida and Aronson 1986). Similarly, *Acacia tortilis* subsp. *raddiana* sporadically enters certain MTEs of northern Africa, but is essentially of Saharan distribution (Quézel 1978).

Although large, conspicuous taxa of nitrogen-fixing legumes are absent, smaller growth forms are very well represented in the shrublands and other open formations around the Mediterranean. Apart from the numerous genera of papilionoid annuals (e.g. *Astragalus, Lathyrus, Medicago, Melilotus, Ononis, Pisum, Trifolium* and *Vicia*) abandoned fields in early stages of succession are colonized by (among other taxa) papilionoid or caesalpinoid shrubs and subshrubs of the genera *Astragalus, Bonjeania, Calycotome, Coronilla, Crotolaria, Cytisus, Dorycnium, Genista* and *Psoralea*. One or more of these annuals or perennials is often quite abundant locally, and they probably play important parts in the nutrient cycling of most terrestrial ecosystems in the Mediterranean region. Functional roles of these species have yet to be tested and quantified in field situations. Since both annual and inter-annual nutrient cycling patterns change with succession (Odum 1969), it is not surprising that prevailing strategies and mechanisms for nitrogen acquisition, utilization and transfer also shift with succession (Gutschick 1981; Lamont 1982). The relative frequency, abundance, growth-form spectra and dominant leaf phenotypes of nitrogen-fixing legumes (and other taxa) must also vary with succession.

Rundel (1989) drew attention to recurrent shifts in relative dominance of certain growth forms and leaf phenotypes among legumes occurring along elevational gradients in both warm desert and chaparral ecosystems of southern California. With this in mind, we sought to evaluate the relative frequency of legume species along the macro-gradient of the Mediterranean Basin, and to compare their growth form and leaf phenotype spectra. We analyzed four regional floras of comparable size and heterogeneity - two from more humid portions of the Mediterranean Basin, and two from the more arid portions. From wettest to driest, these are the Hérault administrative region of southern France, Cataluña (= Catalonia) in northeastern Spain, Tunisia and Israel. Approximately 2000 to 2500 species occur in each of these floras (Loret and Barrandon 1876; Cuénod 1954; Zohary and Feinbrun 1966-1986; Pottier-Alapetite 1979-1981; De Bolos et al. 1990; Salabert and Gastesoleil 1991; Le Floc'h, unpubl. data). Only the native taxa in the MTEs of these regions were included in our analysis. Hyper-arid and exclusively alpine-Mediterranean taxa were also excluded. The relative frequency of legumes in the regional floras is strongly influenced by overall aridity. Legumes contribute only 9.1 and 9.3% of the MTE floras of Hérault and Cataluña, respectively (184 out of 2031 and 230 out of 2464), whereas they account for 12.5 and 14.9% in those of Tunisia and Israel. The proportion of annual legume species increases dramatically toward the warmer end of the gradient, providing 142/807 = 17.6% and 163/210 = 22.8% of the legumes of Tunisia and Israel respectively, compared with only 13.7% and 14.4% of those in Hérault and Cataluña, respectively. Similar trends are apparent for other families across the gradient, but these are not as marked. It is also noteworthy that all four of the regional floras include considerably higher proportions of legume taxa than the South African fynbos (7.3%) or the California chaparral (6-7%), both of which are close to the estimated world average of 7% (Rundel 1989). They are more similar in this regard to the Sonoran Desert of North America, which has about 11.6%, or central Australia, with about 13% (Rundel 1989). Herbaceous perennials when combined with the annuals represent between 70 and 91% of all legume species in the four regional floras studied, with trends similar to those described above occurring along the regional aridity gradient (Fig. 2.6). Accordingly, they contribute about 22% to their respective floras and can be presumed to figure prominently in nitrogen cycling (and community interactions?) of the ecosystems in which they occur. By contrast, the woody legumes represent only 1-3% of their overall floras (Fig. 2.6) and most of the largest ones are not nitrogen-fixers. Nevertheless, important distinctions can be made among life-forms in the woody legumes that may have significance in nutrient and water fluxes at the community level. As opposed to trees and tall shrubs, shrub and subshrub legumes contribute from 12.9-16.1% to the total number of species in these particular life-forms in the four floras. It is striking that Cataluña has about three times as many chamaephyte legume species as Tunisia or Hérault, and more than six times as many as Israel. A high proportion of these species are endemics, on a scale comparable with the shrub flora of the Canary Islands (Shmida and Werger 1992). This contrasts sharply with the situation of trees and tall shrubs throughout the Mediterranean region generally.

Just as important differences were found above in the relative importance of the various types of woody life-forms among the legumes, the relative contribution of five different leaf phenotypes also varied for the woody legumes of the four regional floras, but did not show any clear trends related to climate. Thus, photosynthetically active, green-stemmed species account for as little as 22.2% or as much as 48.6% of the total number of woody legumes of Israel and Cataluña, respectively. Evergreen species among the woody legumes in Cataluña and Tunisia account for 35.7 and 17.2%, respectively, but 38.9% in Israel and 51.9% in Hérault. In Tunisia, deciduous species predominate among woody legumes, and in Israel they contribute about as many as do the evergreen species. To what extent are these differences related to climate, soils and/or

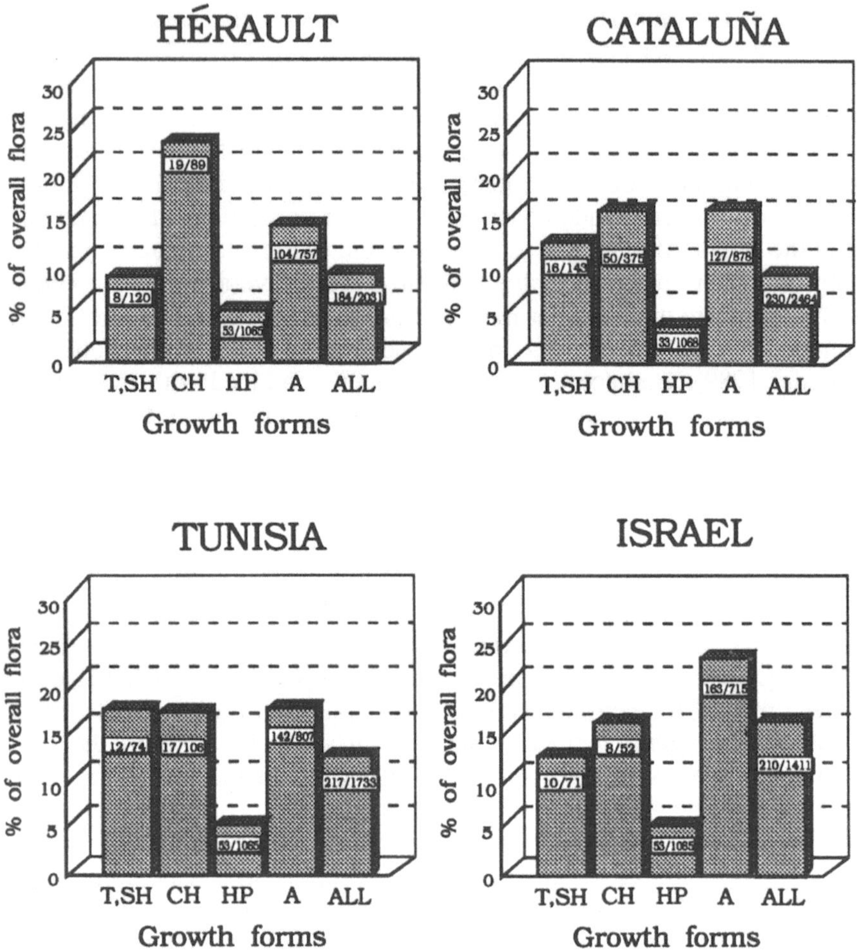

Fig. 2.6. Growth form spectra of legumes in four regional floras of the Mediterranean Basin. *Figures inside columns* indicate the untransformed species numbers of legumes/overall flora for each growth form. **T,SH** trees and tall shrubs; **CH** chamaephytes (= subshrubs); **HP** herbaceous perennials; **A** annuals

historical differences? It makes sense that in more arid, seasonally contrasted regions to the east and south of the Mediterranean Basin, drought-adapted, deciduous species would predominate; we hesitate to go further at this point.

2.7.1.2 Grasses: Growth-Form Spectra and Relative Floristic Contribution

Grasses are also abundant in early to mid-stages of succession, and appear to comprise more than one functional group. For example, Joffre (1990) and Garnier (1992) showed recurrent differences in the nutrient uptake and growth patterns of annual vs. perennial grasses in selected Mediterranean ecosystems. We know of no studies that compared the relative contribution of annuals and perennials across gradients or in different stages of succession. As a first step in this direction, we compared the growth-form spectra and relative contribution of grasses for the same four regional Mediterranean floras as were used for the legumes. Species are fairly evenly divided between herbaceous perennials and annuals (including biennials and facultative annuals), except in Israel where annuals dominate (Fig. 2.7). Note also that the relative contribution of all grasses to the overall floras is rather constant (9-11%), compared with the more widely varying contribution of legumes (9-15%). The contribution of grasses to the total annual flora is unusually high for a single family (87/757=11.5% for Hérault; 114/878=13% for Cataluna; 113/807=14% for Tunisia; and 106/715=14.8% for Israel). The species/genus ratios of grasses and legumes of the four regional floras are quite high in relation to the overall floras (Table 2.2). These two families, therefore, contribute more than average to overall diversity of these floras at the specific level. Results for these two families are in the range of species/genus ratios from the floristically very rich Cape Region of South Africa (Bond and Goldblatt 1984). More detailed analysis of these relationships are now underway at the level of the Mediterranean Basin, North Africa and the Saharan region.

2.7.1.3 Interactions Among Mediterranean Legumes and Grasses

The functional importance of the results presented above remains to be elucidated. Competition for resources exists among plants of different growth forms and functional groups such as annual and perennial grasses and legumes, particularly in situations of mineral or water stress (e.g. Wacquant et al. 1989). However, some herbaceous legumes also apparently form mutualistic associations with grasses in certain grassland ecosystems (Turkington et al. 1977). If such "neighbour relations" do exist, soil-borne microorganisms should play an important mediating role in such associations (Haystead et al. 1988; Thompson et al. 1990; Read 1993), which points to strong links between biodiversity and ecosystem functions related to co-occurring species or functional groups in these two families. Local conditions of soil fertility, soil type and microclimate will also influence

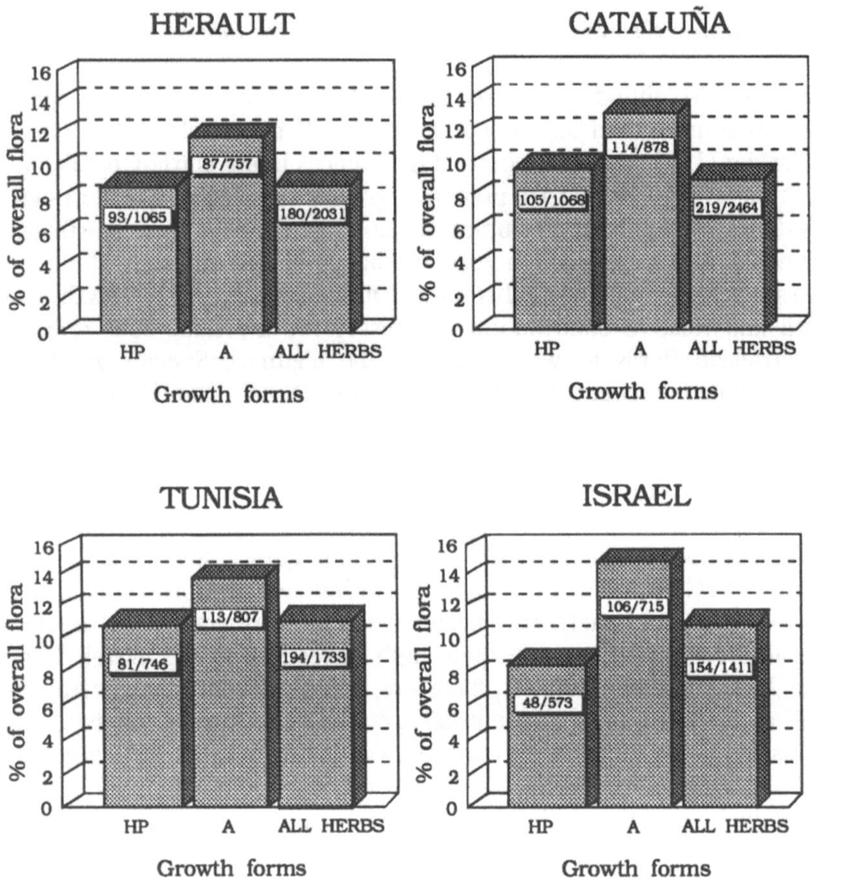

Fig. 2.7. Growth form spectra of grasses in four regional floras of the Mediterranean Basin. *Figures inside columns* indicate the untransformed species numbers of grasses/overall flora for each growth form. **HP** herbaceous perennials; **A** annuals

Table 2.2. Species to genus ratios for legumes, grasses and total floras in four regional floras of the Mediterranan Basin, as compared to figures for South Africa's Cape region

	Legumes	Grasses	Total
Hérault	6.3	3.5	3.2
Cataluña	5.7	2.5	2.7
Tunisia	4.9	4.6	2.6
Israel	5.3	6.2	2.6
Cape Floristic Region	16.9	3.0	9.0

the relative abundance and the functioning of a given taxon at a given time and place.

Table 2.3 lists some of the most common N_2-fixing legume species occurring along an altitudinal gradient in the Hérault region, southern France, together with the seral stage at which they are most common. We then indicate some possible grass associates for each legume. We suggest that both annual and perennial *Medicago* spp., *Psoralea bituminosa, Dorycnium hirsutum* and certain green-stemmed shrubs of the *Genista/Teline/Cytisus* group all appear to be candidates for the term keystone species, in certain sites and in early to middle stages of succession. In later stages of succession, legumes of any life-form are virtually absent from Mediterranean woodlands. Thus, the figures in Table 2.3 tend to support the contention of Read (1993, Fig. 8) that biological nitrogen fixation plays its most important role in intermediate stages of plant community succession, prior to the establishment of full tree cover, but after the initial colonization of disturbed sites.

2.7.2 Avian Frugivores and Bird-Dispersed Plants in Mediterranean Shrublands

The ecological and evolutionary significance of seed dispersal by frugivorous birds in ecosystem structure and function has been much discussed (Van der Pijl 1972), especially in the tropics (Snow 1971; McKey 1975; Howe and Smallwood 1982), but also in the Mediterranean region (Herrera 1982a,b, 1984; Debussche and Isenmann 1989, 1992). However, studies dealing with one or a few plant and bird species often overlook the larger meaning of plant-disperser interactions in the context of ecosystem function, partly because birds usually eat fruits of many species, and plants often depend on several species for dispersal of their seeds (Jordano 1987). In MTEs, as elsewhere, the functional meaning of the links between bird-dispersers and bird-dispersed plant assemblages must be approached at the community level (e.g. Herrera 1982a,b, 1984; Debussche and Isenmann 1989, 1992). To begin such an analysis, the functional attributes of species of both sets of partners of the system should exhibit characteristics allowing us to predict functional links between them. Assuming "diffuse coevolution" (Janzen 1980) between plants and their dispersers, plants that are dispersed by frugivores should produce a moderate number of large, high-quality fleshy fruits with bright colours, conspicuous location at the extremity of stems, high fat and protein contents, large seeds, and long and constant periods of availability (Herrera 1984; Wheelwright 1985). Concurrently, bird-dispersers are expected to be small or moderate-sized species living in the same habitats as the plants

Table 2.3. Most common N_2-fixing legume species along an altitudinal gradient in the Hérault administration region, southern France

	Growth form[a]	Pheno-type[b]	Seral stage[c]	Alt. zone[d]	Possible grass associates[e]
Astragalus spp.	A,HP	E	2	1-2	?
Bonjeania hirsuta	Ch	E	2	1-2	*Bromus erectus*; *Brachypodium phoenicoides*
Dorycnium pentaphyllum	Ch	E	2	1-3	?
Genista scorpius	Ch	GS	2	1-3	*Bromus erectus*; *Brachypodium phoenicoides*
Genista pilosa	Sh	GS	2	1-3	?
Hippocrepis comosa	A	E	2	1-3	?
Lotus corniculatus	HP	E	2-3	1-3	*Dactylis glomerata*
Medicago spp.	A, HP	E	2	1-3	*Brachypodium ramosum*; *Dactylis glomerata*
Psoralea bituminosa	HP	D	2-3	1-2	*Bromus erectus*
Spartium junceum	Sh	GS	2	1-2	?
Sarothamnus scoparius	Sh	GS	2	1-3	*Holcus mollis*; *Festuca* spp.
Trifolium spp.	A, HP	E	2	1-3	Many

[a] Sh = shrubs 1 to 3 m; Ch = chamaephytes (subshrubs) < 1 m; HP = herbaceous perennials, including geophytes and parasites; A = annuals, including biennials and facultative annual-perennials.
[b] E = evergreen and semi-evergreen; GS = photosynthetic stems (often aphyllous); D = drought- or winter-deciduous.
[c] 1- early; 2 - middle; 3 - advanced; 4 - late.
[d] 1 = 0 to 200 m; 2 = 200 to 600 m; 3 = > 600 m.
[e] Sources: Braun-Blanquet 1952; M. Debussche pers. comm., pers. observ.

they are supposed to disperse during the period of fruit availability. These two guilds have been shown to co-occur in Mediterranean shrublands during the ripening period of fruits, from October to January (Herrera 1982a,b, 1984; Debussche and Isenmann 1989, 1992). Fleshy fruit-bearing species represent a large proportion of woody plants in Mediterranean shrublands (76.5% of cover and 65.5% of woody species in a Spanish scrubland) and more than half of them rely entirely on birds for seed dispersal (Herrera 1989; Debussche and Isenmann 1992). In a Spanish lowland shrubland, the three most common species (*Pistacia lentiscus, Myrtus communis* and *Smilax aspera*), which make up 63% of the fruit-producing cover, have a relatively long fruiting period lasting 2.2 to 3.5 months (Herrera 1984). Although the pulp of most species has low protein (2-8%) and lipid (<5%) contents, several species have remarkably fatty fruits (up to 58%), especially *Pistacia lentiscus, P. terebinthus, Viburnum tinus* and *Olea europaea* (Herrera 1984). The highest levels of lipid "profitability" (dry matter yield x lipid content pulp) are found in winter-fruiting species (*Viburnum tinus, Olea europaea, Pistacia lentiscus*) that are thus the most important food plants for birds (Herrera 1982b). Most species of plants display their conspicuous black, red or blue fruits on thin vertical stems which make them readily available for small birds. Thus, all the requirements are met for offering a large food supply for potential frugivorous birds.

The assemblage of frugivorous birds species usually includes 12-16 species of small size (12-18 g body mass) amounting to 66-70% of the total winter bird communities (Blondel 1969; Herrera 1984; Debussche and Isenmann 1992). Most of the individuals (80-90%) are migrant or over-wintering birds that spend from late summer to early spring in the Mediterranean region (Blondel 1969; Moreau 1972). Among these, seven species (*Erithacus rubecula, Sylvia atricapilla, S. melanocephala, S. communis, S. borin, Turdus philomelos* and *T. merula*) are important frugivores (Herrera 1982b,1984; Debussche and Isenmann 1989, 1992), removing 89-100% of all fruits produced (Herrera 1984). Although several species concentrate on one species, *Pistacia lentiscus* (which is very rewarding in terms of lipids and proteins) most of them usually eat the fruits of several species over short time periods. This is presumably because the usually poor and imbalanced composition of fruit pulp (Herrera 1984) makes a mixed fruit diet necessary. Furthermore, the gape width of bird dispersers sets an upper size limit to the fruits they can ingest. Hence, the fraction of the fruit crop not removed by birds significantly increases with increasing fruit diameter (Herrera 1984).

A low number of bird species with differential abundances and frugivory patterns associated with a large number of plants makes the system very asymmetrical (Jordano 1987). Examples are *Lonicera implexa, Rubia peregrina* and *Smilax aspera*, all of which are strongly dependent on the blackcap (*Sylvia atricapilla*) for seed dispersal, but are not an important food resource for that bird (Herrera 1984). On the other hand, one plant species, *Pistacia lentiscus*, may be very important for all the bird species in an area, with the result that it has highly variable and, probably, highly extended seed shadows. To our knowledge, there is not a single case of bird-plant pairs in the Mediterranean Basin wherein the dependence is similarly important for both partners. Therefore, it is very unlikely that the local extinction of one partner would seriously affect the local survival of the other. Thus, many species of both partners of the system may be seen as ecological equivalents. This does not mean, however, that species loss would not have long-term harmful effects because species diversity is important for buffering fragmented multi-species systems against long term changes (Richardson and Cowling 1993). This evolutionary potential means greater security for the future. Nevertheless, it is striking that in Mediterranean habitats, there is not a single example of close bird-plant interactions such as, for instance, that of the Eurasian nutcracker (*Nucifraga caryocatactes*) which relies on the seeds of the Swiss Stone pine (*Pinus cembra*) for 95% of the food, with the pine relying, in turn, on the bird for the dispersal of its seeds in the Alps (Crocq 1990).

Some authors (e.g. Snow 1971) have interpreted the much later fruiting season in the Mediterranean region, relative to that in temperate regions (Stiles 1980; Sorensen 1981), as being a result of co-evolutionary processes between plants and bird-dispersers that are mostly winter visitors. Alternatively, the late-fruiting season of Mediterranean woody plants may be a result of physiological factors related to evergreenness and rainy, mild winters (Debussche and Isenmann 1992; Fuentes 1992). Winter fruiting, which is precluded in more northern habitats because of frost and the dominance of deciduous plants, may also be an adaptation in Mediterranean habitats to escape pests and insect predators (Herrera 1982b), since most potential predators in the region are more active during the warm months of the year. This does not mean, however, that the system does not result from co-evolutionary processes. More or less tight or diffuse co-evolutionary links between fruit plants and birds presumably evolved since the early Tertiary at a time when both plant and bird species were different. The present vegetation includes many taxa that evolved during the Tertiary whereas most extant bird species evolved during the Quaternary (Brodkorb 1971). Therefore, fruits of these plant species have evolved dispersal patterns in connection with bird assemblages other than present-day dispersers, so that present interactions are relatively recent. However, current co-evolution processes may occur to some extent. For example, the close agreement between fruit size and disperser gape width may result from bird selection against large fruits in plants they visit regularly. Many extinction and speciation events occurred in both groups without any breakdown of a mutualistic association. This must therefore be seen as an opportunistic reassortment, over time, of different sets of lineages, rather than as co-evolution occurring among extant species assemblages. Many of the lineages that initiated the process a long time ago might be the ancestors of lineages that occur today in tropical Africa because their ancestors left the Palearctic region during the Miocene-Pliocene climatic deterioration.

In summary, this system includes two inter-related functional groups, each of which includes several important species but no obvious keystone species. Many species are ecological equivalents so that, considering seed dispersal and the food supply for birds only, there is a large amount of redundancy in the processes involved (Howe and Smallwood 1982; Jordano 1987; Debussche and Isenmann 1989). However, some species within a functional group may be more important than others, at a given place and time, depending on the frequency, intensity and scale of environmental disturbances. For instance, cold spells may dramatically decrease population sizes of Mediterranean bird dispersers (e.g. the resident Sardinian warbler, *Sylvia melanocephala*) during several years, so that most of "their" function must be assumed by other members of the functional group. Therefore, in this case, the functional redundancy of most partners of the mutualistic system is probably important in overall ecosystem function because, when faced with environmental shifts or rare events, only redundancy can provide "insurance" for the maintenance of ecosystem function (Lawton and Brown 1993).

2.7.3 Species Equivalence and Density Compensation

In the situation described in Sect. 2.7.2, it is not apparent how to test the hypothesis of ecological redundancy without eliminating birds and/or plants on a sufficiently large scale (several hundreds of hectares or more) as to be meaningful for the birds. As this is technically difficult and ethically indefensible, we must look elsewhere for opportunities to study redundancy in the field. Theoretically, at least, the extent to which some species may be lost without any detectable loss of ecosystem function may be investigated in depauperate biota such as insular ones. MacArthur et al. (1972) emphasized that summed population densities on depauperate islands may be comparable to, or higher than, those in richer communities in comparable habitats on the nearby mainland. This is because resource sharing is achieved by more individuals belonging to a reduced number of species. Such mechanisms as density compensation, niche enlargement, and behaviourial shifts in depauperate island biota raise the problem of ecological equivalence among species belonging to the same guilds. For example, bird faunas on the island of Corsica are impoverished by 30% in comparison with bird faunas in areas of similar size on the nearby mainland. Such a reduction in species numbers on the island apparently did not result in significant changes in the number of functional groups within communities (Blondel et al. 1988). The number of guilds within an insular community of a mature *Q. ilex* forest was the same as that in a community of a similar forest (on the grounds of vegetation structure, Table 2.4) on the mainland but there were fewer species per guild. However, the total biomass of birds was similar in the two communities (Table 2.4). Species that occur on the island achieve very high population densities, presumably because they exploit resources that are shared by more species on the mainland. Changes in populations and communities on the island include: (1) similar or even higher summed population densities, (2) higher species-specific densities, (3) wider habitat-niche breadth, and (4) lower individual body size (Table 2.4). These shifts in the organization of island biota characterize the so-called insularity syndrome (Blondel 1986), but do not seem to induce substantial changes in ecosystem processes. A reduced number of species for achieving similar ecosystem functions may make them more vulnerable to loss of biodiversity. Therefore, island biota may provide natural experiments (natural species removals) that can be utilized to investigate to what extent major ecosystem processes and energetic pathways differ depending on which species are missing and which are present in depauperate (true) islands or in mainland "habitat-islands".

2.7.4 Earthworms: Keystone Species at Risk

Since Darwin (1859), the contribution of earthworms to physical, chemical and biological soil processes has received much attention (Bouché 1972; Edwards and Lofty 1977; Lee 1985; Lavelle 1988; Granval and Muys 1992). Earthworms are assigned to four main groups on the basis of their location in the soil (Bouché 1981; Lavelle 1981). These may be tentatively considered as functional groups: (1) *epigeic* (acid-tolerant species living in the litter they feed upon), (2) *anecic* (burrowers which feed upon the litter and are sensitive to acidity), (3) *endogeic* (deep-dwelling soil species which actively bioturbate and mix soil particles through continuous intestinal transit), and (4) *epianecic* (deep burrowers active in rich forest soils, being more acid tolerant than real anecics). This categorization is largely according to habitat use (on a vertical gradient), food resources (litter vs. soil) and sensitivity to acidity. However, each group presumably contributes differently to such major ecological processes as organic matter recycling,

Table 2.4. Habitat characters and the structure of bird communities in two matched *Quercus ilex* forest habitats, one on an island (Corsica), and one on the nearby mainland. The complexity index of the eight-layered vegetation was calculated as the sum of the horizontal and vertical foliage diversities using Cody's (1975) method (see Blondel 1981)

	Mainland	Island
Height of vegetation (m)	20	25
Complexity index of vegetation	577	571
No. of bird species	24	17
Summed bird density (pairs per 10 ha)	61.3	63.9
Average species-specific density	2.67	3.55
Habitat-niche breadth	1.98	3.81
Mean body weight of the species (g)	29.0	19.8
Summed bird biomass (g)	2 029	1 967

water drainage and bioturbation. Spatial segregation on a vertical soil gradient and geographical variation of these groups in relation to soil type may be the most important cues to assess their relative contribution in soil biology.

Earthworms may constitute a dominant part of the life in soils with a total biomass amounting up to 1000 kg ha^{-1} in rich soils of temperate Europe (up to 4000 kg ha^{-1} in permanent grazed pastures) (Lavelle 1988). The soil processed by them may amount to 230 tons ha^{-1} a^{-1}. The main aspects of earthworm contribution to soil activity are (Granval and Muys 1992): (1) *physical effects* (soil porosity, aeration, water drainage, bioturbation), (2) *biogeochemical effects* (litter decomposition, organo-mineral migration, P and N recycling, balance in pH and C/N equilibrium), (3) *forest diversity* (higher diversity of trees and herbaceous plants through better germination and growth conditions), and (4) source of proteins for many earthworm predators (up to 200 species of birds and mammals, that is around one third of the total richness of these two groups at the scale of Europe; Granval 1988). The stability of earthworm communities depends on historical (Bouché 1983) and anthropogenic (Owen and Galbraith 1989) factors. The much higher species diversity in the Mediterranean region (150 species in southern France) than in northern Europe (30 species) has been attributed to the Pleistocene glaciations which destroyed most communities north of the Mediterranean region (Bouché 1972). In recent times, the increasing impoverishment of earthworm communities has been attributed to nutrient export by logging, litter removal, fire, acid rains, heavy metal pollution and conifer plantations

(Bonneau et al. 1979; Granval and Muys 1992). The combination of soil acidification and earthworm decrease results in a decrease of biological activity of soils, formation of raw humus, soil compaction, reduced natural regeneration of trees, fragilization of trees through superficial rooting, and acceleration of erosion (Hildebrand 1987). Current levels of heavy metal soil contamination may have important consequences for major physical and biogeochemical ecosystem processes. Large-sized species of the genus *Sclerotheca* which are active ground feeders and play an important role in water filtration are much more sensitive to heavy metal pollution (Cu, Zn, Pb, Mn, Cd, Fe) than any other co-occurring group of earthworms (Bouché et al. unpubl. data). All members of the genus have been exterminated from many polluted areas of southern France. Given their ecological function in ecosystems, especially with respect to physical properties of soils, Al Addan (1990) and Bouché et al. (unpubl. data) suggested that their extinction contributed to an increase in soil erosion and in the occurrence of severe flooding that periodically plagues the Mediterranean region, especially in agricultural areas, such as vineyards, where copper is widely used as a fungicide.

Some restoration experiments have demonstrated an increased primary productivity after earthworm introduction in depauperate soils (Van Rhee and Nathans 1961), and an increase in forest productivity by more than 50% after enrichment of soils with fertilizers in the presence of anecic earthworms (Toutain et al. 1988). Moreover, liming and/or fertilizing degraded forest soils is more efficient in the presence of endogeic and anecic earthworms (Lang and Beese 1985; Muys 1989; Granval and Muys 1992). In most degraded forest soils, earthworm reintroduction is the only way for promoting bioturbation and the regeneration of soil activity (Brun et al. 1987), as shown by experiments carried out in temperate Europe and in New Zealand (Huhta 1979; Stockdill 1982).

2.7.5 The Spatial Dynamics of Predatory Ants

The contribution of insects in ecosystem function is of overwhelming importance, as exemplified by such essential functions as pollination (see Sect. 2.6.4). Some groups may be important as predators and parasites in controlling demographic outbreaks of herbivorous and folivorous insects that severely defoliate trees. One example is provided by predatory ants. Ants contribute a major part of the overall insect biomass in most terrestrial ecosystems and may include more than half of the individual insects (Wilson 1992). Therefore, their role in ecosystems at both scales of species-specific interactions and functional groups must be important through many shared functions: seed dispersal (myrmecochory), soil bioturbation, and predation, to say nothing of some extraordinary species-specific interactions with butterflies (*Maculinea* spp.), the caterpillars of which are transported by ants in their nests, where the latter raise them (Thomas et al. 1989). In a typical Mediterranean landscape including all the vegetation belts along an altitudinal gradient (420 - 1880 m) at Mont-Ventoux, southern France, ant species diversity amounted to 64 species (Du Merle 1978). A careful analysis in 58 study sites of ant species richness and species-specific distributional patterns revealed a significant discrimination of community structure in relation to vegetation belts and vegetation units. The richest communities occurred in thermo-, meso-, and supra-mediterranean belts and then declined with increasing altitude. Several species are strongly associated with particular plant species, for instance *Aphaenogaster gibbosa* with *Thymus*, and *Formica gagates* with the Downy oak (*Quercus pubescens*).

An experiment designed to assess one important function, i.e. the predation pressure of ants on eggs of other insects, has shown that many species move seasonally across different habitat patches, resulting in different ant assemblages through time (Benois et al. 1978; Du Merle et al. 1978). Predation was studied in a small mosaic of three habitat patches, a clearing (942 m^2), a forest edge (478 m^2), and a forest patch (952 m^2). At the scale of the three habitats combined, 14 species were recorded (13 in the clearing, 10 in the forest edge and 6 in the forest) and 175 ant nests were found. More than half of these species (9) are insect egg-predators. The egg-eating activity by ants was measured using traps supplied with eggs of the Mediterranean flour moth (*Anagasta kuehniella*). A very high proportion of the moth's eggs were eaten by ants in the three habitats (up to 76%) and important between-habitat exchanges of ants occurred (Du Merle et al. 1978). The relatively low number of predatory ants resident in the forest was compensated by high seasonal migration rates from the two other habitats, especially the clearings. This resulted in marked fluctuations in richness, species densities, community structure and predation pressures by ants according to season. In particular, there was an intense exploitation, especially in mid-summer, of the forest habitat by ant colonies invading this habitat from colonies established in the two nearby habitats. Species involved in this dispersal were *Leptothorax unifasciatus*, *Pheidole pallidula*, and *Myrmica specioides*. Such between-habitat migrations of predatory ants suggest that clearings contribute to controlling populations of insects including foliage-eating insects harmful for tree foliage. Thus, the patchy geographical configuration of the landscapes plays an important role in the dynamics of insect populations at the scale of the system "clearing-ecotone-forest". This is because many forest insects and their predators carry out parts of their life cycle in habitats other than forests. This mosaic structure allows some highly mobile species (e.g. *Leptothorax unifasciatus*) to find, at any time of the year, optimal environmental conditions for breeding. Similarly, such predator species that exploit forest resources from neighbouring habitats may reach higher population sizes, and hence, achieve higher controlling effects on insect populations at the scale of a combination of several habitats than at the scale of only one habitat. This example illustrates the importance of neighbouring effects in some important ecosystem processes at the scale of a mosaic of habitat patches.

2.8 Ecological and Evolutionary Consequences of Patchiness

Scaling Effects, Harlequin Effects and Fragmentation

The substitution of deciduous trees by sclerophyllous evergreen ones and the increase of habitat patchiness over time may affect genetic diversity at the population level if selection regimes become different in these habitat mosaics from those which operated before these changes. It is noteworthy in this context that Gouyon et al. (1986) found higher chemotypic diversity in *Thymus vulgaris* populations in disturbed areas than in relatively more stable environments. This diversity is maintained by greater frequencies of males in such young populations which in turn lead to higher outcrossing rates. Different selection regimes may thus operate depending on the spatial configuration of landscapes in relation to the scale on which organisms disperse. Directional or disruptive selection may lead to new sets of fitness-related life history traits, which could lead to new species assemblages and altered ecosystem processes.

One example has been provided by investigating the relationships between the tree foliage, phyllophagous insects, mainly caterpillars, and the feeding and breeding biology of an insectivorous bird species, the blue tit (*Parus caeruleus*), which mainly feeds upon caterpillars. These relationships have been studied in three habitats: two are in the same landscape on the mainland, one dominated by summergreen trees (*Q. pubescens*), one dominated by evergreen trees (*Q. ilex*), and the third is an evergreen habitat (*Q. ilex*) isolated on the island of Corsica (Blondel et al. 1992ab, Dias et al. 1994). Evergreen oak populations differ from deciduous ones in three main features: (1) evergreenness involves the yearly renewal of about 30% of the foliage instead of 100% in deciduous trees (Floret et al. 1989a); (2) because a large part of the energy available in spring must be allocated to old leaves already active in early spring, the production of new leaves is delayed and occurs some 3 weeks later in evergreen than in summergreen trees; (3) food resources for the tits, especially caterpillars, are available later in the season because they do not start to grow until young leaves are developed. Moreover, caterpillar abundance is lower in evergreen trees because these larvae eat only young and fresh leaves of the year (Blondel et al. 1992a,b). Therefore, in terms of food availability for insectivorous birds, evergreen habitats are much poorer for tits than summergreen ones.

The mean laying date of the blue tit was found to be well correlated with budbreak in the deciduous *Quercus pubescens* but not with that of the evergreen *Q. ilex* (Clamens 1990). In the rich mainland deciduous habitat, blue tits start to breed early (10 April) and have a high clutch-size (9.4 eggs on average) whereas in the poorer sclerophyllous habitat of Corsica, Blue Tits start to breed about 3 weeks later and have a much smaller clutch size (6.5 eggs) (Fig. 2.8). Blue tits isolated on Corsica breed late and have a low fecundity as a response to a late and poor food supply (Blondel et al. 1991). In the mainland evergreen habitat, which closely matches the Corsican one with regards to the phenology and abundance of caterpillars, but which is included within a mosaic of summergreen and evergreen habitat patches, blue tits started to lay as if they were in the rich summergreen compartment of the mosaic. They lay too early and too many eggs in relation to the seasonal availability of food. As a result of this mistiming, many nestlings die from starvation and overall breeding success is low (Blondel et al. 1992b). One explanation of this mismatching is the gene flow hypothesis among sub-populations living in discrete habitat patches of different quality (Blondel et al. 1992a,b).

The mosaic of rich and poor habitat patches is hypothesized to operate as a "source-sink" system (Pulliam 1988). Poor sink habitats, where birds are misadapted because they immigrate from rich source habitats to which they are well adjusted, produce few recruits so that reproduction cannot balance local mortality. Populations in such habitats can persist only if they are permanently restocked by immigrants from more productive source habitats. Because the Corsican blue tit population has been isolated from mainland ones for a very long time, in habitats where there are very few summergreen trees within its range of dispersal, life history traits of the bird have evolved in such a way that they fairly well match local environmental conditions. Such an evolution occurred during the last few millennia because, here again, summergreen trees have been replaced by sclerophyllous ones as a result of human action (Reille 1984). On the other hand, micro-evolution in the poor evergreen mainland habitats is prevented by gene flow from rich, nearby, summergreen ones.

Such scaling and harlequin effects have consequences at both the population and community levels. At the population level, living in habitats of different quality may result in different selection regimes, this in turn leading to complicated patterns of adaptation/maladaptation. In landscapes that include different habitat patches, the contribution of each patch to the

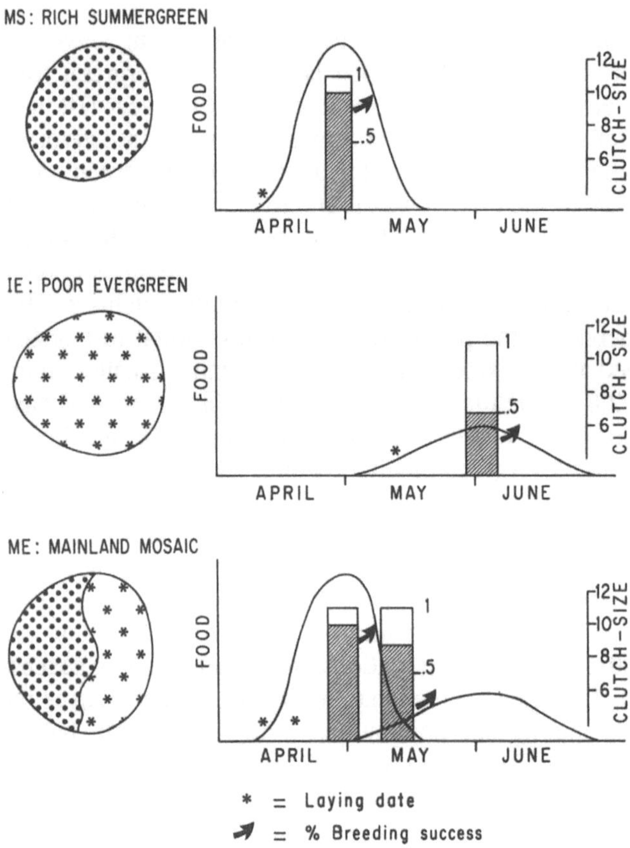

Fig. 2.8. Relationship between breeding traits of the tits and the seasonal variation of food supply (*curves*). *Filled parts of the columns* indicate clutch size, and their *location on the horizontal axis* shows the extent to which breeding time matches the peak of caterpillar abundance (*curves*). There is good matching for populations breeding in both the rich mainland summergreen (*upper*) and the island poor evergreen (*middle*) habitats, but a clear mismatching for the population breeding in the mainland evergreen (*lower*) habitat. *Stars* indicate laying date. For a clutch size of ten eggs, birds will fairly well match the food resources if they lay on average 34 days before the peak date of caterpillars (10 days for completing the clutch because birds lay one egg per day + 14 days incubating + 10 days because the maximum demand of food by the young occurs when they are about 10 days old). *Arrows* indicate breeding success (= number of fledglings/number of eggs laid, total height of the column= 100% breeding success). (after Blondel and Dias 1994)

total population depends on the relative size of each of them. In turn, patch attributes may have consequences at the community level. Most bird species are more sensitive to the structure of the vegetation than to its floristic composition (Cody 1981). Hence, they can be found in a variety of habitats. If the dispersal range of a species is larger than the size of habitat patches, the persistence of local populations in patchy environments and, thus, related components of community diversity, may depend on habitat-specific demographic rates in other habitats. Accordingly, this case study emphasizes the need to consider neighbouring habitats in order to understand community structure and function, because the diversity in any given habitat may depend on both the regional diversity of habitats and the diversity and abundance of resources in each local habitat. Modelling and monitoring the evolution of genetic and demographic systems for any group of organisms in such situations may be useful for landscape management decisions related to the conservation of diversity and the elucidation of the biodiversity/ecosystem function relationship.

2.9 The Future

2.9.1 Regional and Global Changes

Global anthropogenic changes around the Mediterranean Basin have had major impacts on the dynamics and the maintenance of biodiversity over the past several millennia. This includes rates of habitat change, ecosystem xerification through deforestation and soil erosion, habitat fragmentation, nutrient loading etc. These changes will presumably be even more important in the decades to come because of the dramatic changes in land-use practices, human population growth and, eventually, global climatic changes. The future of biodiversity in the Mediterranean region strongly depends on these global changes.

Population trends in the region are very different in the North and the South of the Mediterranean sea. In the European Mediterranean countries, local populations are more or less stabilized, even if seasonal tourist influxes will probably continue to increase in the decades to come, especially along the coasts of both mainland and islands where real estate development increased by 15% between 1965 and 1985 (Le Houérou 1990a). On the other hand, the local populations in the Afro-Asiatic Mediterranean countries are still increasing at a rate of approximately 3% per year (Table 2.1). The consequences of these contrasted trends are important for landscapes and biotic communities. For example, forested areas are decreasing at a rate of 2% per year in the Afro-Asiatic Mediterranean countries whereas they are increasing at a rate of 1-1.5% per year in the Euro-Mediterranean ones, mainly as a result of rural exodus. Concurrently, rangelands and cultivated lands in the latter have decreased between 1965 and 1985 by 7 and 9%, respectively (Le Houérou 1990a).

Thus, there are simultaneous trends of intensification of human pressures in the southern and eastern parts of the Mediterranean Basin, and extensification in the northern and especially northwestern parts. Accordingly, the most important problems Mediterranean people must face are the discrepancies between the north and the south. This makes of the Mediterranean Basin a striking microcosm of current problem of worldwide importance, i.e. the numerous gaps that separate the North from the South in other ways than mere geographic distance. As one high-ranking Tunisian civil servant put it: "the northern (Mediterranean) countries can lean back upon rich and green plains whereas we have only the Desert to lean upon. Where is sustainable regional development to come from when you have huge natural resources and over-industrialisation on one side, and sparse resources and over-population on the other?"

If the current models of global circulation of the atmosphere that predict a doubling of the present day concentration of CO_2 and other "greenhouse gases" (CFCs, HCH, N_2O) by the year 2050 prove to be right, these changes could result in a temperature increase of about 3 °C. Although there is a large amount of uncertainty in these estimates and although the seasonal and geographical manifestations of climatic changes are far from known, possible consequences of these changes in the Mediterranean could be (1) a large increase in evapotranspiration, (2) an increase in primary production, (3) a rise in sea level, (4) an altitudinal and latitudinal shift of vegetation belts, and (v) an increase in the desertification of the Afro-Asiatic parts of the Mediterranean (Le Houérou 1990a). Thus, the future of ecosystems in the Mediterranean Basin over the next half century is highly uncertain at the most fundamental levels.

2.9.2 Restoration, Rehabilitation and Reallocation

Rundel (1979) has pointed out that "enlightened" fire management policies for MTEs require an ability to predict accurately the consequences of varying types of natural and prescribed fires not only on post-fire species composition at a given site, but also on nutrient cycling. Yet fire is just one of the pressing environmental problems facing ecosystems and landscapes around the Mediterranean today. Land-use policies reconciling agricultural development, tourism and nature conservation must be sought. International measures for the protection of endangered species as well as new approaches to ecosystem rehabilitation are needed.

Reafforestation programmes started long ago in many parts of the Mediterranean Basin. For instance, the dramatic destruction of forests by wood cutting, fire and overgrazing produced such catastrophic erosion of soils that important efforts at reafforestation started in Mediterranean France as early as the end of the last century and are still in progress. A well-known example is that of the Mont-Ventoux area (Maury 1960). As a result, many Mediterranean mountains which were almost bare of woody vegetation a century ago are now covered with dense forests. Although artificial, many of these reconstituted forest stands have not only halted soil erosion, but have also re-established suitable conditions for recolonization by various native plants and animals. The result today is more or less diverse and stable communities and, presumably, ecosystems (Blondel 1976; Thinon 1978).

Yet we argue that much more can be done than reforestation of the classic kind if multidisciplinary efforts can be organized to elucidate the links between biodiversity and ecosystem function. Aronson et al. (1993a) argued that for ecosystems subjected to long periods of anarchic human perturbations, three alternatives to continued degradation (or complete abandonment) can be defined: restoration, rehabilitation and reallocation. Restoration sensu stricto aims at the complete (though, admittedly, virtually impossible) return of a site to a pre-existing state in taxonomic terms, i.e. the re-establishment of all indigenous species and the extirpation of all aliens. By contrast, rehabilitation, in our sense, concentrates on repairing damaged or blocked ecosystem functions, with the primary goal of raising ecosystem productivity as fast as possible without sacrificing sustainability. Rehabilitation projects tend to intervene actively and massively and often resort to techniques designed to "jumpstart" the recovery process, such as the reintroduction of nitrogen-fixing plant-microorganism "couples". However, restoration and rehabilitation both aim at recreating self-sustaining ecosystems characterized by succession in plant and animal communities and sufficient resilience to repair themselves following natural or (moderate) human perturbations. Prevailing edaphic and climatic conditions will determine the potential and optimal rates at which both restoration and rehabilitation should proceed. The third response to ecosystem degradation, reallocation, usually disregards the indigenous ecosystem while imposing new uses on a site. Considerable ongoing inputs of energy and nutrients are generally required to keep such reallocated sites productive in the long term.

In the past two or three millennia, most land use in the Mediterranean Basin, apart from the simplest form of pastoralism, has consisted of reallocation of one form or another. Of the more positive forms, at least in terms of conserving biodiversity, agroforestry and sylvo-pastoralism remain among the best-suited systems for sustaining productivity while combatting land degradation in the Mediterranean Basin. As Le Houérou (1990b) stated, the reintroduction of these methods would constitute using "old approaches to new problems". We suggest that the time has come to undertake more experimental research aimed at facilitating restoration and rehabilitation in the Mediterranean region (and other similarly degraded areas) as well as sustainable reallocation.

For example, in the sub-arid portion of the Mediterranean climate region of Tunisia, where three distinct stages of degradation can be clearly identified (Floret and Pontanier 1982), experimental attempts to restore some of the lost biodiversity and, above all, former levels of productivity in the area are now underway. A primary goal of this programme is the reintroduction of nitrogen-fixing legumes of the entire growth-form spectrum indigenous to the region. However, a crucial preliminary stage is to establish sensitive indicators of changing ecosystem function (and structure) as a result of any interventions that might be undertaken.

Table 2.5 shows selected "vital ecosystem attributes" that change in the course of degradation following departure from a pre-disturbance "ecosystem of reference" (Aronson et al. 1993b) in the region of southern Tunisia where restoration and rehabilitation experiments are underway. The tall tree legume *Acacia tortilis* subsp. *raddiana* along with compatible rhizobial inoculants is being experimentally reintroduced, along with a number of other native legume-*Rhizobium* couples, and the effects on ecosystem attributes (those listed in Table 2.5 and many others) will be followed closely. We believe that compiling data for such criteria as these can help evaluate the results of interventions and also be useful in elucidating changes in ecosystem functioning in any and all stages of development.

In the southern part of the Mediterranean Basin, restoration and rehabilitation experiments seem essential and urgent on socio-economic grounds in view of the demographic pressures being applied to already overstrained land ecosystems. In the north, by contrast, there is less urgency as natural regeneration of many ecosystems is now occurring as a result of reduced land use. Nevertheless, where fire control, eco-tourism and environmental protection are emerging as important concerns, and rural exodus is freeing many areas from ongoing agricultural and pastoral exploitation, excellent opportunities for ecosystem research exist in less degraded examples of Mediterranean ecosystems. It remains to be seen if the economic and political incentives will be sufficient to motivate such studies there.

Table 2.5. Vital ecosystem attributes, in late spring, of the typical steppe, 50 km northwest of Gabes, Tunisia, in various stages of degradation. (Floret and Pontanier 1982; Aronson et al. 1993b)

	Vegetation Type[a]			
Vital attribute	RS_3	RS_2	RS_1	SP_1
Number of annual species	13	13	6	2
Number of perennial species	13	8	3	1
Total plant cover (%)	35	20	2	0.5
Average aboveground biomass in spring (kg dry matter ha^{-1} a^{-1})	1 800	600	200	100
Organic matter content in upper soil layers (%)	0.42	n.a.[b]	n.a.	0.1

[a] RS_3 *Rhanterium suaveolens*-dominated shrub-steppe on deep sandy sierozems in relatively undisturbed condition.
RS_2 *Rhanterium suaveolens*-dominated shrub-steppe on deep sandy sierozems somewhat degraded through overgrazing and woodcutting.
RS_1 *Rhanterium suaveolens*-dominated shrub-steppe on deep sandy sierozems badly degraded and with truncated soils.
SP_1 *Stipagrostis pungens*-dominated grass-steppe on aeolian sand deposits.
[b] Not available.

2.9.3 Positive Feedback Switches in Mediterranean Ecosystems

In order to achieve the restoration or rehabilitation of degraded ecosystems, it is not enough to have an inventory of all the species occurring in those systems. Rather, it appears essential to understand the feedback mechanisms that keep ecosystems "running", so to speak, along their traditional trajectory lines. As a corollary, once an ecosystem has become seriously "re-routed" by prolonged human use or other causes, we need to know how to effect a "switch" of some sort that will nudge it back into its former functional lines. Given the poor state of knowledge on virtually all questions concerning ecosystem function and biodiversity, and the predicted changes in human demography and climatic conditions over the next 50 years, there is urgent need to expand research at broad levels. Following Wilson and Agnew (1992), we suggest that studies of possible positive or negative feedback cycles at local or regional levels may be a profitable approach.

In Fig. 2.9, we present what appear to be the switches that have already taken place in typical Mediterranean woodlands on calcareous substrata of southern Europe as a result of multiple stresses or uses imposed by hu-

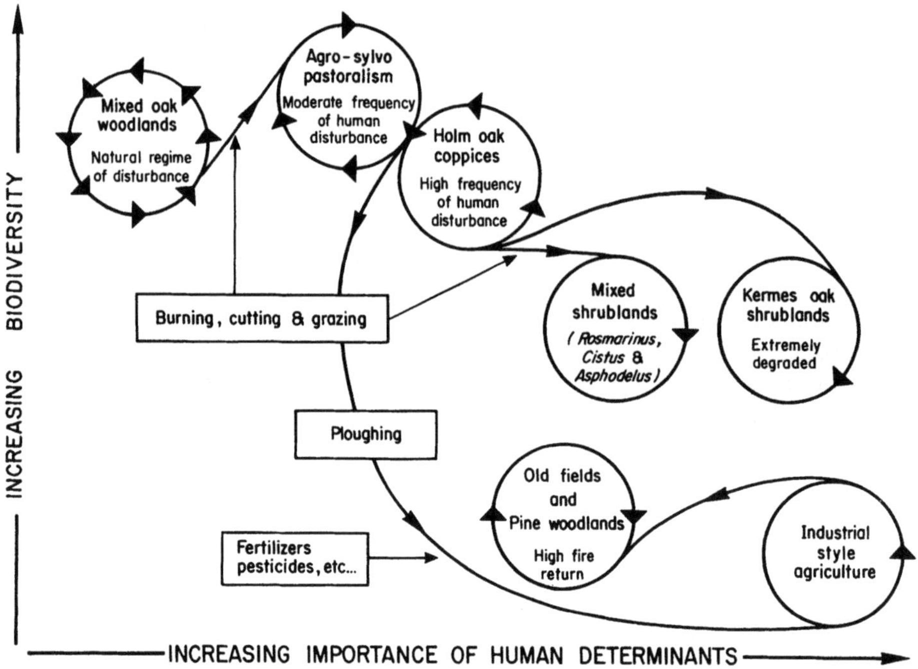

Fig. 2.9. Tentative schematic representation of some human-induced changes in a mixed oak woodland (0-400 m.a.s.l.) in Mediterranean western Europe. At certain thresholds of disturbance, the systems shift from less to more degraded forms, especially through burning, wood-cutting and grazing (but see text). This results in a decrease of energetic pathways and species diversity. *Numbers of arrows on the circles* indicate the richness of ecosystem dynamics (inspired from Woodward 1993), as indicated by "ecological diversity" (i.e. number of functional groups, number of interspecific interactions etc.)

mans. What need to be elucidated here are the changes in nutrient cycles, energy storing, community interactions and other major ecosystem functions concurrent with these shifts and, of course, the importance of the three kinds of biodiversity (intraspecific, interspecific and functional groups). Contrary to common belief, it may be that the highest diversity in Mediterranean ecosystems, in terms of both species richness and ecosystems dynamics and function, did not occur in a so-called climactic mixed oakwood (see Fig. 2.9), but rather in agro-sylvo-pastoral systems such as those that prevailed in many parts of the Mediterranean Basin from the Middle-Age to the middle of the 20th century. In addition, loss of diversity as systems shift from one state to a more degraded one does not necessarily involve all the compartments of ecosystems to the same extent. For example, in habitats dominated by *Q. coccifera, Rosmarinus, Cistus,*

Asphodelus and *Pinus* spp., vertebrate, and, to a lesser extent, insect and plant communities may be heavily impoverished, but soil communities may remain very rich (F. di Castri, pers. comm.).

Above all, we call for better coordination at local, regional and higher levels, of all human activities affecting interlocking ecosystems and landscapes. Throughout this chapter, we have referred repeatedly to the mosaic, harlequin, patchwork, or kaleidoscopic nature of Mediterranean landscapes. To us, these images all suggest living, dynamic, functional systems, and it is precisely this which somehow must be preserved in the Mediterranean region. We need to stop the anarchic exploitation of resources that leads to the destructuring of landscapes and the breakdown of their component parts (see Hobbs et al., Chap. 1 Fig. 1.6, this Vol.).

Acknowledgements. We thank Mireille Cartan-Son, George Davis, Max Debussche, Francesco di Castri, Paul Du Merle, Philippe Granval, Byron Lamont, Roselyne Lumaret, Dave Richardson, Jacques Roy, Christophe Thébaud, John Thompson and two anonymous reviewers for their comments and help in various ways with successive drafts of the manuscript.

References

Al Addan F (1990) Biophysique du sol. Etude quantitative de la régulation, par le travail Lombricien, des propriétés structurales en milieu Méditerranéen. PhD USTL, Montpellier

Allen ON, Allen EK (1981) The Leguminosae. A source book of characteristics, uses and nodulation. University of Wisconsin Press, Madison

Andersen R, Fuentes E, Gadgil M, Lovejoy T, Mooney H, Ojima D, Woodmansee B (1991) Biodiversity from communities to Ecosystems. In: Solbrig OT (ed) From genes to ecosystems: a research agenda for biodiversity. IUBS-SCOPE-UNESCO, Cambridge, Mass, pp 73-82

Arnolds E (1991) Decline of ectomycorrhizal fungi in Europe. Agric Ecosyst Environ 35: 209-244

Aronson J, Floret C, Le Floc'h E, Ovalle C, Pontanier R (1993a) Restoration and rehabilitation of degraded ecosystems, I A view from the south. Restor Ecology 1: 8-17

Aronson J, Floret C, Le Floc'h E, Ovalle C, Pontanier R (1993b) Restoration and rehabilitation of degraded ecosystems. II Case studies in southern Tunisia, central Chile and northern Cameroon. Restor Ecology 1: 168-187

Aronson J, Shmida A (1992) Plant species diversity along a Mediterranean-desert gradient and its correlation with interannual rainfall fluctuations. J Arid Environ 23: 235-247

Auffray JC, Vanlerberghe F, Britton-Davidian J (1990) The house mouse progression in Eurasia: a palaeontological and archaeozoological approach. Biol J Linn Soc 41: 13-25

Axelrod DI (1973) History of the Mediterranean ecosystems in California. In: di Castri F, Mooney HA (eds) Mediterranean-type ecosystems: origin and structure. Ecological Studies 7, Springer, Berlin Heidelberg New York, pp 225-283

Axelrod DI (1975) Evolution and biogeography of Madrean-Tethyan sclerophyll vegetation. Ann Mis Bot Gard. 62: 280-334

Bagnouls F, Gaussen H (1953) Saison sèche et indice xérothermique. Bull Soc Hist Nat Toulouse 83: 193-239
Baker HG, Stebbins GL (1965) The genetics of colonizing species. Academic Press, New York
Barbault R (1991) Dynamique et fonctions de la biodiversité: éléments de réflexion pour un programme de recherches. Environnement 3: 11-12
Barbour MG, Minnich RA (1990) The myth of chaparral convergence. Isr J Bot 39: 453-463
Beaulieu JL (1977) Contribution pollenanalytique à l'histoire tardiglaciaire et Holocène de la végétation des Alpes Méridionales Françaises. Thèse, Fac Sci Techn St Jerôme, Marseille
Belsky AJ, Amundson RG, Duxbury JM, Riha SJ, Ali AR, Mwonga (1989) The effects of trees on their physical, chemical and biological environments in a semi-arid savanna in Kenya. J Appl Ecol 26: 1005-1024
Benois A, Du Merle P, Lafont JP, Marro JP (1978) L'activité oophage de la myrmécofaune dans différents milieux du Mont Ventoux (Vaucluse). Ann Zool Ecol Anim 10: 205-219
Bergström G (1978) Role of volatile chemicals in *Ophrys*-pollinator interactions. In Heyborne JB (ed) Biochemical aspects of plant and animal coevolution. Academic Press, New York, pp 207-232
Beug HJ (1967) On the forest history of the Dalmatian coast. Rev Palaeobot Palynol 2: 271-279
Beug HJ (1975) Changes of climate and vegetation belts in the mountains of Mediterranean Europe during the Holocene. Bull Geol 19: 101-110
Beug HJ (1977) Vegetationsgeschichtliche Untersuchung im Küstenreich von Istiren. Flora 166: 357-381
Billes G, Cortez J, Lossaint P, Lafont F, Prioton J (1975) Étude comparative de l'activité biologique des sols sous peuplements arbustifs et herbacés de la garrigue méditerranéene. I Minéralisation du carbone et de l'azote. Rev Ecol Biol Sol 12: 115-139
Blondel J (1969) Synécologie des passereaux résidents et migrateurs dans un échantillon de la région méditerranéenne française. Thèse Doct Etat, CRDP, Marseille
Blondel J (1976) L'influence des reboisements sur les communautés d'oiseaux, l'exemple du Mont-Ventoux. Ann Sci For 33: 221-245
Blondel J (1981) Structure and dynamics of bird communities in Mediterranean habitats. In: di Castri F, Goodall DW, Specht RL (eds) Mediterranean-type shrublands, Ecosystems of the World 11. Elsevier, Amsterdam, pp 361-385
Blondel J (1985) Mediterranean bird faunas in the light of anthropic pressure since the Neolithic. In: Acta XVIII Congr Intern Ornithol, vol II. Moscow, pp 594-607
Blondel J (1986) Biogéographie évolutive. Masson, Paris
Blondel J (1987) From biogeography to life history theory: a multithematic approach. J Biogeogr 14: 405-422
Blondel J (1988) Biogéographie évolutive à différentes échelles: l'histoire des avifaunes méditerranéennes. In: Ouellet H (ed) Acta XIX Congressus Internationalis Ornithologici 1986, Ottawa, pp 155-188
Blondel J (1991) Invasions and range modifications of birds in the Mediterranean Basin. In: Groves RH, di Castri F (eds) Biogeography of Mediterranean invasions. Cambridge University Press, Cambridge, pp 307-321
Blondel J, Dias PC (1994) Summergreenness, evergreenness and life history variation in Mediterranean Blue Tits. In: Arianoutsou M, Groves RH (eds) Plant-animal interactions

in mediterranean-type ecosystems. Kluwer Academic Publishers, The Netherlands, pp 25-36

Blondel J, Farré H (1988) The convergent trajectories of bird communities in European Forests. Oecologia 75: 83-93

Blondel J, Vigne J-D (1993) Space, time, and man as determinants of diversity of birds and mammals in the Mediterranean region. In: Ricklefs RE, Schluter D (eds) Species diversity in ecological communities: historical and geographical perspectives. Chicago University Press, Chicago, pp 135-146

Blondel J, Vuilleumier F, Marcus LF, Terouanne E (1984) Is there ecomorphological convergence among Mediterranean bird communities of Chile, California and France? Evol Biol 18: 141-213

Blondel J, Chessel D, Frochot B (1988) Bird species impoverishment, niche expansion and density inflation in Mediterranean island habitats. Ecology 69: 1899-1917

Blondel J, Dervieux A, Maistre M, Perret Ph (1991) Feeding ecology and life history variation of the blue tit in Mediterranean mainland and island habitats. Oecologia 88: 9-14

Blondel J, Isenmann P, Maistre M, Perret Ph (1992a) What are the consequences of being a downy oak (*Quercus pubescens*) or a holm oak (*Q. ilex*) for breeding blue tits (*Parus caeruleus*)? Vegetatio 99-100: 129-136

Blondel J, Perret Ph, Maistre M, Dias P (1992b) Do harlequin Mediterranean environments function as source sink for blue tits (*Parus caeruleus* L.)? Landscape Ecol 6: 213-219

Bond P, Goldblatt P (1984) Plants of the Cape Flora. A descriptive catalogue. J S Afr Bot Suppl 13: 1-455

Bond WJ (1993) Keystone species. In: Schulze E-D, Mooney HA (eds) Biodiversity and Ecosystem Function. Ecological studies 99. Springer, Berlin Heidelberg New York, pp 237-253

Bonneau M, Brethes A, Lelong F, Levy G, Nys C, Souchier B (1979) Effets de boisements résineux purs sur l'évolution de la fertilité du sol. Rev For Fr 31: 196-207

Bottner P, Couteaux MM, Vallejo VR (1994) Soil organic matter in Mediterranean type ecosystems and global climatic changes. A case study: the soils of the Mediterranean Basin. In: Morello JM, Oechel WC (eds) Anticipated effects of a changing global environment on mediterranean-type ecosystems. Proc Workshop, Valencia, Spain, 14-17 Sept 1992 (in press)

Bouché MB (1972) Lombriciens de France, Ecologie et Systématique. Ann Zool Ecol Anim, No spéc 72 (2). INRA, Paris, 670 pp

Bouché MB (1981) Contribution des lombriciens aux migrations d'éléments dans les sols tempérés. In: Migrations organo-minérales dans les sols tempérés. Colloq Int CNRS 303: 145-153

Bouché MB (1983) The establishment of earthworm communities. In: Satchell JE (ed) Earthworm ecology from Darwin to vermiculture. Chapman and Hall, London, pp 431-448

Braun-Blanquet J (1936) La Forêt d'Yeuse Languedocienne: *Quercion ilicis*. Comm SIGMA 5, Mem Soc Hist Nat Nimes, Montpellier

Braun-Blanquet J (1952) Les groupements végétaux de la France méditerranéenne. CNRS, Paris

Brodkorb P (1971) Origin and evolution of birds. In: Farner DS, King JR (eds) Avian biology, vol I. Academic Press, New York, pp 19-55

Brun JJ, Cluzeau D, Trehen P, Bouché MB (1987) Biostimulation: perspectives et limites de l'amélioration biologique des sols par stimulation ou introduction d'espèces lombriciennes. Rev Ecol Biol Sol 34: 685-701

Cheylan G (1979) Recherches sur l'organisation du peuplement de vertébrés d'une montagne méditerranéenne: la Sainte-Victoire, Bouches du Rhône. PhD Dissertation, Paris

Cheylan G (1991) Patterns of Pleistocene turnover, current distribution and speciation among Mediterranean mammals. In: Groves RH, di Castri F (eds) Biogeography of Mediterranean Invasions. Cambridge University Press, Cambridge, pp 227-262

Cheylan M, Poitevin F (1993) Conservation of reptiles, frogs and toads. In: Monbailliu X (ed) Environmental management of Mediterranean islands and coasts. Medmaravis, Saint-Maximin

Clamens A (1990) Influence of oak (*Quercus*) leafing on blue tits (*Parus caerulus*) laying date in Mediterranean habitats. Acta Oecol 11: 539-544

Cody ML (1968) On the methods of resource division in grassland bird communities. Am Nat 102: 107-147

Cody ML (1974) Competition and the structure of bird communities. Princeton University Press, Princeton, NJ

Cody ML (1975) Towards a theory of continental species diversities: bird distributions over Mediterranean habitat gradients. In: Cody ML, Diamond JM (eds) Ecology and evolution of communities. Cambridge University Press, London, pp 214-257

Cody ML (1981) Habitat selection in birds: the roles of vegetation structure, competitors, and productivity. BioScience 31: 107-113

Cody ML, Mooney HA (1978) Convergence versus nonconvergence in Mediterranean-climate ecosystems. Annu Rev Ecol Syst 9: 265-231

Cowling RM, Campbell BM (1980) Convergence in vegetation structure in the mediterranean communities of California, Chile and South Africa. Vegetatio 43: 191-197

Crocq C (1990) Le Casse-Noix Moucheté. Lechevallier et Chabaud, Paris

Cuénod A (1954) Flore de Tunisie: cryptogrames vasculaires, gymnospermes et monocotylédones. Imprimerie SEFAN, Tunis

Daget P (1977) Le bioclimat méditerranéen, caractères généraux, modes de caractérisation. Vegetatio 34: 1-20

Daget P (1980) Un élément actuel de caractérisation du monde méditerranéen: le climat. Naturalia Monspeliensia 237: 101-126

Darwin C (1859) On the origin of species by means of natural selection, 6th edn (1972). Murray, London

Daubenmire R (1968) Plant communities. A text book of plant synecology. Harper and Row, New York

De Bolos 0, Vigo J, Masalles RM, Ninot JM (1990) Flora Manual dels Països Catalans. Editorial Portic, SA, Barcelona

De Bonneval L (1990) D'un taillis à l'autre. La déshérence d'un patrimoine forestier communal (Valliguières, Gard) 1820-1990. INRA, Unité d'Ecodéveloppement, Montfavet, France

Debussche M, Isenmann P (1989) Fleshy fruit characters and the choices of bird and mammal seed dispersers in a Mediterranean region. Oikos 56: 327-338

Debussche M, Isenmann P (1992) A mediterranean bird disperser assemblage: composition and phenology in relation to fruit availability. Rev Ecol (Terre et Vie) 47: 56-80

De Faria SM, Lewis GP, Sprent JI, Sutherland JM (1989) Occurrence of nodulation in the Leguminosae. New Phytol 111: 607-619

Delaugerre M, Cheylan M (1992) Batraciens et reptiles de Corse. Parc Naturel Régional de Corse et Ecole Pratique des Hautes Etudes, Ajaccio, Montpellier
De Lillis E (1991) An ecomorphological study of the evergreen leaf. Braun-Blanquetia 7: 1-127
Delmas J (1986) La truffe et sa culture. INRA, Paris
De Martonne E (1927) Traité de géographie physique, vol 1. A Colin, Paris
Diamond JM (1992) Twilight of the pygmy hippos. Nature 359: 15
Dias PC, Meunier F, Beltra S, Cartan-Son M (1994) Blue tits in Mediterranean mosaics. Ardea (in press)
di Castri F, Hansen AJ, Debussche M (eds) (1990) Biological invasions in Europe and the Mediterrranean Basin. Kluwer, Dordrecht
Du Merle P (1978) Les peuplements de fourmis et les peuplements d'acridiens du Mont Ventoux, II Les peuplements de fourmis. La Terre et la Vie, suppl 1: 161-218
Du Merle P, Jourdheuil P, Marro JP, Mazet R (1978) Evolution saisonnière de la myrmécofaune et de son activité prédatrice dans un milieu forestier: les interactions clairière-lisière-forêt. Ann Soc Entomol Fr 14: 141-157
ECE/FAO (1990) Forest fire statistics. Genève, ECE/TIM/51
Edwards AC, Lofty JR (1977) Biology of earthworms. Chapman and Hall, London
Ehrlich PR, Ehrlich AH (1981) Extinctions: the causes and consequences of the disappearance of species. Random, New York
Elton CS (1958) The ecology of invasion by animals and plants. Methuen, London
Emberger L (1930) La végétation de la région Méditerranéenne. Essai d'une classification des groupements végétaux. Rev Gén Bot 42: 641-662
Emberger L (1943) Les limites de l'aire de végétation Méditerranéenne en France. Bull Soc Hist Nat Toulouse 78: 159-180
Etienne M (1991) Protection of mediterranean forests against fire: an ecological approach for redevelopment. In: Ravera O (ed) Terrestrial and aquatic ecosystems, perturbations and recovery. Ellis Norwood, New York, pp 493-498
Flahault C (1937) La distribution géographique des végétaux dans la région Méditerranéenne française. Encyclop. Biologique, vol 18. Lechevalier, Paris
Floret C, Pontanier R (1982) L'aridité en Tunisie présaharienne. Climat, sol, végétation et aménagement. Travaux et Documents de l'ORSTOM 150, Paris
Floret C, Galan MJ, Le Floc'h E, Leprince F, Romane F (1989a) Description of plant annual cycles (France) In: Orshan G (ed) Plant pheno-morphological studies in mediterranean-type ecosystems. Geobotany 12. Kluwer, Dordrecht, pp 9-97
Floret C, Galan MJ, Le Floc'h E, Orshan G, Romane F (1989b) Growth forms and phenomorphology traits along an environmental gradient: tools for studying vegetation? J Veg Sci 1: 71-80
Florschultz F, Ménendez Amor J, Wijmstra TA (1971) Palynology of a thick Quaternary succession in southern Spain. Palaeogeogr Palaeoclimatol Palaeoecol 10: 233-264
Fuentes M (1992) Latitudinal and elevational variation in fruiting phenology among western European bird-dispersed plants. Ecography 15: 177-183
Garnier E (1992) Growth analysis of congeneric annual and perennial grass species. J Ecol 80: 665-675
Gaussen H (1954) Théorie et classification des climats et microclimats. VII Congr Int Bot, Paris 7, pp 125-130
Gomez-Campo C (1985) The conservation of Mediterranean plants: principles and problems. In: Gomez-Campo C (ed) Plant conservation in the Mediterranean area. Junk, Dordrecht, pp 3-8

Gouyon PH, Vernet Ph, Guillerm JL, Valdeyron G (1986) Polymorphisms and environment: the adaptive value of the oil polymorphisms in *Thymus vulgaris* L. Heredity 57: 59-66
Granval P (1988) Approche écologique de la gestion de l'espace rural: des besoins de la bécasse à la qualité des milieux. PhD Dissertation, Université de Rennes, Rennes
Granval P, Muys B (1992) Management of forest soils and earthworms to improve woodcock (*Scolopax* sp.) habitats: a literature survey. Gibier Faune Sauvage 9: 243-255
Groves RH, Burdon JJ (eds) (1986) The ecology of biological invasions: an Australian perspective. Cambridge University Press, Cambridge
Grubb PJ (1977) The maintenance of species richness in plant communities and the importance of the regeneration niche. Biol Rev 52: 107-145
Gutschick VP (1981) Evolved strategies in nitrogen acquisition by plants. Am Nat 118: 607-637
Hanski I (1982) Dynamics of regional distribution: the core and satellite species hypothesis. Oikos 38: 210-221
Haystead A, Malajczuk N, Groves TS (1988) Underground transfer of nitrogen between plants infected with vesicular arbuscular mycorrhizal fungi. New Phytol 108: 417-423
Heim de Balsac H (1936) Biogéographie des mammifères et des Oiseaux de l'Afrique du Nord. Presses Universitaires de France, Paris
Herrera CM (1982a) Defense of ripe fruits from pests: its significance in relation to plant-disperser interactions. Am Nat 120: 218-241
Herrera CM (1982b) Seasonal variation in the quality of fruits and diffuse coevolution between plants and avian dispersers. Ecology 63: 773-785
Herrera CM (1984) A study of avian frugivores, bird-dispersed plants, and their interaction in Mediterranean scrublands. Ecol Monogr 54: 1-23
Herrera CM (1989) Frugivory and seed dispersal by carnivorous mammals and associated fruit characteristics, in undisturbed Mediterranean habitats. Oikos 55: 250-262
Herrera CM (1992) Historical effects and sorting processes as explanations for contemporary ecological patterns: character syndromes in Mediterranean woody plants. Am Nat 140: 421-446
Hildebrand EE (1987) Die Struktur von Waldböden - ein gefährdetes Fliessgleichgewicht. Allgemeine Forst Zeitschrift 16/17: 424-426
Howe HF, Smallwood J (1982) Ecology of seed dispersal. Annu Rev Ecol Syst 13: 201-228
Hubert B, Bellon S, Chassany JP, Guérin G, Martinand P, Prévost F (1989) Intégrer les activités pastorales et forestières dans la gestion de l'espace méditerranéen. For Méditerr 11: 238-251
Huhta V (1979) Effects of liming and deciduous litter on earthworm Lumbricidae populations of a spruce forest, with an inoculation experiment on *Allolobophora caliginosa*. Pedobiologia 19: 340-345
Huntley B, Birks HJB (1983) An atlas of past and present pollen maps for Europe: 0-13000 years ago. Cambridge University Press, Cambridge
INED (1993) Tous les pays du monde. Population et sociétés. INED 282, Paris
Jaeger J-J (1975) The mammalian faunas and hominid fossils of the Middle Pleistocene of the Maghreb. In: Butzer KW, Isaac GL (eds) After the Australopithecines. Mouton, The Hague
Jaeger J-J, Coiffait B, Tong H, Denys C (1987) Rodent extinctions following Messinian faunal exchanges between western Europe and Northern Africa. Mém Soc Géol Fr 150: 153-158
Jaenkie J (1991) Mass extinction of European fungi. Trends Ecol Evol 6: 174-175

Jalut G (1976) Les débuts de l'agriculture en France, les défrichements. In: La Préhistoire française. CNRS, Paris, pp 180-185
Janzen DH (1980) When is it coevolution? Evolution 34: 611-612
Jarrell WM, Virginia RA (1990) Soil cation accumulation in a mesquite woodland: sustained production and long-term estimates of water use and nitrogen fixation. J Arid Environ 18: 51-58
Joffre R (1990) Plant and soil nitrogen dynamics in mediterranean grasslands: a comparison of annual and perennial grasses. Oecologia 85: 142-149
Joffre R, Vacher J, de los Llanos C, Long G (1988) The dehesa: an agrosilvopastoral system of the Mediterranean region with special reference to the Sierra Morena area of Spain. Agrofor Systems 6: 71-96
Johnson DL (1977) The California ice-age refugium and the Rancholabrean extinction problem. Quatern Res 8: 154-178
Jordano P (1987) Patterns of mutualistic interactions in pollination and seed dispersal: connectance, dependence asymmetries and coevolution. Am Nat 129: 657-677
Kolodny EY (1974) La population des Iles de la Grèce, essai de géographie insulaire en Méditerranée Orientale. Edisud, Aix-en-Provence
Körner Ch (1993) Scaling from species to vegetation: the usefulness of functional groups. In: Schulze E-D, Mooney HA (eds) Biodiversity and ecosystem function. Ecological studies 99. Springer, Berlin Heidelberg New York, pp 117-142
Kruger FJ (1983) Plant community diversity and dynamics in relation to fire. In: Kruger FJ, Mitchell DT, Jarvis JUM (eds) Mediterranean-type ecosystems. The role of nutrients. Ecological studies 43. Springer, Berlin Heidelberg New York, pp 466-472
Kuhnholtz-Lordat G (1938) La terre incendiée. Essai d'agronomie comparée. Maison carrée, Nîmes
Kurten B (1968) Pleistocene mammals of Europe. Weidenfeld & Nicolson, London
Lamont BB (1982) Mechanisms for enhancing nutrient uptake in plants, with particular reference to Mediterranean South Africa and Western Australia. Bot Rev 48: 597-689
Lamont BB (1984) Specialized modes of nutrition. In: Pate JS, Beard JS (eds) Kwongan: plant life of the Sandplain. University of Western Australia Press, Nedlands, Western Australia, pp 236-245
Lamont BB (1992) Functional interactions within plants, the contribution of keystone and other species to biological diversity. In: Hobbs RJ (ed) Biodiversity of mediterranean ecosystems in Australia. Surrey Beatty, Chipping Norton NSW, pp 95-127
Lang E, Beese F (1985) Die Reaktion der mikrobiellen Bodenpopulation eines Buchenwaldes auf Kalkungsmassnahmen. Allg Forst Zeitschrift 43: 1166-1169
Lavelle P (1981) Stratégies adaptatives globales chez les vers de terre. les types écologiques. Acta Oecol/Oecol Gen 2: 119-133
Lavelle P (1988) Earthworm activities and the soil system. Biol Fertil Soils 6,3: 237-251
Lawton JH, Brown VK (1993) Redundancy in ecosystems. In: Schulze E-D, Mooney HA (eds) Biodiversity and ecosystem function. Ecological studies 99. Springer, Berlin Heidelberg New York, pp 255-270
Lee KE (1985) Earthworms: their ecology and relationships with soils and land use. Academic Press, Sydney
Le Houérou HN (1968) La désertisation du Sahara septentrional et des steppes limitrophes. Ann Algér Géogr 3: 2-27
Le Houérou HN (1980) L'impact de l'homme et de ses animaux sur la forêt méditerranéenne. For Méditerr 2: 31-44;155-174

Le Houérou HN (1981) Impact of man and his animals on Mediterranean vegetation. In: di Castri F, Goodall DW, Specht RL (eds) Mediterranean-type shrublands. Ecosystems of the world, vol 11. Elsevier, Amsterdam, pp 479-517

Le Houérou HN (1990a) Global change: vegetation, ecosystems and land use in the southern Mediterranean Basin by the mid twenty-first century. Isr J Bot 39: 481-508

Le Houérou HN (1990b) Agroforestry and sylvopastoralism to combat degradation in the Mediterranean Basin: old approaches to new problems. Agric Ecosyst Environ 33: 99-109

Le Houérou HN (1991) La Méditerranée en l'an 2050: Impacts respectifs d'une éventuelle évolution climatique et de la démographie sur la végétation, les écosystèmes et l'utilisation des terres. Météorologie 36: 4-37

Leon C, Lucas G, Synge HK (1985) The value of information in saving threatened Mediterranean plants. In: Gomez-Campo C (ed) Plant conservation in the Mediterranean area. Junk, Dordrecht, pp 177-196

Leonardi S, Rapp M (1990) Production de phytomasse et utilisation des bioéléments lors de la reconstitution d'un taillis de chêne vert. Acta Oecol 11: 819-834

Leopold A (1949) A Sand County almanac. Oxford University Press, Oxford

Loret H, Barrandon A (1876) Flore de Montpellier. 2 vols. Coulet, Paris

Lumaret R (1988) Cytology, genetics, and evolution in the genus *Dactylis*. CRC Crit Rev Plant Sci 7(1): 55-91

MacArthur RH, Karr JR, Diamond JM (1972) Density compensation in island faunas. Ecology 53: 330-342

Macdonald IAW, Kruger FJ, Ferrar AA (1986) The ecology and management of biological invasions in Southern Africa. Oxford University Press, Cape Town

Main AR (1981) Ecosystem theory and management. J R Soc West Aust 4: 1-4

Marchand H (1990) Les forêts Méditerranéennes. Enjeux et Perspectives. Les Fascicules du Plan Bleu 2. Economica, Paris

Margaris NS, Vokou D (1982) Structural and physiological features of woody plants in phryganic ecosystems related to adaptive mechanisms. Ecol Mediterr 8: 449-459

Martin PS (1984) Prehistoric overkill: the global model. In: Martin PS, Klein RG (eds) Quaternary extinctions. University of Arizona Press, Tucson, pp 354-403

Maury P (1960) Le reboisement de la forêt de Bédoin et son enseignement. Ann Ecol Nat Eaux For 17: 119-149

May RM (1973) Stability and complexity in model ecosystems. Princeton University Press, Princeton

May RM (1977) Thresholds and breakpoints in ecosystems with a multiplicity of stable states. Nature 269: 471-477

McKey D (1975) The ecology of coevolved seed dispersal systems. In: Gilbert LE, Raven PH (eds) Coevolution of animals and plants. University of Texas Press, Austin, pp 159-191

McNaughton SJ (1993) Biodiversity and function of grazing ecosystems. In Schulze E-D, Mooney HA (eds) Biodiversity and ecosystem function. Ecological studies 99. Springer, Berlin Heidelberg New York, pp 361-408

Medus J, Pons A (1980) Les prédécesseurs des végétaux méditerranéens actuels jusqu'au début du Miocène. Naturalia Monspeliensia 237: 11-20

Ménendez-Amor J, Florschultz F (1964) Results of the preliminary palynological investigation of samples from a 50 m boring in southern Spain. Bol R Soc Esp Hist Nat 62: 251-255

Mooney HA (ed) (1977a) Convergent evolution in Chile and California. US/IBP Synthesis Ser 5, Dowden, Hutchinson and Ross, Stroudsburg, PA
Mooney HA (1977b) The carbon cycle in mediterranean-climate evergreen scrub communities. In: Environmental consequences of fire and fuel management in mediterranean-climate ecosystems. US For Serv Gen Tech Rep WO-3 Washington DC, pp 107-115
Mooney HA (1983) Carbon-gaining capacity and allocation patterns of mediterranean-climate plants. In: Kruger FJ, Mitchell DT, Jarvis JUM (eds) Mediterranean-type ecosystems. The role of nutrients. Ecological studies 43. Springer, Berlin Heidelberg New York, pp 103-119
Mooney HA, Drake JA (eds) (1986) Ecology of biological invasions of North America and Hawaii. Springer, Berlin Heidelberg New York
Mooney HA, Dunn EL (1970) Convergent evolution of mediterranean shrubs. Evolution 24: 292-303
Mooney HA, Hamburg SP, Drake JA (1986) The invasions of plants and animals into California. In: Mooney HA, Drake JA (eds) Ecology of biological invasions of North America and Hawaii. Ecological studies 58. Springer, Berlin Heidelberg New York, pp 250-272
Moreau RE (1966) The bird faunas of Africa and its Islands. Academic Press, New York
Moreau RE (1972) The Palaearctic-African bird migrations systems. Academic Press, London
Mourer-Chauviré C (1975) Les oiseaux du Pléistocène Moyen et Supérieur de France. Thèse d'Etat, Université Claude Bernard, Lyon
Muys B (1989) Evaluation of conversion of tree species and liming as measures to decrease soil compaction in a beech forest on loamy soil. In: FAO/ECE/ILO (eds) Seminar on the impact of mechanization of forest operations to the soil. Louvain-la-Neuve, pp 341-355
Nahal I (1981) The Mediterranean climate from a biological viewpoint. In: di Castri F, Goodall DW, Specht RL (eds) Mediterranean-type shrublands. Ecosystems of the world. vol 11. Elsevier, Amsterdam, pp 63-86
Naveh Z (1967) Mediterranean ecosystems and vegetation types in California and Israel. Ecology 48: 445-459
Naveh Z (1975) The evolutionary significance of fire in the Mediterranean region. Vegetatio 29: 199-208
Naveh Z, Dan J (1973) The human degradation of Mediterranean landscapes in Israel. In: di Castri F, Mooney HA (eds) Mediterranean-type ecosystems: origins and structure. Ecological studies 7. Springer, Berlin Heidelberg New York, pp 373-390
Niklewski J, Van Zeist W (1970) A late quaternary pollen diagram from Northwestern Syria. Acta Bot Neerl 19: 737-754
Odum EP (1969) The strategy of ecosystem development. Science 164: 262-270
Oppenheimer HR (1960) Adaptation to drought: xerophytism. In: Arid zones Res XV Plant water relationships in arid and semi-arid conditions. Unesco, Paris, pp 105-138
Orshan G (ed) (1989) Plant pheno-morphological Studies in Mediterranean-type ecosystems. Geobotany 12. Kluwer, Dordrecht
Owen RB, Galbraith WJ (1989) Earthworm biomass in relation to forest types, soil and land use: implications for woodcock management. Wildl Soc Bull 17: 130-136
Pagnol J (1973) La truffe, 3rd edn. Aubanel, Paris
Paine RT (1966) Food web complexity and species diversity. Am Nat 100: 65-75
Paine RT (1992) Food-web analysis through field measurement of per capita interaction strength. Nature 355: 73-75

Palamarev E (1989) Paleobotanical evidences of the tertiary history and origin of the Mediterranean sclerophyll dendroflora. Plant Syst Evol 162: 93-107

Parks CR, Wendel JF (1990) Molecular divergence between Asian and North American species of *Liriodendron* (Magnoliaceae) with implications for interpretation of fossil floras. Am J Bot 77: 1243-1256

Pate JS, Hopper SD (1993) Rare and common plants in ecosystems, with special reference to the South-west Australian Flora. In: Schulze E-D, Mooney HA (eds) Biodiversity and ecosystem function. Ecological studies 99. Springer, Berlin Heidelberg New York, pp 293-326

Péguy Ch (1970) Précis de climatologie. Masson, Paris

Pickett STA, White PS (eds) (1985) The ecology of natural disturbance and patch dynamics. Academic Press, New York

Pignatti S (1978) Evolutionary trends in Mediterranean flora and vegetation. Vegetatio 37: 175-185

Pignatti S (1983) Human impact on the Mediterranean Basin. In: Holzner W et al (eds) Man's impact on vegetation. Junk, The Hague, pp 151-162

Pons A (1964) Contribution palynologique à l'étude de la flore et de la végétation pliocènes de la région rhodanienne. Ann Soc Nat Bot Paris 5: 499-722

Pons A (1981) The history of the Mediterranean shrublands. In: di Castri F, Goodall DW, Specht RL (eds) Mediterranean-type shrublands. Ecosystems of the world. vol 11. Elsevier, Amsterdam, pp 131-138

Pons A, Quézel P (1985) The history of the flora and vegetation and past and present human disturbance in the Mediterranean region. In: Gomez-Campo C (ed) Plant conservation in the Mediterranean area. Junk, Dordrecht, pp 25-43

Pons A, Suc JP (1980) Les témoignages des structures actuelles de végétation méditerranéenne durant le passé antérieur à l'action de l'homme. In: Denizot M, Sauvage Ch (eds) La mise en place, l'évolution et la caractérisation de la flore et de la végétation circum-méditerranéennes. Naturalia Monspeliensia 237: 69-78

Poplin F (1979) Origine du Mouflon de Corse dans une nouvelle perspective paléontologique: par marronnage. Ann Gén Sél Anim 11: 133-143

Pottier-Alapetite G (1979-1981) Flore de la Tunisie: Angiospermes-Dicotylédones, 2 vol. Publ Scientifiques Tunisiennes, Tunis

Prodon R, Fons R, Athias-Binche F (1987) The impact of fire on animal communities in mediterranean area. In: Trabaud L (ed) The role of fire in ecological Systems. SPB Academic Publ, The Hague, pp 121-157

Prudhomme P (1986) Dynamique de l'interface Elevage-Espace des Garrigues de la région d'Aumelas. LECSA-INRA, Montpellier, France

Pulliam HR (1988) Sources, sinks, and population regulation. Am Nat 132: 652-661

Quézel P (1965) La Végétation du Sahara. Fischer, Stuttgart

Quézel P (1978) Analysis of the flora of Mediterranean and Saharan Africa. Ann Mo Bot Garden 65: 479-534

Quézel P (1985) Definition of the Mediterranean region and origin of its flora. In: Gomez-Campo C (ed) Plant conservation in the Mediterranean Area. Junk, Dordrecht, pp 9-24

Quézel P, Gamisans J, Gruber M (1980) Biogéographie et mise en place des flores Méditerranéennes. Naturalia Monspeliensia 237: 41-51

Ramade F (1990) Conservation des écosystèmes Méditerranéens. Plan Bleu 3. Economica, Paris

Rapp M, Lossaint P (1981) Some aspects of mineral cycling in the garrigue of southern France. In: di Castri F, Goodall DW, Specht RL (eds) Mediterranean-type shrublands. Ecosystems of the world 11. Elsevier, Amsterdam, pp 289-301

Raunkiaer C (1934) The life-forms of plants and statistical plant geography. Oxford University Press, Oxford

Raven PH (1973) The evolution of Mediterranean flora. In: di Castri F, Mooney HA (eds) Mediterranean-type ecosystems: origin and structure. Ecological studies 7. Springer, Berlin Heidelberg New York, pp 213-224

Read DJ (1993) Plant-microbe mutualisms and community structure. In: Schulze E-D, Mooney HA (eds) Biodiversity and ecosystem function. Ecological studies 99. Springer, Berlin Heidelberg New York, pp 181-209

Rebière J (1982) La truffe en Périgord, 3rd edn. Editions Fanlac, Périgueux

Reille M (1975) Contribution pollenanalytique à l'histoire holocène de la végétation de la montagne Corse. PhD Dissertation, Fac Sci Aix-Marseille III

Reille M (1977) Contribution pollenanalytique à l'histoire holocène de la végétation des montagnes du Rif (Maroc septentrional). In: Recherches françaises sur le Quaternaire. INQUA (1977) Supp Bull AFEQ, 50: 53-76

Reille M (1984) Origine de la végétation actuelle de la Corse sud-orientale; analyse pollinique de cinq marais côtiers. Pollen Spores 26: 43-60

Reille M, Triat H, Vernet JL (1980) Les témoignages des structures actuelles de végétation méditerranéenne durant le passé contemporain de l'action de l'homme. Naturalia Monspeliensia 237: 79-87

Reumer JWF, Payne S (1986) Notes on the Soricidae (Insectivora, Mammalia) from Crete, II, The shrew remains from Minoan and Classical Kommos. Bonn Zool Beitr 3: 173-182

Richardson DM, Cowling RM (1993) Biodiversity and ecosystem processes: opportunities in Mediterranean-type ecosystems. Trends Ecol Evol 8: 79-80

Richardson DM, Macdonald IAW, Holmes PM, Cowling RM (1992) Plant and animal invasions. In: Cowling RM (ed) The ecology of fynbos: nutrients, fire and diversity, Oxford University Press, Cape Town, pp 271-308

Romane F (1987) Efficacité de la distribution des formes de croissance des végétaux pour l'analyse de la végétation à l'échelle régionale. PhD Dissertation, Université de Droit, d'Economie et des Sciences d'Aix-Marseille

Root RB (1967) The niche exploitation pattern of the blue-gray gnatcatcher. Ecol Monogr 37: 317-319;331-349

Roy J (1981) Intraspecific variation in the physiological characteristics of perennial grasses of the Mediterranean region. Ecol Mediterr 8: 435-448

Roy J, Lumaret R (1987) Associated clinal variation in leaf tissue water relations and allozyme polymorphism in *Dactylis glomerata* L. populations. Evol Trends Plants 1: 9-19

Roy J, Sonié L (1992) Germination and population dynamics of *Cistus* species in relation to fire. J Appl Ecol 29: 647-655

Rundel P (1979) Ecological impact of fires on mineral and sediment pools and fluxes. In: Agee JK (ed) Fire and fuel management in mediterranean-climate ecosystems: research priorities and programmes. MAB Technical Notes 11. UNESCO, Paris, pp 17-22

Rundel P (1989) Ecological success in relation to plant form and function in woody legumes. In: Stirton CH, Zarucchi JL (eds) Advances in legume biology. Missouri Botanical Garden, St Louis, Missouri, pp 377-398

Saint-Girons H (1983) Régimes et rations alimentaires des serpents. Bull Soc Zool Fr 108: 431-437

Saint-Quirin (1904) Les verriers du Languedoc (1290-1790). Reissued by Association L Reveillée, Montpellier, France (1985)
Salabert J, Gastesoleil J (1991) Contribution à l'inventaire de la flore de l'Hérault. Mond Plant (Toulouse) 442: 16-18
Sauvage C (1963) Le quotient pluviothermique d'Emberger, son utilisation et la représentation géographique de ses variations au Maroc. Ann Serv Phys Globe Météo Inst Sci Chérifien 20: 11-23
Schluter D (1986) Tests of similarity and convergence of finch communities. Ecology 67: 1073-1083
Schulze E-D (1982) Plant life forms as related to plant carbon, water and nutrient relations. In: Lange OL, Nobel PS, Osmond CB, Ziegler H (eds) Encyclopedia of plant physiology. vol 12B. Springer, Berlin Heidelberg New York, pp 646-652
Schulze E-D (1989) Ökosystemforschung - die Entwicklung einer jungen Wissenschaft. In: Gerwin R (ed) Wie die Zukunft Wurzeln schlug. Springer, Berlin Heidelberg New York, pp 55-64
Schulze E-D, Mooney HA (eds) (1993) Biodiversity and ecosystem function. Ecological studies 99. Springer, Berlin Heidelberg New York
Schulze E-D, Hall AE, Lange OL, Evenari M, Kapen L, Buschbom U (1980) Long-term effects of drought on wild and cultivated plants in the Negev desert, I Maximal rates of net photosynthesis. Oecologia 45: 11-18
Shmida A (1981) Mediterranean vegetation in California and Israel: similarities and differences. Isr J Bot 30: 105-123
Shmida A, Aronson J (1986) Sudanian elements in the flora of Israel. Annals Mo Bot Gard 73: 1-28
Shmida A, Werger MJA (1992) Growth form diversity on the Canary Islands. Vegetatio 102: 183-199
Simberloff D, Dayan T (1991) The guild concept and the structure of ecological communities. Annu Rev Ecol Syst 22: 115-143
Snow DW (1971) Evolutionary aspects of fruit-eating by birds. Ibis 113: 194-202
Sorensen AE (1981) Interactions between birds and fruit in a temperate woodland. Oecologia 50: 242-249
Soulier A (1993) Le Languedoc pour héritage. Presses du Languedoc, Montpellier, France
Sousa WP (1984) The role of disturbance in natural communities. Annu Rev Ecol Syst 15: 353-391
Stebbins GL (1970) Transference of function as a factor in the evolution of seeds and their accessory structures. Isr J Bot 19: 59-70
Stebbins GL (1982) Perspectives in evolutionary theory. Evolution 36: 1109-1118
Stebbins GL, Zohary D (1959) Cytogenetic and evolutionary studies in the genus *Dactylis*. I Morphology, distribution and interrelationships of the diploid subspecies. Univ Calif Berkeley Publ Bot 31: 1
Stiles EW (1980) Patterns of fruit presentation and seed dispersal in bird-disseminated woody plants in the eastern deciduous forests. Am Nat 116: 670-688
Stockdill SMJ (1982) Effects of introduced earthworms on the productivity of New Zealand pastures. Pedobiologia 24: 29-35
Suc JP (1978) L'étude palynologique du Pliocène du sud de la France, méthode d'approche et résultats. Ann Mines Belg 6: 120-126
Suc JP (1980) Contribution à la connaissance du Pliocène et du Pléistocène inférieur des régions méditerranéennes d'Europe occidentale par l'analyse palynologique des dépôts

du Languedoc-Roussillon (sud de la France) et de la Catalogne (nord-est de l'Espagne). PhD Dissertation, Univ Sci Techn Languedoc, Montpellier

Suc JP (1984) Origin and evolution of the Mediterranean vegetation and climate in Europe. Nature 307: 409-432

Tchernov E (1984) Faunal turnover and extinction rate in the Levant. In: Martin PS, Klein RG (eds) Quaternary extinctions. University of Arizona Press, Tucson, pp 528-552

Thébaud C, Debussche M (1991) Rapid invasion of *Fraxinus ornus* L. along the Hérault River system in southern France: the importance of seed dispersal by water. J Biogeogr 18: 7-12

Thinon M (1978) Quelques aspects floristiques et pédologiques de l'incidence écologique des reboisements du Mont Ventoux. Terre Vie suppl 1: 67-109

Thirgood JV (1981) Man and the Mediterranean forest. Academic Press, New York

Thomas JA, Elmes GW, Wardlaw JC, Woyciechowski M (1989) Host specificity among *Maculinea* butterflies in Myrmica ant nests. Oecologia 79: 452-457

Thompson JD, Turkington R, Holl FB (1990) The influence of *Rhizobium leguminosarum* biovar. *trifolii* on the growth and neighbor relationships of *Trifolium repens* and three grasses. Can J Bot 68: 296-303

Tomaselli R (1976) La dégradation du maquis méditerranéen. In: Notes techniques du MAB, 2 Forêts et Maquis Méditerranéens: écologie, conservation et aménagement. UNESCO, Paris, pp 35-76

Toutain F, Diagne A, Le Tacon F (1988) Possibilités de modification du type d'humus et d'amélioration de la fertilité des sols à moyen terme en hêtraie par apport d'éléments minéraux. Rev For Fr 40: 99-107

Trabaud LV (1984) Man and fire: impacts on mediterranean vegetation. In: di Castri F, Goodall DW, Specht RL (eds) Mediterranean-type shrublands. Ecosystems of the world. vol 11. Elsevier, Amsterdam, pp 523-537

Trabaud LV (1987) Dynamics after fire of sclerophyllous communities in the Mediterranean Basin. Ecol Med 13: 25-37

Trabaud LV, Christensen NL, Gill AM (1993) Historical biogeography of fire in temperate and mediterranean ecosystems. In: Crutzen PJ, Goldammer JG (eds) Fire in the environment: its ecological and atmospheric importance. John Wiley, New York, pp 277-295

Triat H (1978) Contribution pollenanalytique à l'histoire Tardi- et Post-glaciaire de la végétation de la Basse Vallée du Rhône. PhD Dissertation, Université Aix-Marseille III, France

Triat H (1979) Histoire de la forêt provençale depuis 15000 ans d'après l'analyse pollinique. For Méditerr 1: 19-24

Turkington R, Cavers PB, Aarssin LW (1977) Neighbor relations in grass-legume communities. I Inter-specific contacts in four grassland communities near London, Ontario. Can J Bot 55: 2701-2711

UNESCO (1977) Atlas of the world water balance. UNESCO, Paris

Valverde JA (1957) Aves del Sahara Espanol. IDEA, Madrid

Valverde JA (1967) Estructura de Una Communidad de Vertebrados Terrestres. CSIC, Madrid

Van Campo M, Guinet Ph, Cohen J (1968) Fossil pollen from late Tertiary and middle Pleistocene deposits of the Kurkur oasis. In: Butzer KW, Hansen CL (eds) Desert and river in Nubia. University of Wisconsin Press, Madison, pp 515-520

Van der Pijl L (1972) Principles of dispersal in higher plants, 2nd edn. Springer, Berlin Heidelberg New York

Van Rhee JA, Nathans S (1961) Observations on earthworm populations in orchard soils. Neth J Agric Sci 9: 94-100
Van Zeist W (1980) Aperçu sur la diffusion des végétaux cultivés dans la région méditerranéenne. Naturalia Monspeliensia 237: 129-145
Van Zeist W, Bottema S (1977) The pleniglacial, late glacial and early postglacial vegetation of Zeribar and their present-day counterparts. Palaeohistoria 19: 19-85
Van Zeist W, Woldring H, Stapert D (1975) Late quaternary vegetation and climate of Southwestern Turkey. Palaeohistoria 17: 55-143
Vaurie C (1959-1965) The birds of the Palearctic fauna, vol I: non Passeriformes. Witherby, London
Vernet JL (1972) Contribution à l'étude de la végétation du sud-est de la France au Quaternaire. Etude des macroflores de charbon de bois principalement. PhD Dissertation, Univ Sci Techn Languedoc, Montpellier
Vernet JL (1973) Etude sur l'histoire de la végétation du sud-est de la France au Quaternaire d'après les charbons de bois principalement. Paléobiol Continent 4: 1-90
Vigne J-D (1988) Les Mammifères post-glaciaires de Corse, étude archéozoologique. XXVIe suppl Gallia Préhistoire. CNRS, Paris
Vigne J-D (1992) Zooarchaeology and the biogeographical history of the mammals of Corsica and Sardinia since the last ice age. Mammal Rev 22: 87-96
Vigne J-D, Marinval-Vigne MC (1983) Méthode pour la mise en évidence de la consommation du petit gibier. Br Archaeol Rep Int Ser 163: 239-242
Vilette J (1983) Avifaunes de la fin du Pléistocène Supérieur et de l'Holocène dans le Sud de la France et en Catalogne. Systématique, Paléoenvironnement, Paléoethnologie. Thèse, Univ Cl Bernard, Lyon
Vitousek PM, Hooper DU (1993) Biological diversity and terrestrial ecosystem biogeochemistry. In: Schulze E-D, Mooney HA (eds) (1993) Biodiversity and ecosystem function. Ecological studies 99. Springer, Berlin Heidelberg New York, pp 4-14
Vogel P, Maddalena T, Schembri PJ (1990) Cytotaxonomy of Shrews of the genus *Crocidura* from Mediterranean islands. Vie Milieu 40: 124-129
Voous KH (1960) Atlas of European birds. Nelson, Edinburgh
Vuilleumier F (1991) Invasions in the Mediterranean avifaunas of California and Chile. In: Groves RH, di Castri F (eds) Biogeography of Mediterranean invasions. Cambridge University Press, Cambridge, pp 326-358
Wacquant JP (1990) Biogeographical and physiological aspects of the invasion by *Dittrichia (ex-Inula) viscosa* W Greuter, a ruderal species in the Mediterranean Basin. In: di Castri F, Hansen A, Debussche M (eds) Biological invasions in Europe and the Mediterranean Basin. Kluwer, Dordrecht, pp 353-364
Wacquant JP, El-Chahatha H, Jacquard P (1989) Effect of mineral stress on competition and associated growth of a grass (*Lolium italicum* L.) and a legume (*Trifolium pratense* L.). In: Wright EE (ed) Proc Symp on Plant Physiology and Herbage Production. Brit Grassland Soc, Maidenhead, pp 231-234
Walter H (1983) Vegetation of the Earth and ecological systems of the Geo-Biosphere. Springer, Berlin Heidelberg
Walter H, Lieth H (1960) Klimadiagramm Weltatlas. Fischer, Jena
Wheelwright NT (1985) Fruit size, gape width, and the diets of fruit-eating birds. Ecology 66: 808-818
Wijmstra TA (1969) Palynology of the first 30 m of a 120 m deep section in Northern Greece. Acta Bot Neerl 18: 511-527
Wilson EO (1988) Biodiversity. Nat Acad Press, Washington, DC

Wilson EO (1992) The diversity of life. Belknap, Harvard University, MA
Wilson JB, Agnew ADQ (1992) Positive-feedback switches in plant communities. Adv Ecol Res 23: 264-336
Woodward FI (1993) How many species are required for a functional ecosystem? In: Schulze E-D, Mooney HA (eds) (1993) Biodiversity and ecosystem function. Ecological studies 99. Springer, Berlin Heidelberg New York, pp 271-291
Woodwell GM, Smith HH (eds) (1969) Diversity and stability in ecological systems. Brookhaven National Laboratory, Upton, New York
Zohary D (1969) The progenitors of wheat and barley in relation to domestication and agricultural dispersal in the Old World. In: Ucko PJ, Dimpley GW (eds) The domestication of plants and animals. Aldine, Chicago, pp 47-66
Zohary D, Spiegel-Roy P (1975) Beginnings of fruit-growing in the Old World. Science 187: 319-327
Zohary M (1973) Geobotanical foundations of the Middle East. 2 parts, Fischer, Stuttgart
Zohary M, Feinbrun N (1966-1986) Flora Palaestina. Parts 1-4, The Israel Academy of Sciences and Humanities, Jerusalem

3 Biodiversity and Ecosystem Functioning in Mediterranean-Climate California

J.E. Keeley and C.C. Swift

3.1 Introduction

Rapid population growth and land development in California have already contributed, largely through habitat destruction, to a significant loss of biodiversity (Jensen et al. 1990). These obvious alterations in our flora and fauna have led to increasing concern about further losses of species diversity resulting from fragmented populations and from the introduction of non-native species and concern about the impact of losses in genetic diversity arising from reduced population sizes (Keeley 1993). In addition, predicted climatic changes are likely to alter landscape patterns of diversity with unknown consequences on regional ecosystem processes.

We believe this is a matter of concern for all citizens because of our dependence upon the proper functioning of natural ecosystems. Not only do these ecosystems have aesthetic, recreational, and scientific value, they play fundamental roles in regional landscape processes that are required for the air we breath, water we drink and food we eat. However, it is unclear exactly how these known and anticipated changes in biodiversity might alter or have altered the functioning of Californian ecosystems; specifically, the transfer of water, nutrients and carbon. The reasons for this include a depauperate body of knowledge on the basic population biology of most native species, the lack of full understanding of the mechanisms underlying the assemblage of species in communities, and the paucity of studies that have attempted to relate population biology or community ecology to ecosystem function.

In this chapter we address the question of how biodiversity affects the functioning of natural ecosystems in California. *Ecosystem function is defined as the set of processes which maintain natural systems through the transfer of energy, matter and information.* These processes are regulated by species and the biotic interactions that characterise ecological systems (see Hobbs et al., Chap. 1, this Vol.). We know relatively little of how individual taxa affect ecosystem processes and far less of how their role in the system affects the overall functioning of the system. While this region has a relatively rich source of scientific studies on the flora and fauna, nearly all have focused on one or a few species and a single level of

G.W. Davis and D.M. Richardson (Eds.)
Ecological Studies, Vol.109

organisation such as either populations or communities. Some of these studies provide clues about the effect of individual species on certain ecosystem processes, however, such studies have not addressed the question of how biodiversity per se, i.e. the numbers and kinds of species assembled at one time in one place, affects ecosystem functioning. Thus, any attempt to answer the question posed here must rely upon induction and extrapolation. We will briefly review past and present patterns of biodiversity in the State. The extent to which species affect ecosystem processes that control functioning is explored and the evidence for functional groups of ecological equivalents is analyzed. Critical to answering this question is whether species are redundant in the roles they play and the extent to which seemingly minor "players" are linked to ecosystem processes. Lastly, we will focus on evidence that diversity affects ecosystem functions such as carbon gain, water balance and nutrient cycling.

3.2 Patterns and Determinants of Diversity

Diversity encompasses a hierarchy of scales, and each level of diversity, from genetic diversity to gamma diversity, produces effects at other levels. Thus, processes affecting one level will invariably alter other levels of diversity. Climatic and topographic change, particularly over the past million years, has created our particular pattern of landscape (gamma and beta) diversity. Today, both landscape and community (alpha) diversity of plant and animal species is influenced by disturbances such as fire, urbanisation and invasive organisms. For example, the anthropogenic impact on fire frequency has altered the balance of species composition within communities and the distribution of communities across the landscape. Urbanisation has had a profound impact on the fragmentation of habitats, and invasives are potential threats to native species. These changes are of concern because the persistence of populations, particularly in the face of global climate change, may ultimately be determined by existing patterns of diversity. In other words, communities are a storehouse of diversity, some of which may not be tied to the proper functioning of ecosystems now, but could be critical under future conditions.

3.2.1 Historical Determinants of Floristic Diversity

The California Floristic Province, which encompasses the mediterranean-climate region to the west of the Sierra Nevada and other interior ranges, extends over 1500 km from Oregon to Baja California (Fig. 3.1).

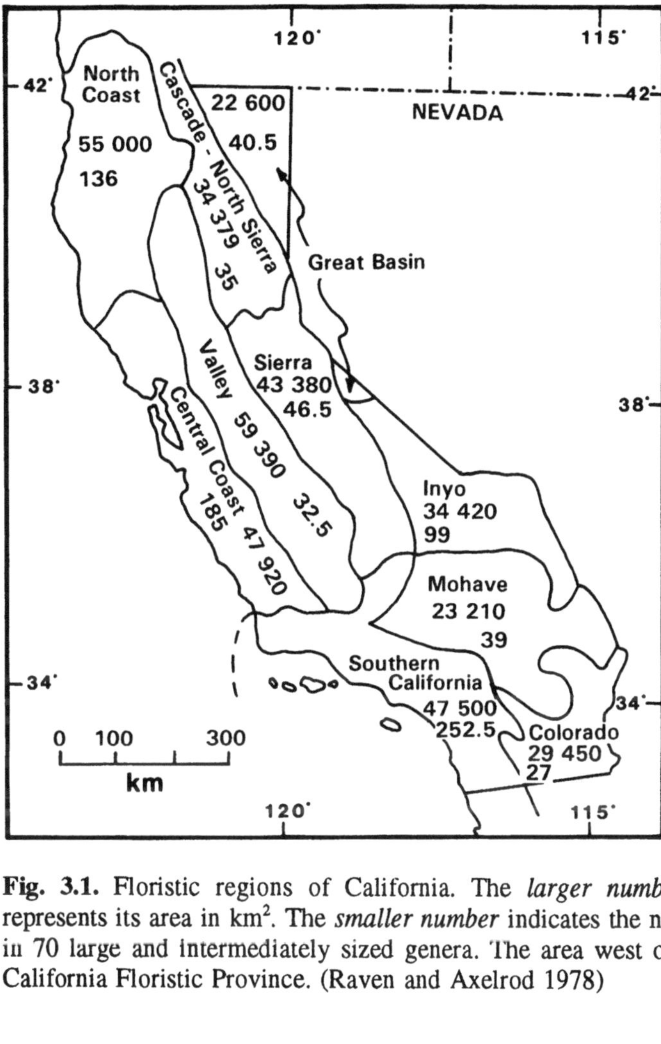

Fig. 3.1. Floristic regions of California. The *larger number* in each region represents its area in km^2. The *smaller number* indicates the number of endemics in 70 large and intermediately sized genera. The area west of *heavy line* is the California Floristic Province. (Raven and Axelrod 1978)

Historically, this broad latitudinal gradient comprised an ecotone between the Arcto-Tertiary geoflora to the north and the Madro-Tertiary geoflora to the south (Axelrod 1958; Raven and Axelrod 1978). During the Miocene (25-15 Ma B.P.) central California included both xeric adapted Madro-Tertiary and mesic adapted Arcto-Tertiary species. These elements segregated in a mixture of community types with Arcto-Tertiary species restricted to cooler habitats, such as high elevations and north-facing slopes or in southern California in riparian habitats (Sauer 1988). During Pleistocene shifts between warm-dry and cool-wet climates, elements of both assemblages mixed as climate forced a cycle of latitudinal migration and retreat (Axelrod 1981, 1986). Beginning in the Late Pliocene, increasing topographic relief, generated by the uplift of the Sierra Nevada and other ranges, had profound impacts on local climates and created many novel combinations of environmental characteristics (Raven and Axelrod 1978; Axelrod 1981). Other geological features of the region, such as unusual substrates like serpentine, have further increased the number of habitats (Raven and Axelrod 1978; Axelrod 1981). Thus, modern plant assemblages represent unique mixtures of species derived from both geofloras.

The present mediterranean climate of California is the result of a climate shift of diminishing summer rains and warmer, drier conditions beginning in the early Pleistocene (Axelrod 1981, 1986; Raven and Axelrod 1978). This, coupled with the glacial/interglacial shifts in climate, resulted in an ecological disequilibrium, which has been the driving force behind the rapid speciation in many characteristic Californian taxa such as *Ceanothus, Arctostaphylos* and *Phacelia* (Stebbins 1976; Raven and Axelrod 1978). In addition, the spread of the summer-drought climate affected wildfire frequency, which has had an impact on speciation patterns in some taxa (e.g. Wells 1969; Keeley and Keeley 1988). Concomitantly, with the shift from a summer-rain to a summer-drought climate, many taxa were eliminated, leading to a marked decline in diversity at the generic level (Axelrod 1973, 1975). The summer-drought conditions would have eliminated many more taxa, but the newly uplifted Sierra Nevada provided mountaintop refugia for taxa such as *Picea, Abies* and *Sequoiadendron*, which were eliminated from lower altitudes and lower latitudes (Warter 1976; Johnson 1977a; Axelrod 1981). Due to the moderate winter temperatures of the mediterranean climate, California also has been a refuge for previously widespread taxa such as *Sequoia*, which were eliminated from colder parts of western North America (Stebbins and Major 1965; Raven and Axelrod 1978).

During the warming of the Xerothermic (8000 - 4000 B.P.) regional climate changes of as little as 1°C had additional profound effects on present-day species distributions (Johnson 1977b; Raven and Axelrod 1978; Axelrod 1981). Taxa migrated north and west into previously cool wet areas and remain today as endemic taxa or northern extensions of a broader southern distribution (Stebbins and Major 1965; Raven and Axelrod 1978; Axelrod 1981). Such climatic patterns may have played a role in the wide disjunctions observed in the distribution of some taxa; e.g. populations of the madrone tree, *Arbutus menziesii*, in southern California are interpreted as relicts of a former widespread distribution pattern, which were left when the species retreated northward due to increasing aridity.

In summary, the flora of California has undergone periods of rapid speciation as well as recombinations of species assemblages. These factors account for the preponderance of endemic plant taxa in the different floristic regions of California (Fig. 3.1) and are certainly important determinants of our present patterns of diversity at all levels. Clearly, large-scale changes in landscape diversity have occurred, with marked shifts in patterns of diversity at other levels. The effect of such changes on ecosystem function are unknown.

3.2.2 Historical Determinants of Vertebrate Diversity

Near the end of the Pleistocene, during a brief period between 11200 and 10800 B.P., over 70% of the genera of large mammalian herbivores in North America became extinct (Owen-Smith 1987; Diamond 1992). In California, this included four genera of megaherbivores (>1000 kg) and many smaller herbivores such as cervids, bovids, antilocaprid, camelid and equid species. In addition, this period also saw the extinction of numerous carnivores, including canids, ursids, and felids, and a diverse raptorial and scavenging aviafauna (Stock 1930; Lundelius et al. 1983).

Climate change cannot account entirely for these extinctions, since massive extinctions did not accompany prior glacial/interglacial periods (Graham 1992). Several other arguments can be made against the climatic explanation for this massive extinction event. Based on Pleistocene fossil floras, it appears that suitable climates did not disappear but rather moved northward and in the process there were few extinctions of plant species on which these megaherbivores depended. Additionally, many of these mammalian taxa were very widespread across North America (Lundelius et al. 1983), suggesting tolerance of a broad climatic regime.

The coincidence of these extinctions with the populating of North America by humans skilled in hunting large prey is used as evidence to support the hypothesis that humans were directly responsible (Martin 1984). Simultaneous changes in climate may have hastened the extinction process (Johnson 1977b), since escalating summer droughts would have forced species to congregate near water, making them easy targets for increasingly larger populations of humans 11000 B.P. (Johnson 1977a). However, the human overkill hypothesis fails to account for the massive extinction of many species which were not likely targets of human predation.

Owen-Smith (1989) has proposed the "keystone herbivore" hypothesis to explain the extinction of the diverse fauna not directly a target of human exploitation. Based on the principles derived from studies of megaherbivores in Africa, he suggests that these Pleistocene mammals were the source of massive disturbances, which would have maintained widespread landscape diversity. The mosaic of habitats thus created may have been critical to the survival of other herbivore species. Today, megaherbivores in Africa are postulated to be an important link in nutrient cycling, acting to keep nutrients near the soil surface where they are easily reincorporated into primary production (Botkin et al. 1981). Thus, Pleistocene megaherbivores would have had a massive impact on nutrient cycling and regrowth after grazing and trampling and this in turn may have been critical to the production of suitable habitats for other species (Owen-Smith 1987; Edwards 1992). With the extinction of the megaherbivores, many of these habitats would have disappeared and been replaced by unsuitable habitat comprising a greater abundance of evergreen woody plants that were less nutritious or toxic (Guthrie 1984). Additionally (as observed today in African ecosystems), with the demise of the major disturbance-creating megaherbivores, shrublands and woodlands would have expanded, leading to more widespread catastrophic wildfires (Owen-Smith 1988).

Since the arrival in southern California of the first Europeans, large mammals, the grizzly bear (*Ursus horribilis*) and the pronghorn antelope (*Antilocarpra americana*) have been eliminated from cismontane California. Many other large carnivores and herbivores have undoubtedly shrunk in numbers due to habitat destruction and lack of corridors between habitat fragments. More recently, due to stream channelisation in the early part of this century, salmon (*Salmo* spp.) spawning has been eliminated from southern Californian rivers.

In conclusion, over a geologically brief period of time there has been a dramatic decline in faunal diversity in California. Based on analogous contemporary megaherbivore-dominated systems, it seems certain that this loss of biodiversity had a profound impact on those ecosystems' capacity for transfer of water, nutrients and energy, and these extinctions had a domino effect, causing further extinctions. The remaining fauna is not only a small subset of the Pleistocene fauna, but it is also a subset of the smallest members of that fauna. Thus, it appears that these changes in ecosystem functioning did not impact all growth forms in a like manner. Apparently, much of the contemporary fauna was represented in the Pleistocene fauna and is likely to have undergone shifts in distribution and abundance. Almost certainly the patterns of natural disturbance have changed, changing also the landscape diversity and distribution of plant communities. Loss of the megaherbivores would have resulted in greatly altered patterns of nutrient cycling and, since sclerophylly is a functional response to decreased nutrient availability (Gulmon and Chu 1981; Mooney 1986), may have reinforced the climatically induced trend toward the dominance of sclerophyllous shrubs and trees.

3.2.3 Present Patterns of Plant Diversity

The vascular flora of mediterranean-climate California includes approximately 5050 species, 30% of which are endemic and 15% are naturalised non-native species (Mooney et al. 1986a). The annual growth form represents a sizable proportion (27%) of the flora and nearly half are endemic to the region (Raven and Axelrod 1978). Diversity of the total flora is highest in mountainous areas but growth forms show different patterns (Richerson and Lum 1980). Trees are proportionately more diverse in the northern coastal areas, herbaceous perennials are most diverse in northern coastal and montane regions, shrub diversity is highest in central and southern coastal areas, whereas annuals are most diverse in the Central Valley and southern coast ranges. Many of these annual species are temporally sporadic in occurrence, arising after wildfires in chaparral or after heavy rainfall in the desert. Due to their ephemeral nature, communities dominated by annuals show marked seasonal changes in diversity.

Present day gamma diversity patterns of Californian vegetation including the total flora and its various life-forms and taxonomic subdivisions can be predicted by climatic and topographic variables, and thus serve as a reasonable test of proposed environmental hypotheses of diversity regulation (Richerson and Lum 1980). The actual relationship between environment and diversity patterns was more complex than might be supposed. Topographic heterogeneity had a strong effect, however, weather variables accounted for the bulk of the diversity patterns in the models used. The results were unexpected in that favourableness indices (mean weather variables) were generally more important than indices of seasonality or irregularity. These results suggest that favourableness hypotheses deserve more theoretical and empirical attention.

There are a number of large (>50 spp.) endemic-rich genera in the state; including woody taxa (*Arctostaphylos, Ceanothus*), suffrutescent or herbaceous perennial taxa (*Eriogonum, Penstemon, Lupinus, Astragalus, Carex*), and mostly annual taxa (*Phacelia, Cryptantha, Mimulus*). Many of the species in these genera show remarkable morphological similarity with congeners, suggesting some level of redundancy or ecological equivalency.

Cody (1986) has addressed this possibility by looking at species turnover as a function of distance in the two largest shrub genera in California. In the genera *Ceanothus* and *Arctostaphylos*, about a quarter of the species in the former genus are restricted to one or two counties, and 40% of the species in the latter genus are local endemics (with ranges less than 50000 km^2). In both genera the percentage of species turnover among 1/2° x 1/2° latitude-longitude quadrats (about 50 x 50 km) along the coast is nearly linear (Fig. 3.2), with a 4-5% turnover per unit shift in distance over a 500 km interval. However, species turnover from coastal points (Santa Cruz,

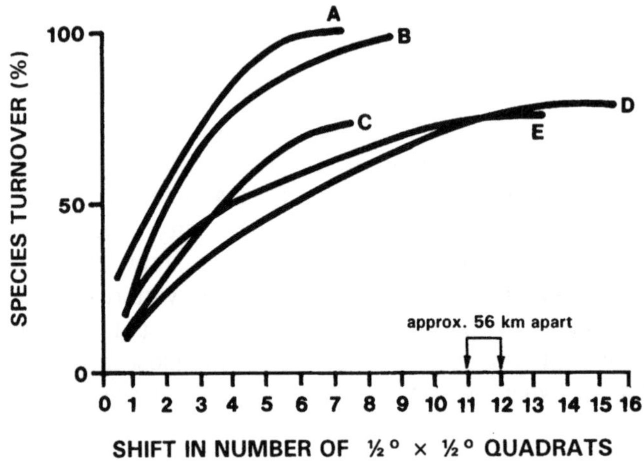

Fig. 3.2. Species turnover rates in *Ceanothus* and *Arctostaphylos* in 0.5° x 0.5° latitude-longitude quadrats, expressed as a percentage of the species that the sites have in common. Santa Barbara/Santa Cruz to inland transects illustrate coast to interior changes (**A**, **B**, **C**), and coastal transects illustrate latitudinal changes (**D**, **E**). (Cody 1986)

Santa Barbara) inland, over a steeper climatic gradient together with topographical changes, is much steeper – about 20% per unit shift in distance. The steeper gradient reflects greater sediment heterogeneity and climate heterogeneity which comes about through local topographic effects on fog distribution. These patterns suggest finely tuned morpho-physiological types adapted to fairly specific ecological conditions, dictated by sediment and climate, and would argue against the notion of redundancy in these large genera.

Even within the same community there is evidence that pairs of morphologically similar annual species of *Cryptantha* and *Lotus* are separated by micro-topographic niche differences, one species of each pair is restricted to gaps and the other to beneath the chaparral canopy (Shmida and Whittaker 1981). Regional differences in shrub composition of coastal sage scrub was suggested as a factor accounting for herb species diversity as herb species seem to prefer associating with different shrub species (Westman 1987).

3.2.4 Present Patterns of Insect Diversity

For natural ecosystems, there is relatively limited data on diversity and abundance of most invertebrate groups. Powell and Hogue (1979) estimate that there are 28000 insect species in California but, outside of agricultural systems, relatively few of these have been studied in any detail. Flying insect communities in chaparral appear to follow marked seasonal patterns with greatest biomass and diversity in late spring (Atkins 1977). Summer drought appears to be a major limiting factor to these invertebrates, as the lowest diversity is observed at the end of the summer drought in October (Atkins 1977). This would also be supported by the report that flying insect abundance on the pole-facing slopes peaks later in the season than on equator-facing slopes (Cody 1974). Clear microsite differences are seen in the distribution of aphid galls (Sholes and Beatty 1987) and are evident by the restriction of biting-flies to humid ravines, a phenomenon well-known to chaparral field biologists.

Ground-dwelling insects in chaparral reach peak diversity and biomass peak much later, in September (Saiz 1977), suggesting that the limiting factors operating in the understorey are different than those in the canopy. In Saiz's study, which lasted 2 years, only 118 species were reported, primarily beetles. It is unclear how closely this approximates the total invertebrate ground-dwelling fauna. In California, it is estimated that there are 206 ant species (Snelling pers. comm. reported in Cody et al. 1977) and Hunt (1977) noted marked elevational differences in ant diversity, with more than double the diversity at 1000 m than at either coast or montane (2000 m) sites. Some of this diversity, however, may be tied to greater plant community diversity sampled at the mid-elevation site (Force 1990).

Caterpillars, moths and butterflies make up the major component of the foliage consuming trophic level in California and are relatively well known. Powell (1992a) has been conducting a census of the lepidopteran fauna in coastal central California at the Big Creek Reserve near Monterey. In total, 810 species of lepidoptera (60 butterflies and 750 moths) have been recognised, and it is estimated that this represents approximately two-thirds of the total Lepidopteran fauna. Adults of 17% of the microlepidoptera at Big Creek, and 9% of all moth species are diurnal, and these are primarily pollinators.

There is some evidence that lepidopteran diversity is tied to plant diversity. For example, Brown (1987) has shown that butterfly diversity is highly correlated with floral diversity in California and Baja California. He suggests that this is because these macrolepidopterans are highly host-specific. Powell (1992a) has made an intensive examination of the larval food plants for a community of microlepidoptera (Fig. 3.3). There are confirmed host plants for 171 insect species at his study site in Monterey and, based on other studies, the host is inferred for 96 additional species. Thus, host plants for about 65% of the recorded microlepidopteran species in the fauna are documented, and known detritivores (34) raise the total number of insect taxa accounted for to 73%. Powell (1992a) found that these species use only 67 native and 3 alien genera, or about 32% of the

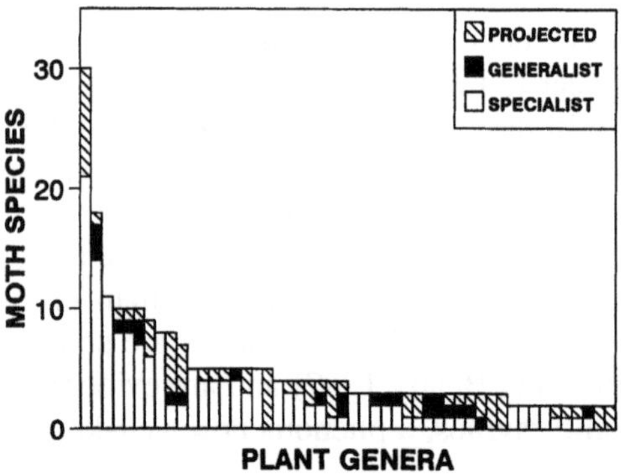

Fig. 3.3. Number of moth species recorded feeding or projected to feed on genera of vascular plants at Big Creek Reserve, California. Each *vertical column* represents one genus, from *left to right*: *Quercus* (31 moth species), *Ceanothus* (18 spp.), *Salix* (11), *Baccharis, Gnaphalium, Arbutus* (10 each), *Arctostaphylos* (9), *Alnus* and *Lupinus* (8), *Artemisia* (7), etc. Remaining 19 plant genera are host to a single species. (Powell 1992a)

native plant genera at Big Creek. It is possible that the remaining moth species that are unaccounted for use many of the other 70% of the flora. However, in a comparable study at the Antioch National Wildlife Refuge in Contra Costa County the pattern is similar. Antioch is a much simpler community with about 110 native plant species in 87 genera, and Powell (1992a) could account for 86% of the microlepidoptera fauna, but they utilize only about 48% of the native plant genera.

Detailed studies of a single shrub species *Eriogonum fasciculatum* in chaparral revealed that as many as 100 genera of insects were recorded from daytime collections and the species composition of nighttime collections was quite different (Swisher 1979).

Bees represent the primary pollinators in California. There are over 1200 native species and it is estimated that they pollinate 95% of the insect-pollinated plants in California (Force 1990). Bees are particularly abundant in chaparral, and Moldenke (1976) suggests that this can be attributed to the fact that most are ground-nesting and the understorey of chaparral, devoid of low ground cover but protected from wind, is the ideal environment. Force (1990) suggests that wind poses an obstacle to pollinating insects and accounts for the decrease in insect pollinators near the coast. Not surprisingly, pollinator diversity, as well as insect abundance in general, rises with increasing biomass of the floral resources (Cody et al. 1977).

Several studies have documented successional changes in insect communities. In mature chaparral insect diversity is generally greater than in the immediate postfire years (Moldenke 1977; Force 1981) and there are marked successional shifts in insect guilds (Force 1982). For example, predatory insects are particularly abundant after fire and parasitic insects increase over time. Also, pollinators are abundant immediately after fire but are gradually replaced by phytophagous insects. There are at least two groups of insects that are the invertebrate equivalent of postfire annual plants. These fire-following insects include species of beetles (*Melanophila*) (Barr and Linsley 1947) and flies (*Microsamia*) (Kessel 1947). Thus, there is reason to believe that insect diversity and trophic structure is tied to plant diversity, climate and disturbance.

3.2.5 Present Patterns of Vertebrate Diversity

There are fewer than 150 species of amphibians and reptiles in California (Laudenslayer et al. 1991) and the best studied are the lizards, which comprise about a third of these taxa. In general, alpha diversity is relatively low, although even so, species sub-divide the habitat by micro-topography (Cody et al. 1977). Relative to Chilean matorral, Californian lizard communities have lower alpha and beta diversity, but greater gamma diversity.

Excluding rare visitors, the Californian avifauna comprises approximately 450 native species, although a large percentage of these are associated with the marine environment (Laudenslayer et al. 1991). Besides the Californian condor (*Gymnogyps californianus*), only the yellow-billed magpie (*Pica nuttalli*) is endemic to the state. The Passiformes or perching birds comprise more than 40% of the total avifauna and nearly 65% of the terrestrial bird fauna. Relatively few of the passerines are restricted to just a single plant community, although exceptions include the Californian gnatcatcher (*Polioptila californica*) which is restricted to coastal sage scrub (Atwood 1993) and the wrentit (*Chamaea fasciata*) and Californian thrasher (*Toxostoma redivivum*) which occur only in chaparral (Cody 1973). The narrow restricted distribution of the latter species inspired Grinnell (1917) to coin the now famous term "ecological niche" to describe the thrasher's role in the chaparral community.

Focussing largely on the Passiformes, Cody et al. (1977) showed a very marked turnover of species across the landscape with a peak in species turnover per habitat in mid-elevation chaparral (Fig. 3.4). In addition, alpha diversity changed markedly in different plant communities, reaching a peak in plant communities with the greatest biomass (Fig. 3.4). The association between avian diversity and plant biomass is tied to different foraging opportunities provided by communities of greater stature. For example, Dutton (1981) found that during the breeding season in chaparral 27 bird

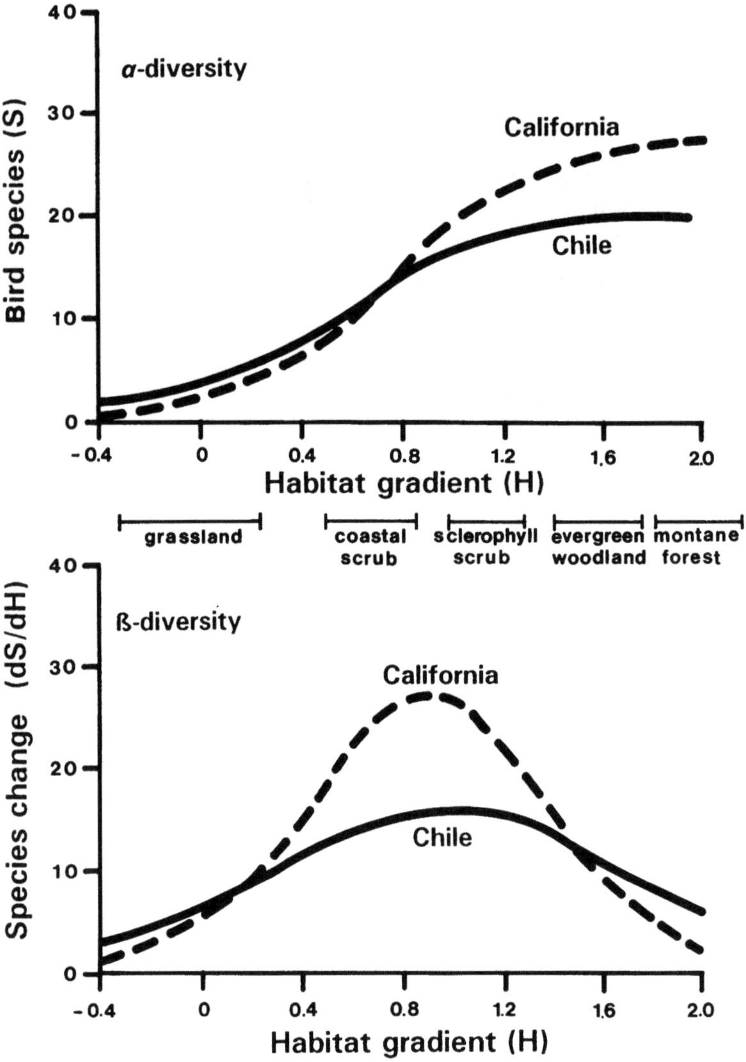

Fig. 3.4. Bird diversity in California and Chile. *Above* Alpha diversity plotted as a function of position on the habitat gradient. *Below* Beta diversity plotted as a function of position on the habitat gradient (Cody et al. 1977)

species partitioned resources primarily by differential foraging height in the shrub canopy.

The California gnatcatcher is an extreme habitat specialist restricted to coastal sage scrub vegetation (Atwood 1993). Due to rapid urbanisation and destruction of this community, the gnatcatcher is a federally listed endangered species. Not only is it restricted to coastal sage scrub, but there is evidence that it is restricted to particular sub-associations. For example, it is apparently absent from *Salvia mellifera*-dominated sites (Atwood 1993) and even within communities there is evidence of shrub preferences.

Temporal changes in avifauna diversity have been reported following wildfires in chaparral. In recent burns, Wirtz (1979) showed that breeding bird species diversity increased over unburned chaparral. Five years after fire, species composition of burned chaparral was still decidedly different from mature stands (Yoder 1976), and as noted above for insects, there were marked changes in the trophic guilds of birds present in burned and mature chaparral. However, bird density may be higher in burns due to a decrease in territory size resulting from increasing food resources such as seeds and insects. In fact, Alten (1981) observed that initial postfire bird diversity was correlated with insect diversity, but as the stand aged this did not hold.

There are approximately 210 species and subspecies of native terrestrial mammals in California of which rodents (Rodentia) comprise 57% (Laudenslayer et al. 1991). Evidence indicates that environmental complexity is an important and general determinant of animal diversity in the state. Additionally, small mammal diversity may vary seasonally or annually. For example, rodent communities in coastal sage scrub experience fluctuations in the relative abundance of species and the variable presence of rare species results in temporal fluctuations in species diversity (McCloskey 1972).

Unlike birds, which are capable of more efficient utilisation of different levels in the canopy, mammal diversity does not increase with increasing plant biomass in California (Cody et al. 1977). Rather, much of the small mammal diversity in California is accounted for by local differences between habitats (beta diversity) and thus gamma or landscape diversity is relatively low.

3.2.6 Diversity Patterns of Fungi

For the more conspicuous fungi it is estimated that there are perhaps 4000 species of Basidiomycetes, 300 species of slime molds and 1000 species of lichen (Jensen et al. 1990). According to Armstrong and Platt (1993), there has been a dramatic decline in the lichen flora of southern California, due largely to air pollution.

Microscopic fungi in our systems have not been studied in great detail but some generalisations are possible. Any given site may have a dozen or more genera represented in the soil and, as with other taxonomic groups, there are widespread generalist species as well as highly specialised species.

In a study of different vegetation types on Mount Diablo in northern California it was observed that the genus *Penicillium* could be divided into species that were nearly ubiquitous in time and space and species that were highly specialised to different vegetation types and even to different types of chaparral, dependent upon the shrub dominant (Fordham 1977). The effect wildfires have on fungi and other soil microbes is not exactly clear. Apparently there is an initial dieback but they may come back in greater mass early in succession (Dunn et al. 1985).

3.2.7 Patterns of Genetic Diversity

Life history mode has had a profound influence on patterns of genetic diversity and this has most likely been important in adapting to the radical climatic shifts since early Pleistocene. Speciation in plant genera such as *Arctostaphylos* and *Ceanothus* was driven by increasing aridity and increasing probability of fire. Wells (1969) suggested that adaptive radiation of these genera is tied to the evolutionary loss of the resprouting mode, and this is supported by the observation that the majority of the species are obligate seeders, which do not resprout after fire. Obligate seeders have greater numbers of sexually produced generations, resulting in greater genetic recombination, which in turn contributes to more rapid speciation. A consequence of this life history mode is the potential for greater fine-tuning to the aridity of the summer-drought climate. Thus, it is not surprising that obligate seeders are far better able to tolerate drought stress, both as seedlings and mature shrubs, than resprouting taxa (Keeley and Keeley 1988). This may also explain why several resprouting species of *Arctostaphylos* produce obligate seeding populations on the most arid borders of their distribution (Keeley, unpubl. data).

The differences in life history mode have observable effects on patterns of genetic variation. Within the genus *Arctostaphylos*, populations of obligate seeding species tend to be morphologically homogeneous relative to populations of resprouting species. Each new generation of obligate seeders arising after fire is the product of genetic mixing, recombination and intense selection. Populations of resprouters, on the other hand, are far more heterogeneous and appear to represent a collection of morphologically distinct clones, many of them potentially quite ancient (Keeley pers. observ.).

Related perhaps to the rapidity of evolution, as well as the shifting climatic pattern which maintains ecological disequilibrium, woody taxa such as *Arctostaphylos* and *Ceanothus* have not evolved fertility barriers (Raven 1973). These shrubs are largely outcrossing and where species distributions overlap, hybrid swarms are common (Dobzhansky 1953;

Gottlieb 1968; Keeley 1976; Ellstrand et al. 1987) and there is evidence that hybrids and backcrosses may segregate out in the landscape in different microhabitats (Ball et al. 1983). Also, such hybrid swarms may represent a reservoir of genetic variation poised to colonise newly available habitat which is provided by disturbance such as fire. Genotypes of individuals on the margins of populations, where the potential for hybridisation exists, are likely to differ from those in the center (Bazzaz 1986). Patterns of variation in *Arctostaphylos* strongly support this idea and it undoubtedly accounts for the taxonomic difficulty of the group. For example, the very widespread resprouting *A. glandulosa* has had more than a dozen subspecific epithets assigned, apparently because of previous hybridization events (Keeley, unpubl. data). They persist as clones and are not swamped out by genetic recombination.

On the other hand, some clonal species may be genetically depauperate. For example, the very rare southern Californian *Nolina interrata* is a rosette-forming subshrub represented by only a few widely scattered populations. Genetic analysis suggests that one such population of 2000 rosettes is apparently composed of just a single clone (Bauder 1993), a particular problem for this dioecious species.

In the herbaceous flora, annuals are the obligate seeders and perennials are analogous to resprouters; however, the annual turnover in generations has led to internal barriers to hybridization (Raven 1973). Thus, we find that annuals are less variable genetically (Karron 1991). For example, genetic identities among seven populations of the annual *Pentachaeta lyonii* range between 1.00 and 0.990, and mean heterozygosity ranged from 0.000 to 0.058 (Arias et al., unpubl. data). This endangered species is endemic to the Santa Monica Mountains of southern California and has been reported from fewer than 50 locations (California Natural Heritage Program Natural Diversity Data Base 1991). The lack of variability within this species does not necessarily reflect decreasing population sizes, but may instead be related to its life history strategy. *Pentachaeta lyonii* appears to expand preferentially into disturbed areas on a temporary basis, persisting until resources become limiting as a result of increasing competition (Keeley and Keeley 1993). Such a life history, which restricts reproduction to periods of favourable conditions is similar to that exhibited by chaparral annual species (Swift 1991) and desert annuals (Mulroy and Rundel 1977) and would increase reproductive isolation of populations responding to localized changes in resource availability. Relative to long-lived species, annuals are more likely to have locally adapted populations and are replaced more rapidly as we move from one habitat to the next, i.e. they exhibit greater beta diversity.

An appreciable portion of theoretical population biology is based on the temporal and spatial variations in the frequency of different chromosomal inversions in *Drosophila pseudoobscura*. Thus, the effect of chaparral wildfires on such genetic changes is of some interest. Moore et al. (1979) demonstrated that the gene arrangements of the third chromosome of *D. pseudoobscura* were unchanged in populations of mature chaparral and recently burned chaparral. What is remarkable is that the frequency of various inversions observed by Dobzhansky in the 1950s have remained relatively stable, despite repeated wildfires and colonization by closely related species. Thus, the ecological role of genetic variation in some species is far from resolved.

Smith (1978) studied allele frequencies between populations of the deermouse *Peromyscus californicus* and suggested that the pattern of geographic variation may have been affected by historical events and the potential exists that fragmentation of the central coast region may have contributed to isolation of the northern populations.

3.2.8 Effect of Wildfire Frequency on Landscape Diversity

Wildfires play an important ecological and evolutionary role in many Californian ecosystems and potentially affect biodiversity. It is clear that species are not equally adapted to all fire frequencies and there is evidence that entire ecosystems may be replaced by altering the burning regime. The landscape mosaic predicted by Owen-Smith (1989) for the Pleistocene environment of southern California would have had a very discontinuous fuel load. This landscape heterogeneity predicts a more heterogenous fire environment than that created today by massive stands of contiguous fuels. This, coupled with anthropogenic ignitions, has increased the fire frequency (Wells 1962), with the result that many shrublands have been replaced by non-native annual grasslands (Fig. 3.5), and this has had measurable impacts on vertebrate diversity patterns (Lillywhite 1977). Such increased disturbance has possibly increased landscape diversity and, due to the dynamic nature of shrubland displacement and re-invasion, there is much temporal diversity in the landscape (Fig. 3.6).

In montane regions, apparently decreased fire frequency has likewise altered ecosystem processes and potentially altered community characteristics (Rundel 1971; Kilgore and Taylor 1979).

Predicted climate changes may alter fire frequency, and Parsons (1991) has suggested that with a predicted increase in lightning ignitions (perhaps as much as 26%) the potential exists for profound landscape changes that might accompany such an increased source of fire ignition.

Fig. 3.5. Vegetation mosaic in southern California coastal ranges. Grasslands are dominated by non-native annuals and have displaced native perennial grasslands on the alluvial flats in the *foreground* and have displaced coastal sage scrub on the *slopes*. (Keeley 1990)

3.2.9 Effect of Alien Species' Invasions on Native Flora and Fauna

Invasion and establishment of alien species has had a profound impact on Californian ecosystems and has altered diversity patterns at all levels from genetic to landscape diversity (Mooncy and Drake 1986; Groves and di Castri 1991; D'Antonio and Vitousek 1992). Genetic diversity of native species is often affected through hybridization with invasive taxa, as is the case with southern Californian cacti following the introduction of *Opuntia* from Mexico (Benson and Walkington 1965). Sometimes regional movement of plant material can affect local taxa. For example, the island cherry, *Prunus lyonii*, endemic to offshore islands, is widely planted and seeded in southern California and readily hybridizes with the mainland *Prunus ilicifolia* (Keeley, unpubl. data). Intraspecific effects are possible when locally adapted gene complexes are disrupted by the intentional introduction of foreign ecotypes in community restoration projects (Ledig 1992); however, such introductions could have a beneficial impact on populations that are genetically depauperate due to demographic bottlenecks.

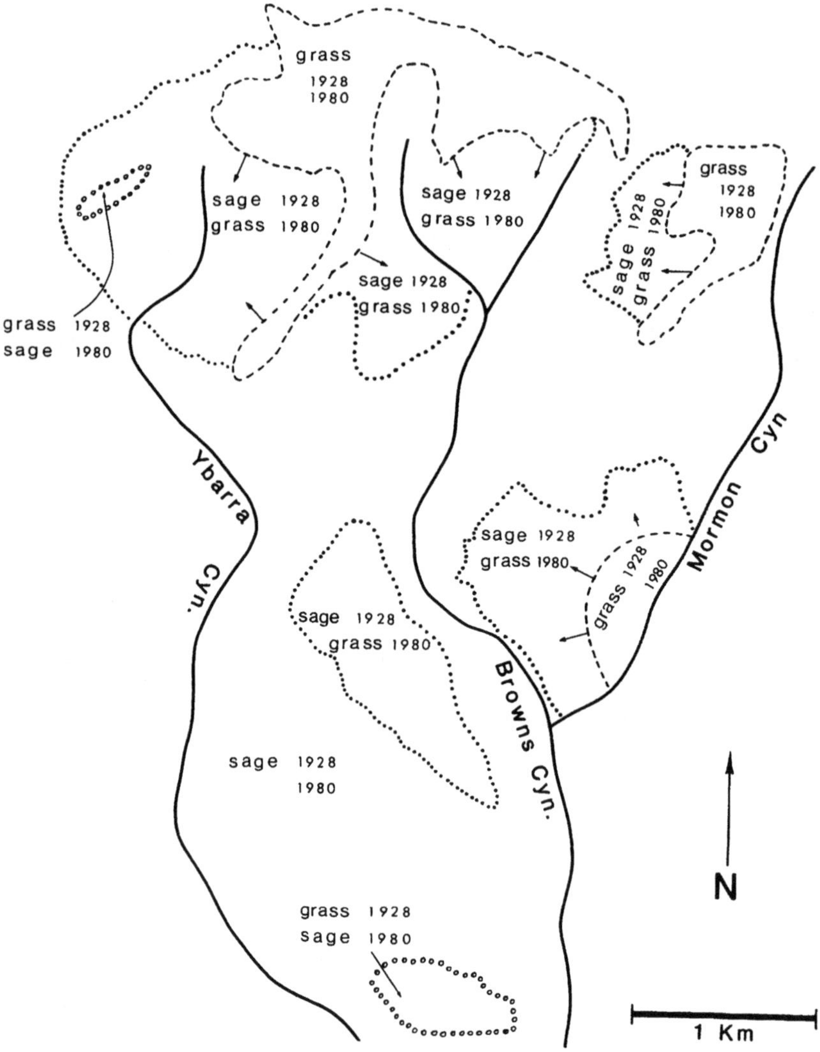

Fig. 3.6. Historical and present-day distribution of grasslands and coastal sage scrub (as of 1980) on a portion of the Oat Mountain Quadrangle, Los Angeles Co. (Freudenberger et al. 1987)

Invasives may also impact community diversity patterns by causing the extirpation of native species. For example, in chaparral the diversity of herbaceous species has been shown to be drastically reduced by the management practice of seeding the alien grass *Lolium multiflorum* (Keeley et al. 1981; Nadkarni and Odion 1986). Likewise, avian community diversity in California may have been altered by the introduction of the European Starling, *Sturnus vulgaris*, which was introduced in 1890 in New York and arrived in California in 1942. It competes with native hole-nesting species such as bluebirds, titmice, nuthatches, swallows, wrens, woodpeckers and nuthatches (Vuilleumier 1991). Although a total of 19 bird species have been introduced to California, nearly all are restricted to urban environments and only the starling has successfully invaded the chaparral (Vuilleumier 1991).

Ecosystem effects of feral sheep and goat introduction on coastal islands of southern California are horrific. The effect of uncontrolled overgrazing and trampling have denuded many of the island ecosystems (Van Vuren and Coblentz 1987) and often invoke the image of a lunar landscape. These introductions into an environment devoid of the necessary predator diversity illustrate clearly the impact of simplified community structure on ecosystem stability.

Invasive species can also affect landscape diversity as illustrated by the relatively rapid and widespread establishment of annual grasses and forbs (e.g. *Avena, Bromus, Hordeum, Lolium, Brassica and Erodium*). Today, annual grasslands persist under a wide range of topographic and edaphic conditions. They have replaced shrublands after repeated fires and replaced perennial grasslands under regimes of overgrazing and drought (Fig. 3.5). Under both conditions the landscape has changed from one dominated by native perennials to non-native annuals.

A consequence of this invasion is the destruction of much of the native grassland community. The invasion of the pristine Californian Valley Grassland illustrates how rapidly and thoroughly an alien flora can dominate another landscape. The process began in earnest in the late 18th century and was nearly complete by the middle of the 19th century. Perhaps what is most uniquc about Californian annual grasslands is the comparative stability of this ecosystem dominated by non-native annuals with relatively little threat of being displaced by natives.

The precise effect of this invasion on grassland biodiversity is unknown. The pristine prairie in California was dominated by perennial bunchgrasses in genera such as *Stipa, Poa, Koeleria* and others although there is evidence that they seldom exceeded 50% ground cover (Keeley 1990). The interstitial spaces between the bunchgrasses would have been filled by native annuals and geophytes. Replacement of the native flora by introduced species has most likely reduced the diversity of life histories and phenologies in Californian grasslands. The earliest flowering non-native species flower later than the earliest natives and the latest-flowering non-natives flower earlier than the summer active natives (Chiariello 1989). Reduced growth form and life history diversity in modern annual grasslands have potentially created a very different environment for foraging animal species. It has been demonstrated in southeastern Arizona that native grasslands invaded by non-native grasses have a significantly lower variety and abundance of indigenous grasses, herbs, shrubs, grasshoppers, rodents and birds than pristine grasslands (Bock et al. 1986).

In California, invasion by non-natives has most likely resulted in a more homogenous pattern of landscape diversity. Most annual grass species have a great deal of ecological and geographical breadth. However, the native grassland community exhibited much diversity across the state with replacement of the widespread *Stipa pulchra* by other native species on different sites (Beetle 1947). For example, moister sites were dominated by *Muhlenbergia rigens* and drier interior sites by *Aristida* and *Oryzopsis*. On rich alluvial soils were rhizomatous species of *Elymus* and *Agropyron*, on gravelly ridges and serpentine, *Sitanion jubatum* dominated, and alkaline flats were dominated by the rhizomatous perennial *Distichlis spicata* and the bunchgrass *Sporobolus airoides*.

Ellstrand (pers. comm.) has suggested that this colonization event may have had a profound impact on the ability of ecological communities to respond to pollution and global climate change. Native perennials as a group naturally harbour high levels of genetic diversity and are largely outcrossers whereas the invasive annuals are genetically less diverse and are typically selfers. However, this analysis is complicated by the greater adaptive potential of the annual life cycle. Predictions for future climate change indicate that the rate of temperature rise in the Central Valley of California may approach 5°C over the next 75 years (Jensen et al. 1990). Almost certainly, perennials will be unable to adapt or migrate rapidly enough to such unprecedented rates of change, and thus these introduced annuals may spread even further.

3.2.10 Effect of Habitat Fragmentation on Patterns of Diversity

Rapid population growth in California has resulted in widespread habitat destruction throughout the state; consequently, remaining natural areas represent habitat islands often without corridors between (Swift et al. 1993). Several studies have investigated the effect of such fragmentation on biotic diversity and have inferred effects on the functioning of these systems.

Soulé et al. (1988) did an island biogeographic study of specialist bird species in chaparral habitat fragments in coastal urban areas of San Diego County, California. These fragments are portions of canyons that have been insularized by encroaching development on canyon rims and in canyon bottoms. An archipelago of "islands" of native chaparral vegetation has been formed with fragments ranging in size from less than 1 hectare to hundreds of hectares and of varying ages up to 80 years.

They found a significant negative partial correlation between species diversity and fragment age (time since insularization) and a positive correlation between a species' density and the number of habitat fragments it currently occupies. They concluded that populations of species that naturally occur at lower density become extinct more rapidly than do those of more abundant species, i.e. species differ in their vulnerability to extinction. Additionally, Bolger et al. (1991) compared the species-area relationship in unfragmented chaparral habitat with that in urban chaparral fragments and confirmed that rapid population extinction of resident bird species had occurred in these fragments. A strong positive correlation between the relative persistence ability of a species and its density remained even after correcting for the sampling effect of area. Bird species present in species-poor fragments were nested subsets of those in species-rich fragments. Once isolated, these fragments degrade with time as a result of invasion by alien plants.

Effect of habitat fragmentation on plant diversity in these same coastal scrub communities has been studied by Alberts et al. (1993). They found that for 25 isolated patches of coastal scrub vegetation in San Diego County, the variance in native plant species number was best accounted for by size and age variables, with larger and more recently isolated patches supporting more species. Fragments with longer perimeters contained more ruderal weeds, whereas older fragments and those with artificially supplemented water sources supported higher numbers of escaped ornamentals.

Human disturbance, including clearing of vegetation, addition of non-natural water supplies, and disruption of fire regimes, has probably contributed to the loss of natives and subsequent invasion of alien plants into fragments. Obviously, any change in the flora could potentially upset the community structure of animal communities. Other factors that could affect these communities include the introduction of "subsidized" predators such as domestic cats (*Felis domestica*) and dogs (*Canis domesticus*) which are effective predators of native bird and mammal populations (Wilcove et al. 1986). Soulé et al. (1988) proposed that in less disturbed habitats large predators such as the coyote (*Canis latrans*) may reduce the populations of medium-sized nest predators such as the opossum (*Didelphis virginianus*). In small, highly disturbed fragments the large predators go extinct and populations of "meso-predators" increase, causing increased nest predation. In this instance the coyote is considered a "keystone" species (see Hobbs et al., Chap. 1 Sect. 1.4 this Vol. for a definition), which, when locally extinct has profound impact on the biotic diversity of other animal guilds. It remains, however, to develop a general conceptual framework that will predict which kinds of organisms play key roles in different kinds of ecosystems. With such knowledge, if extirpation of these keystone organisms cannot be avoided, their roles could potentially be replaced by other species or other forms of ecosystem management.

Despite the lure of Soule's "keystone coyote" hypothesis, not all subsequent studies have supported it. Langen et al. (1991) found that the predation rate of artificial birds' nests was highest in an unfragmented site. Among fragments, predation was higher at moderately disturbed than at highly disturbed sites. These results suggest that nest predator species diversity or density is reduced in disturbed chaparral fragments and that nest predation is probably not the most important cause of the observed loss of chaparral breeding bird diversity in habitat fragments.

All of these studies suggest the important role of habitat size in affecting faunal and flora diversity. Consistent with this is the finding by Sauvajot and Buechner (1993) that for 12 habitat fragments in the Santa Monica Mountains of southern California variation in bird and mammal species richness and relative abundance were not correlated with proximity to development. These studies illustrate how chaparral habitat very close to urban development can have important conservation value.

It is apparent, however, that direct alteration of a site during development has a very destructive effect on biodiversity (Sauvajot and Buechner 1993). However, in a study of central Californian suburban bird populations, Vale and Vale (1976) found that both numbers of species and numbers of individuals increased with time after development, although most of the increase in species diversity is concentrated in the initial years following housing construction. Garden plantings seem most influential in determining the distribution and density of birds, but other factors (e.g. the presence of bird feeders) may also be important. When compared to the pre-suburban habitats adjacent to the suburbs studied, the residential neighbourhoods supported larger numbers of both species and individuals and potentially could visit adjacent habitats and have a detrimental influence on these communities.

3.2.10.1 Fragmentation and Ecosystem Processes

Ecosystem functioning of habitats may be altered by fragmentation and proximity to urban development. For example, pollinator services may be greatly upset by insecticide application in adjoining agricultural fields. In southern California, much of the landscape is dissected by golf courses which have a myriad of indirect impacts on ecosystem processes. For example, increased human activity and noise may reduce raptor and other predator activity in adjoining grasslands and scrub. Loss of these predators could threaten plant species sensitive to uncontrolled rodent and rabbit populations. The threat may be exacerbated by the fact that summer drought acts as a natural check on these predator populations. Golf courses provide a high nutrition food resource during the summer, which contributes to the maintenance of unnaturally high rodent and rabbit populations. Not only are predation rates increased under such conditions but, due to the connectedness of these natural and unnatural systems, these habitats will be nutrient- enriched by movement of animals between the golf courses and the grasslands. In native grasslands, such nutrient enrichment is known to favour non-native weeds at the expense of native species (Keeley 1990).

Theoretical considerations predict a negative impact of subdivision on regional species numbers when species interactions are strongly positive or mutualistic (Gilbert 1980). Conversely, subdivision is particularly likely to increase diversity when strong negative interactions occur among species and otherwise threatened species are able to escape in space and time. Few, if any, empirical tests have examined diversity in otherwise comparable areas differing only in the degree of subdivision. Numerous species-area studies indicate that diversity should be maximized in single contiguous areas, whereas patch models often predict regional co-existence and thus enhanced total diversity, resulting from spatial subdivision of habitat (Soulé and Simberloff 1986).

This problem was investigated experimentally in a Californian annual grassland by isolating experimental areas of 64 m^2, subdivided into 2, 8 and 32 subunits, with each subunit separated by intensive grazing (Quinn and Robinson 1987). Species richness and species evenness of flowering plants increased with increasing subdivision, with approximately 40% more species in the most subdivided treatment than in the least. It was concluded that aboveground interspecific competition appears to be an important causal mechanism. There is evidence of strong "priority effects" in which dominant species pre-empt particular plots and prevent the establishment of competitors. Late-emerging and low-growing species occur primarily on plot edges and the greater relative perimeter of the more subdivided treatments may contribute substantially to their greater diversity.

A far-reaching consequence of habitat fragmentation is the decreased potential for plant migration in response to impending climatic change which could result in large scale changes in gamma diversity. Movement of species between interior ranges and coastal ranges during the Xerothermic most likely involved movement along watercourses (Axelrod 1981). In addition, these watercourses probably acted as localized mini-refugia (particularly in southern California) (Sauer 1988), allowing taxa to persist in California which otherwise might have been lost during warming over the last million years. Large-scale elimination (89% of valley riparian habitat has been lost as a result of habitat conversion) and alteration of such natural corridors (Mooney 1983; Jensen et al. 1990) may have eliminated important migration routes for plant and animal species responding to climatic change with shifts in population distribution.

3.3 Biodiversity and Community Structure and Function

Functional linkages between different trophic levels in a community are common and important to evaluating the degree to which biodiversity controls functioning of entire ecosystems. MacArthur (1955) argued that more complex communities were also more stable. Complexity may be defined by the diversity of species or the diversity of pathways energy takes to reach a particular population. The examples chosen here illustrate the difficulty of lumping species into "functional groups", as species within such a group may be tightly linked to species in quite another functional group (Schulze and Mooney 1993).

3.3.1 Plant Linkages with Other Trophic Levels: Oaks as a Case History

Perhaps more than any other element, oaks (*Quercus* spp., Fagaceae) play a fundamental role as keystone species in Californian mediterranean-type ecosystems (MTEs). They affect plant and animal diversity through impacts on the functioning of food webs and as important functional links between other ecosystems.

Species of oaks are critical food sources for a myriad of fungal and animal species, including aboriginal peoples (Pavlik et al. 1991). Oaks are utilized by a diverse array of herbivores. Most obvious are the granivores, that depend upon acorns as an important part of their diet. Some, such as the scrub jay (*Aphelocoma coerulescens*), in turn play critical roles in the dispersal and caching of acorns. Without their assistance, oaks could not persist on steep slopes, as gravity tends to carry acorns to the bottom of ravines (Grinnell 1936). Studies in other parts of North America reveal that, although jays feed exclusively on acorns during the fall, they are unable to exist entirely on this diet. Consumption of acorns infested with the acorn weevil larvae is critical to counteracting the effects of acorn tannins and this accounts for their apparent subsistence on just acorns during the fall caching season (Johnson et al 1993). Oaks sustain these much-needed dispersers during the interim periods of acorn crops by the array of insect larvae specialized on oaks. One of these is the Californian oak moth (*Phryganidia californica*) that feeds on the coast live oak (*Quercus agrifolia*). Peaks of herbivore damage occur during the summer drought as the oak moth preferentially feeds on the less nutritious mature foliage, as a means of circumventing quantitative plant defenses such as condensed tannins (Mauffette and Oechel 1989). Populations of the oak moth are controlled by a virus which is always present but increases in populations over time, possibly contributing to a cyclic pattern of oak moth infestation (Baker 1965). Mycorrhizal fungi, which proliferate on oak roots, produce mushrooms that are important food sources for insects and vertebrates.

Another granivore is the acorn woodpecker (*Melanerpes formicivorus*), which stores acorns in small holes that it drills in storage trees known as granaries. Trees are sometimes riddled with holes and this provides needed entrance for insects and microbes, which play a critical role in the decomposition process of dead trees. Similar ecosystem services are provided by sapsucker woodpeckers (*Sphyrapicus* spp.) that tap living oak tissues for sap.

All oaks harbor gall-forming wasps, and in the western USA oaks are host to more than 140 species (Brown and Eads 1965). All are in a single family of Hymenoptera, the Cynipidae, and most are restricted to oaks. This appears to represent an ancient association, as black oaks and white oaks have their own separate cynipid fauna. In some species of wasp there is an alternation of generations in which bisexual populations generate asexual populations, separated in time and space, and each generation produces a different gall on a different part of the plant. Gall larvae are often parasitized by other species of wasps, and the galls themselves may harbour other insects long after the original resident has left.

Evergreen and deciduous oak species play different roles in the landscape. Some animals are dependent upon the perennial nature of the foliage in evergreens, both as food and shelter, whereas other species may require the dense litter layers that develop under deciduous species. All oaks play an important role as substrates for lichens, and mistletoes and lichens may be important sources of nesting material for an array of bird species.

In savanna landscapes oaks are linked to the functioning and biodiversity of surrounding grasslands through their impact on the light regime and on nutrient cycling (Gordon et al. 1989; McClaran and Bartolome 1989; Callaway et al. 1991; Welker et al. 1991) In shrubland habitats scrub oaks may play critical roles as shelter for animals that graze the adjacent grasslands (Davis and Mooney 1985; Hobbs and Mooney 1986) as well as facilitating the establishment of more mesic trees such as coast live oaks (Callaway and D'Antonio 1991; Keeley 1992a).

On coastal slopes oaks act as fog-collectors, contributing significant amounts of dew precipitation, which affects the water budget of the site. Seedling recruitment by other species may be dependent upon this added moisture.

Oak communities also play key roles in the linkage of communities. For example, as seasonal food sources for birds that migrate from montane environments during the fall when acorns are dispersing, such as the large flocks of band-tailed pigeons (*Columba fasciata*) that seek out the massive acorn crops that occur in mast years. Phainopeplas (*Phainopepla nitans*) move in winter from their desert environments to obtain mistletoe berries. Mistletoes also provide a critical food source for migratory frugivores such as waxwings (*Bombycilla* spp.).

3.3.2 Vertebrate Communities

MacArthur (1972) hypothesized that competitors in communities of species that depend on a common limiting resource are likely to develop divergent resource utilizations. He concluded that abundances of competitors will equilibrate at a configuration minimizing a quantity which is the difference between the amount of resources produced by an environment and the amount utilized by the community. If too much of one type of resource is left unused, then a new species that uses the resource efficiently can invade the community, thereby reducing the ratio. Similarly, a particular species that depends on a rare resource will be rare or absent. In short, this theory provides a method for predicting species abundances from knowledge of the availabilities and species utilizations of critical resources. The theory has implications for the role of biodiversity in the functioning of communities and ecosystems, and predicts that loss of biodiversity from an ecosystem will prevent complete utilization of resources.

Price and Waser (1984) compared predicted changes in rodent abundances with those observed over the two and one-half years following a fire in a topographically uniform area of California coastal sage scrub vegetation. They used microhabitat-specific capture frequencies as resource-utilization functions for each species, because structural microhabitats appear to be related to underlying resources that limit rodent populations. Resource-availability functions were derived from censuses of vegetation and other structural features on burned and unburned sides of the permanent grid. Predictions from MacArthur's theory were upheld in that (1) abundances of each species on one side of the grid relative to the other correlated significantly with relative K values on the two sides of the grid (K values were calculated as the overlap between microhabitat utilization and availability functions); (2) relative abundances of the five rodent species on the burned side of the grid were correlated with those predicted from relative K values; and (3) the total microhabitat utilization of the rodent community on each side of the grid matched overall microhabitat availability on that side well and matched availability on the other side poorly. These results provide support for a potentially powerful protocol for predicting identities and relative abundances of co-existing species.

One means of evaluating the extent to which redundancy occurs in natural ecosystems is to look at the manner in which animal guilds utilize plant resources in their environment. Studies to date are consistent in demonstrating that co-existence of rodent communities is a function of the manner in which species divide up resources. For example, Meserve (1976a, 1976b) studied the food habits of seven species of rodents in a Californian coastal sage scrub community. Laboratory feeding experiments using the cafeteria method indicated different patterns of food preference from wide acceptance of shrubs, forbs, and grasses (*Dipodomys agilis*) to lower preferences for shrubs and high preferences for grasses (*Reithrodontomys megalotis, Peromyscus eremicus* and *P. californicus*) to relatively narrow acceptance of only grasses (*Perognathus longimembris*). Other dietary axes that act to separate these rodents include a high utilization of insects and *Eriogonum fasciculatum* by *Reithrodontomys*, a generalized diet for *Perognathus maniculatus*, and a specialization on various seeds, flowers, and fruits, by *P. californicus* and to a lesser degree by *P. eremicus*.

Additional factors affecting co-existence include differences in nest sites, e.g. *Dipodomys* burrows in sandy soils whereas some other species do not use burrows. *Peromyscus* species are dependent upon free water while *Perognathus* and *Dipodomys* are independent of free water. Also, species have horizontal and vertical differences in spatial preferences within the coastal sage scrub, which contribute to differences in resource exploitation.

One type of observation that is suggestive of incomplete redundancy is that microhabitat specializations do not disappear entirely when competitors are removed (Price 1978). Additionally, it is unlikely that removal of one species from a community will be exactly compensated for by increased consumptive biomass of another species. For example, Brown and Munger (1985) found that *Dipodomys* spp. removed from a site are not replaced by equivalent consumptive biomass of other small granivores. The reason for this is probably one of efficiency: no two species will have exactly equal efficiency at utilizing a particular food type. Hence, loss of one species means lower net efficiency by the remaining species. For example, less consumption means less rapid biomass recycling, which means less fertile soil, which translates into reduced plant productivity.

3.3.2.1 Rodents and Plant Community Structure

There is tremendous potential and at least a few documented cases in which rodents are linked to processes that would impact plant species diversity. Seed predation is ubiquitous and widespread and by its sheer force would seem to be potentially capable of affecting plant diversity. One study showed that predation of *Arctostaphylos* seeds was significantly greater on one species than another (Keeley and Hays 1976) and this could potentially affect not only community composition but genetic diversity for fruit size since rodents selectively removed the largest seeds. In annual-dominated communities, rodents that selectively remove large-seeded species have been shown to have a profound effect on plant species diversity (Samson et al. 1992).

The bare zone between mature chaparral or coastal sage scrub and grassland, possibly controlled in part by allelopathic inhibition (Muller et al. 1964), owes its existence on some sites to rodent and rabbit predation (Fig. 3.7) (Bartholomew 1970). These animal activities not only control community boundaries but may play roles in nutrient linkages between communities.

Postfire herbivory also has the potential for markedly impacting plant diversity. Studies by Mills (1983, 1986) have shown that small mammals consistently prefer seedlings of *Ceanothus greggii* over *Adenostoma fasciculatum*. When seedlings were protected from herbivory, *Adenostoma* seedlings experienced higher mortality. Quinn (1991) demonstrated that *Ceanothus greggii* dominated a site several years after fire in an area fenced from small mammals; however, in an adjacent unenclosed site this species was replaced by *Adenostoma fasciculatum*. Quinn (1986) has shown a similar phenomenon with *Ceanothus crassifolius*.

The potential exists in mediterranean-climate vegetations for marked control of vegetation structure by rodent guilds. Brown and Heske (1990) studied the more xeric desert scrub habitat which shares many of the same Heteromyidae genera, such as *Dipodomys* and *Perognathus*, with more coastal communities in truly mediterranean-climate vegetations. In their study, 12 years after three species of *Dipodomys* were removed from plots, density of tall perennial and annual grasses increased about three-fold, and rodent species more typical of this community had colonized the site. They concluded that species of *Dipodomys* were a keystone guild: through seed predation and soil disturbance they have major effects on biological diversity and biogeochemical processes. In this case the "keystone" organism was not a single species, but a guild of three taxonomically related and ecological similar *Dipodomys*. In their study there was some evidence of redundancy in that removal of the largest of the three species had significant effects on the distribution of other rodents, but it required the removal of all three species to cause wholesale changes in vegetation. This study was significant because it demonstrated that the eight common and several rare species of rodents that remained after *Dipodomys* had been removed clearly did not play the same keystone role and were not able to prevent the conversion of scrub to grassland.

Fig. 3.7. Bare zone between coastal sage scrub and grassland with rodent-proof exclosure in bare zone in the *foreground*

Changes in important carnivores may have indirect impacts related to foraging strategies of predators and prey. If nocturnal predators such as coyotes are removed and diurnal predators such as hawks remain, this could change the balance between nocturnal prey such as kangaroo rats and diurnal prey such as squirrels. Since the former are granivores and the latter herbivores, we might expect these changes to impact the vegetation composition.

Removal of large carnivores may also have regional impacts in that it reduces the across-habitat connectance, which potentially decreases food-web stability. This is particularly so in southern California, where the juxtaposition of grasslands and woodlands may be needed to sustain raptors that forage in grasslands but require woodlands for shelter (M. Price, pers. comm.).

3.3.3 Invertebrate Communities

3.3.3.1 Pollinator Linkages

In a detailed study of pollinator communities along an altitudinal transect of 15 community types across central California, Moldenke (1975, 1979a,b) concluded that competition between plants for pollinators increases the blooming season of plant species in pollinator-limited environments. This has the effect of increasing the percentage of the flora that is blooming during any given week. It was noted that some of the least important contributors to community floral biomass are the most heavily visited by pollinators and vice versa. In other words, species that have limited control of ecosystem functions such as energy flow, may be strong controllers of ecosystem processes such as pollinator resources.

Competition for pollinators was further suggested by the observation that self-compatible plant species not visited by pollinators and therefore forced to self, are highest at the peak of blooming. Additionally, this competition for pollinators is revealed in the disproportionate occurrence of genetically incompatible annual species flanking the anthesis peak, while a disproportionate abundance of self-compatible perennial species occurs at the anthesis peak.

A mechanism that permits successful outcrossing at the peak of blooming is the exceptionally high number of plants serviced by specialist-feeding bees which allows for pollination even when a plant species is in low density or when competing species may have successfully usurped all the generalist pollinators. Specialized feeding habits of course would not evolve in bees, if it were not competitively forced upon them by competition for their floral resources as well; more species of specialist-feeding bees are in fact active during the peak synchrony of anthesis than at any other period.

Moldenke concluded that while there is a tendency for specialization in most environments, it is only in the most predictable or stable environments that the loss of both behavioural and genetic flexibility, associated with specializing, is a viable strategy.

Extreme specialization has long been noted between all species of the genus *Yucca* and moth species in the genus *Tegeticula* (or *Parategeticula*) (Powell 1992b). It is curious that almost all of the many *Yucca* species from Nevada to Florida are pollinated by a single species of moth, yet each of the three of Californian *Yucca* species is pollinated by a different species of moth. Such extreme specialization illustrates a marked degree of co-evolution between insect and plant.

3.3.3.2 Herbivore Linkages

Factors controlling diversity of insect communities are not well understood. One approach is to determine the degree of niche overlap. This was addressed in a very detailed study of microlepidoptera on its host, the California live oak (Opler 1974). In these herbivores the question is complicated by the fact that species interact as adults as well as larvae. On this host there are 35 species of foliage-feeding microlepidoptera; 18 external feeders and 17 leaf miners (see also Fig. 3.3). These groups segregate out on young and mature foliage, respectively. All of the leaf miners are host-specific, some being restricted to coastal sites and some interior. Opler found relatively little niche overlap in these moths, as species on the same tree exhibited trophic and temporal separation.

Another approach is to separate ecological from biogeographical processes. Regional (gamma) herbivore richness is defined as the total number of herbivore species associated with a host plant in any part of its range and local (alpha) richness is the number of species associated with a host population at a single site. Understanding the mechanisms operating on community diversity requires an understanding of the relationship between these two variables. When local ecological interactions are weak and herbivore species are widespread, the correlation between richness at these two spatial scales should be strong. If local interactions are strong and herbivore species have restricted ranges due to habitat specialization, the correlation should be weak.

Cornell (1979, 1985, 1986) attempted to separate these two factors in a study of Californian cynipid oak gall-forming wasps. Different oak species support significantly different numbers of cynipine species on trees of a given size. Cornell found that within oak species, tree height and the number of cynipine species available to colonize trees explains much of the variation in cynipine richness among trees. However, there is no evidence of vertical or horizontal zonation; upper tree canopies support the same species as the lower canopies. Those species found in smaller trees are either random or non-random subsets of those in larger trees.

On the other hand, the total number of cynipine species associated with each oak species (regional richness) varies widely among five oak species and correlates significantly with cynipine richness on individual trees in multiple regressions with tree size. Path analysis indicates that regional richness is correlated with richness on trees exclusively through increasing the species pool from which the assemblage on individual trees is drawn

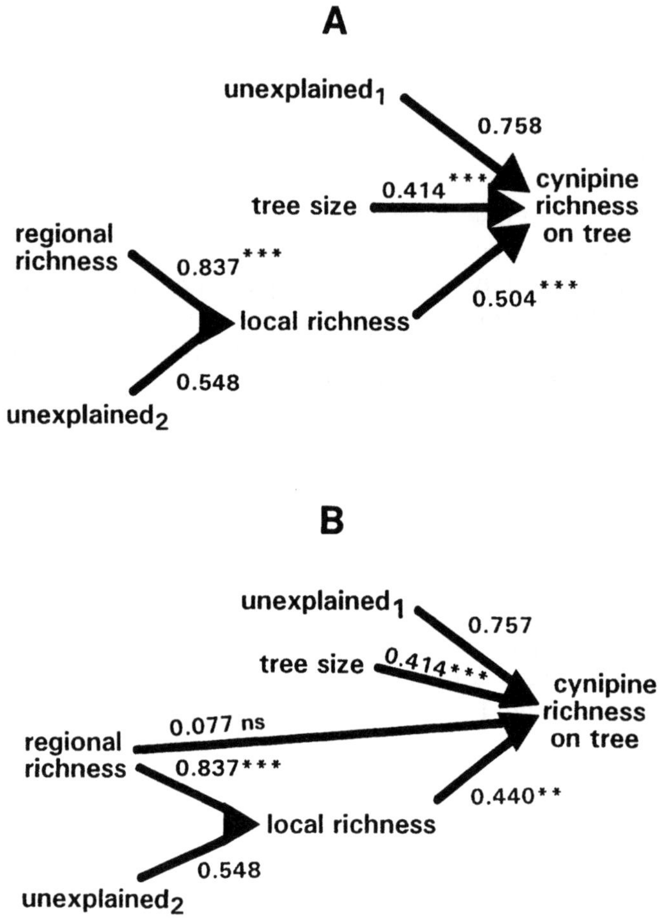

Fig. 3.8. Path analysis to determine the contribution of regional richness of cynipine wasp species to cynipine richness on trees either via its influence on local richness (**A**), or directly (**B**). *Numbers* associated with *arrows* are path coefficients. (Cornell 1985)

(Fig. 3.8). In other words, there is an unabated increase in alpha diversity as gamma diversity increases, suggesting that excess niche space is available on host oaks for additional cynipine species, and that limits to similarity of co-existing cynipines have not been reached. This indicates that sufficient information on the causes responsible for local herbivore richness cannot be gained by studying the local host population in isolation. The spatial dynamics of the system are such that measures of the regional pool from which the local assemblage is drawn must also be taken. However, cynipine habitat specialization and local interactions are not predictable from regional richness, but may influence local richness on certain oak species independently of regional richness.

3.3.3.3 Parasite Linkages

Although widespread in North America, Lyme disease has a unique vector system in Californian chaparral. The western black-legged tick, *Ixodes pacificus*, gives the disease to several hundred people each year. This tick species picks up the disease, borne by a spirochete *Borrelia burgdorferi*, from the dusky-footed woodrat *Neotoma fuscipes*. However, only 1-5% of the *I. pacificus* ticks are infected, a number that epidemiologists believe is too low to maintain a spirochete reservoir in the woodrat population. Recently, Brown and Lane (1992) demonstrated that the disease cycle requires a second tick species, *I. neotomae*, which does not bite humans but is responsible for maintaining high levels of the spirochete in the woodrat population. Thus, Lyme disease can only exist in areas where the spirochete, the woodrat, both ticks, and humans co-exist.

3.4 Biodiversity and Ecosystem Function

Here we explore the hypothesis that different patterns of community diversity affect the efficiency of resource use in two Californian ecosystems. The resources considered are water, light energy and nitrogen, and resource use efficiency is defined as the ratio of the amount of resource taken up to the amount available. There are reasons to believe that communities should converge towards a set of species that produces a higher resource use than any other mixture of species (MacArthur 1972). Loss of biodiversity thus translates into less efficient resource utilization.

3.4.1 The Annual Grassland Ecosystem

Close to one-fifth of the Californian landscape is covered by grasslands, most of which are dominated by non-native annuals (Huenneke 1989). Biodiversity of grasses and forbs plays an important role in efficient exploitation of resources in this ecosystem. Factors affecting diversity patterns included unpredictable annual weather patterns, spatial distribution of resources and disturbance by animals and fire.

3.4.1.1 Disturbance, Predation, Inundation and Grassland Diversity

Grassland community structure is strongly controlled by disturbances. Gopher mounds, for example, have been shown to create special microhabitats within the annual grasslands and result in very different survivorship curves for different species (Hobbs and Mooney 1985). Clearly, species sort out along a disturbance gradient by their very different regeneration niches and some species even require gopher disturbance to enable them to establish and remain within the community. Thus, gopher activity maintains a higher overall community diversity and this is predicted by the model of Hobbs and Hobbs (1987).

Further, the disturbance regime creates a complex pattern of microsites which are most suitable for different species in different rainfall years (Hobbs and Mooney 1991). Domestic livestock grazing has the potential for upsetting this balance since it leads to exclusion of gophers (Hunter 1991), and although it creates bare soil of a different nature (Table 3.1), there is evidence that canopy gaps created by grazers and gopher mounds favour the same species (Peart 1989). The effect of grazers such as rodents and livestock on altering community composition has been documented by Talbot et al. (1939) (Table 3.2).

Harvester ants (*Veromessor andrei*) likewise have the potential for affecting the diversity of grasslands through their preference for seeds of non-dominant species (Hobbs 1985). In addition, ant nests alter the chemical and physical properties of the soil by elevating temperatures and levels of phosphorous, potassium and nitrogen, relative to the surrounding soils (Beattie 1989). Thus, loss of the ant guild could potentially greatly alter the community composition in these grasslands. It is unknown whether other seed-eating guilds, such as the Heteromyidae rodents, that compete with ants, have similar preferences.

In some respects, selection for diverse approaches to resource exploitation has resulted in a sort of co-evolutionary dance with other microbial and animal guilds that, in turn, select for greater diversity. For example, grasshoppers can affect the quality of their own host plants through feeding, potentially influencing biomass, resource allocation and the actual distribution of the individuals in the host plant population (Joern 1989). Perennials and C_4 species may tolerate grazing better than annuals or C_3 species. In addition, different weather patterns may generate different phenological patterns, both in plants and in their predators, and consequently the vulnerable species one year may not be the same in the next year. For example, larvae of the butterfly, *Euphydryas editha bayensis* depend on precise coupling between the host-plant phenology and larval development. Very wet years can uncouple this relationship by retarding larval development and result in localized extinction (Chiariello 1989).

Table 3.1. Mean percentage ground cover (and standard error) in grazed and ungrazed areas at Vina Plains Preserve, Tehama Co., California. (Hunter 1991)

Cover type	Ungrazed (n = 100)	Grazed (n = 50)
Bare soil	9.62 (1.2)	28.60 (3.3)
Gopher mounds	7.05 (0.9)	0.67 (0.5)
Rock	1.25 (0.4)	1.09 (0.3)
Vegetation	81.92 (1.3)	68.03 (3.3)

Table 3.2. Fluctuations in composition of annual grasslands under different treatments. (Talbot et al. 1939)

Plot	Treatment	Year	Total		Four most abundant species			
			Forbs	Grasses	*Erodium* spp.	*Bromus hordeaceus*	*Avena barbata*	*Bromus rigidus*
			Per cent of cover					
1	Rodents	1933	98	2	86	1	T	T
	and	1935	4	96	3	20	72	4
	livestock	1937	1	99	0	10	1	88
	excluded	1938	20	80	1	10	10	60
2	Only	1933	99	1	80	T	0	T
	livestock	1935	15	85	14	75	10	T
	excluded	1937	4	96	2	85	8	T
		1938	20	80	0	55	20	T
3	Grazed by	1933	98	2	82	T	T	T
	livestock	1935	79	21	50	5	T	T
	and	1937	67	33[a]	50	0	0	T
	rodents	1938	43	57[a]	10	24	T	T

T (Trace) = values less than 1%
[a] Includes *Juncus bufonius*

Impacts of herbivory by rodents has the potential for producing substantial changes in vegetation composition. Not only are there qualitative differences in food preferences but peaks in animal abundance may vary from year to year such that peaks may occur in winter one year and in summer the next year (Lidicker 1989). A classical example of herbivory's effect on biodiversity is preferential feeding on the imported Klamath weed (*Hypericum perforatum*) by the introduction of certain *Chrysomela* beetles as biological control agents which resulted in the return of native bunchgrasses (Huffaker 1951). Thus, herbivory has the potential for creating a heterogenous selective landscape that would select for higher biodiversity.

Grasslands are also quite resilient to wildfires and such perturbations play an important role in eliminating invading shrubs and thus maintaining the herbaceous nature of this community. Wildfires also have the potential for altering the balance between different taxa (Larson and Duncan 1982; Parsons and Stohlgren 1989).

Inundation of grassland depressions which are underlain by an impervious hardpan support a unique vernal pool flora adapted to the seasonal alteration of winter inundation and summer drought. These azonal communities support a rich endemic flora and non-natives have not permanently established in most vernal pool communities. Unpredictable rainfall results in pools that remain filled for variable lengths of time from year to year and species appear to segregate out spatially and temporally according to the average water duration period (Fig. 3.9). (Zedler 1987). It could be argued that full exploitation of resources in this community is dependent upon this diverse array of plant strategies. Based on detailed comparison of carbon gain in representatives of this flora it has been argued that co-existence of this diverse community is enhanced by a disequilibrium process that favours some species in very wet years and others in dry years (Keeley and Sandquist 1991).

3.4.1.2 Grassland Diversity and Patterns of Resource Utilization

Biomass production in the annual grassland ecosystem is largely controlled by patterns of precipitation and temperature. Pitt and Heady (1978) described the impact of variation in weather patterns over an 18-year period on standing crop, cover and botanical composition. Typically, standing crop increased as the growing season progressed from March to June, although some species declined during this period and were replaced by other species. They also found that consistent precipitation following fall germination, and again during the period of rapid plant growth in spring, contributes to a "grass year", whereas quite a different set of conditions would favour forbs, e.g. biomass of *Erodium* spp. correlated with dry fall conditions.

Other researchers have likewise found evidence of significant niche differentiation in grassland species that could impact total net production for the community. For example, in laboratory studies, Hufstader (1976) showed marked differences in mean maximum, mean median and mean minimum temperatures necessary for maximum biomass production in two seemingly ecologically equivalent species of *Bromus*. In the field Johnston and Lathrop (1978) showed temporal segregation of *Erodium obtusiplica-*

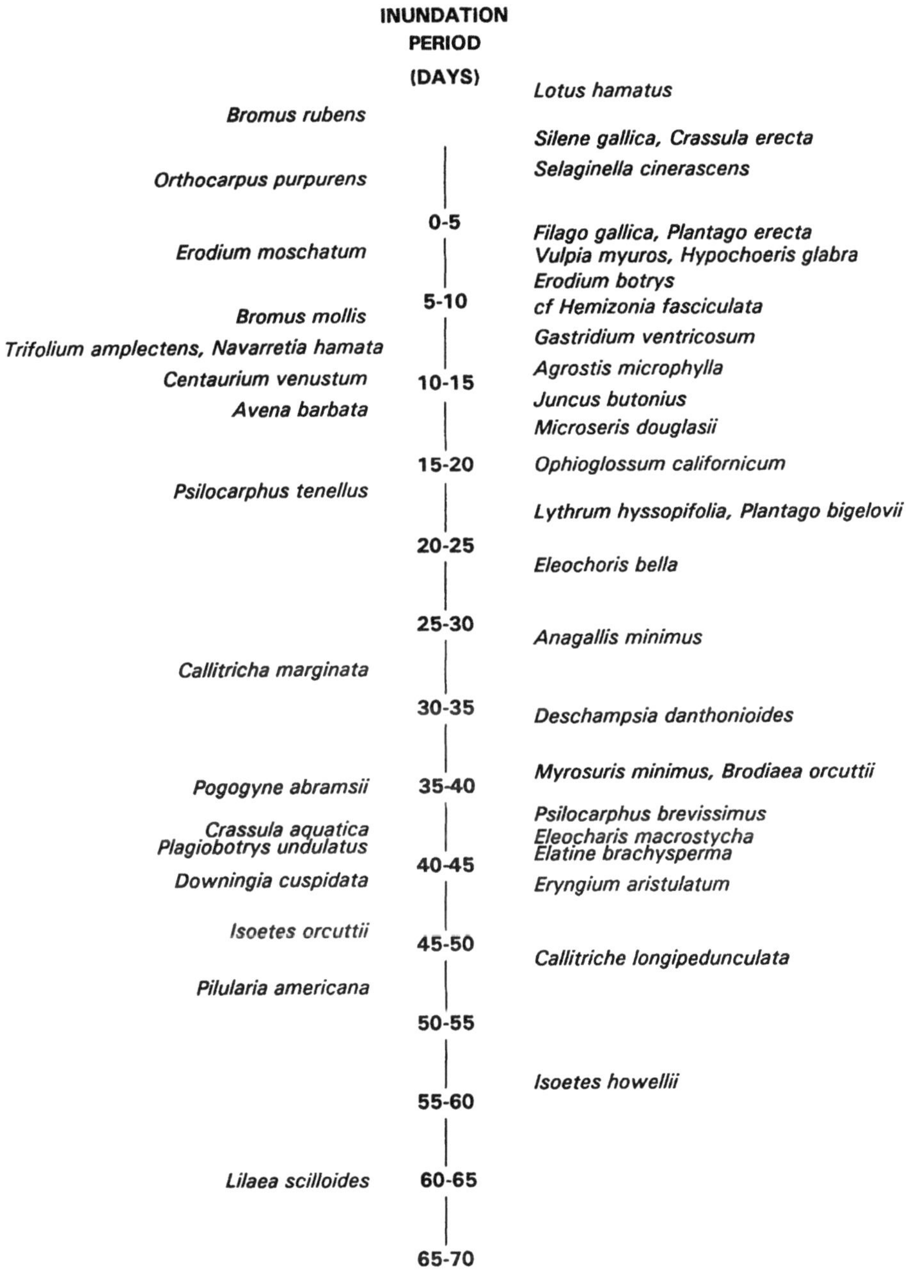

Fig. 3.9. Average water duration class for common vernal pool species for Kearney Mesa, southern California. (Zedler 1987)

tum, Avena barbata and *Holocarpha virgata* (Fig. 3.10). Elongation and phenological development of each species were correlated with temperature, daylength and soil moisture, and manipulated association experiments indicated that competition effects are not totally responsible for temporal phase segregation in the annual grassland. Martin and Harding (1981) showed that total reproductive rates of sympatric populations of two species of *Erodium* were greater than allopatric populations, further supporting the notion that species have evolved finely tuned differences in the manner in which they exploit the environment and that intra-specific competition is generally more intense than inter-specific.

Growth form diversity also leads to greater utilization of resources. Jackson and Roy (1986) showed that different annual patterns of rainfall and timing and duration of summer drought had markedly different effects on annual grasses compared to perennial grasses. It is to be expected that in exceptionally wet years perennial bunchgrasses would predominate and in exceptionally dry years annuals would succeed.

Marked phenological differences in the annual flora have been demonstrated by Gulmon et al. (1983) and Mooney et al. (1986b). Species differ in rooting depth and thus are separated by the portion of the soil profile where water is primarily extracted. This has a profound effect on flowering phenology and acts to extend the period of energy capture (Fig. 3.11). These species can be viewed as comprising two functional groups; the early flowering annuals and the late flowering annuals. Field studies have shown that if one or the other of these functional groups is removed from a site, it will result in decreased productivity (N. Chiariello, pers. comm.).

Marshall and Jain (1969) showed density-dependent differences in exploitation of resources by the nearly indistinguishable, and seemingly redundant species, *Avena fatua* and *A. barbata.* An analysis of the relative fitness of the two species in mixed populations indicated that they could successfully cohabit over the range of densities studied due to frequency-dependent selection favouring *A. barbata* in relative rarity.

Chiariello and Roughgarden (1984) have shown the potential for evolution of intra-genomic diversity as a mechanism of optimizing carbon allocation in ecotypes of an annual grassland plant that are either vernal or autumnal. Their results showed that autumnal ecotypes delay reproduction to later in the year as a means of reducing seed predation and this has transformed into different patterns for storage carbon.

Seed dormancy is one means of avoiding inter-specific competition and enhancing the distribution of different species in time in a manner that increases the exploitation of the environment. Jain (1982) showed that there was much variability in germination behaviour of five grass species and two legume species, and seasonal dormancy of *Avena* and *Bromus* was

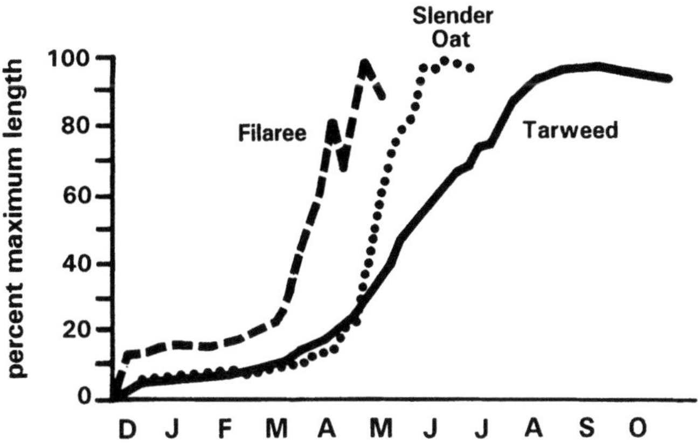

Fig. 3.10. Seasonal change in leaf length in filaree (*Erodium obtusiplicatum*), slender oat (*Avena barbata*) and tarweed (*Holocarpha virgata*). (Johnston and Lathrop 1978)

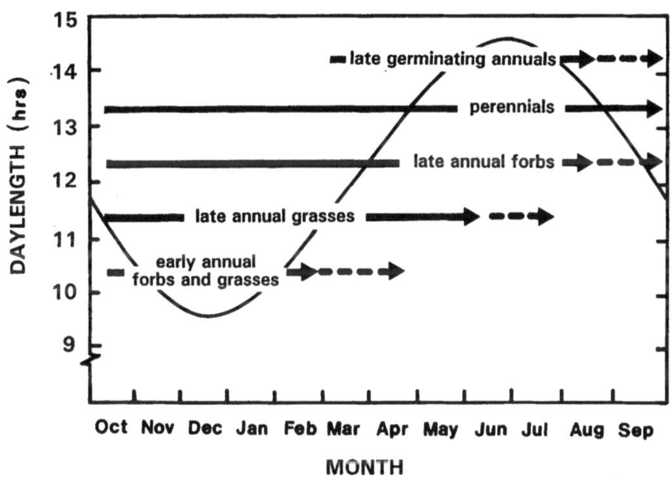

Fig. 3.11. Functional groups of grassland species based on phenological patterns. (Chiariello 1989)

highly correlated with the patterns of a probable rainstorm at different sites during the summer months. There is evidence that some of this variation is carried over between generations as maternal effects (Ewing 1981). In some cases a single species may diversify its germination behaviour. Stamp (1990) showed that *Erodium brachcarpum* seeds produced late in the fruiting period are likely to germinate in the year of seed production, whereas seeds produced early in the season are larger, with a thicker seed coat and are likely to germinate in subsequent years.

Also, in this nitrogen-limited system there is microsite heterogeneity of inorganic-N availability. This arises because some microsites may be nitrate-depleted due to inherent spatial variation in the NH^{4+} pool available to ammonium oxidizers and because in some microsites microbial assimilation may out-compete plant assimilation (Davidson et al. 1990). Growth form diversity may play an important role in the spatial distribution of ammonium due to differences in amounts of standing dead and thus the amount retained against leaching. For example, prostrate forbs disarticulate and possibly decompose more rapidly than annual grasses, which stay erect in dry air. Grazers will contribute even more to this heterogeneity of nutrient pools (Center et al. 1989). Growth form diversity underground will also carry with it different competitive abilities in different seasons. Therefore, differences in nitrogen assimilation rates between species (Hull and Mooney 1990) and between annual and perennial grasses (Joffre 1990) may play important roles in affecting N cycling processes of this ecosystem. Differences between species may change during the year as demand for nutrients changes, and these will be affected by relative growth rates and by photosynthetic rates that determine the carbon available for nitrogen assimilation (Schimel et al. 1989).

The important role of biodiversity in exploiting this heterogenous micro-landscape of nutrient pools is suggested by experiments that show loss of native diversity and invasion by non-native weeds when this heterogeneity is buffered by the more or less even distribution of nutrients (Huenneke et al. 1990). In fact, McCown and Williams (1968) have presented experimental verification that under favourable nutrient conditions *Bromus mollis* becomes the superior competitor for light because of its greater stature and more erect leaf habit, but under nutrient-stressed conditions leaf area is considerably reduced and *Erodium botrys* becomes the superior competitor. Hull and Muller (1976) also demonstrated that *Bromus* species were superior under enhanced nutrients whereas *Avena* species had superior production under well-watered conditions, and Hobbs et al. (1988) showed that this superiority of *Bromus* under high nutrient conditions was over forbs as well.

Species specific differences in degree of mycotrophy may also play a critical role in the exploitation of soil nutrients. It appears that much of this annual flora is mycorrhizal (Chiariello et al. 1982; Hopkins 1987) and there are apparent differences in the dependence upon mycotrophy between different species and different growth forms (St. John 1993). These differences may have important ramifications for successional processes and interactions between ecosystems. Preliminary data suggest that many of these annuals and successional subshrubs such as *Diplacus, Eriogonum* and

Artemisia are facultatively mycorrhizal, whereas chaparral climax shrubs are often obligately mycorrhizal.

3.4.2 The Chaparral Ecosystem

This evergreen shrub vegetation dominates much of the Californian landscape, particularly on mountain slopes below 1500 m. Wildfires are an important disturbance factor in this community and they have a profound impact on many ecosystem processes.

3.4.2.1 Regeneration, Disturbance and Biodiversity

Grubb (1977) has suggested that maintenance of species richness in plant communities may be tied to differences in their regeneration niche. The regeneration niche of chaparral species can be represented by two functional groups; those having "disturbance-dependent" recruitment and those with "disturbance-avoiding" recruitment. In the first group, opportunities for recruitment and population expansion are restricted to a narrow window of time immediately after fire e.g. *Ceanothus* species in the subgenus *Cerastes*. In the latter group are species that survive fires by resprouting, but opportunities for recruitment and population expansion occur in the absence of disturbance e.g. shrub species of *Quercus*. These two functional groups have different patterns of survivorship as chaparral ages (Fig. 3.12).

Although species within each of these functional groups share many life history characteristics, they are not ecologically equivalent; individual species are often displaced geographically, as described above (Cody 1986).

Most communities are a mixture of these two functional groups and the relative success of one group over the other will vary with the pattern of disturbance. When the fire-return interval is very short, many disturbance-dependent taxa are unable to regenerate due to insufficient time for regenerating the soil seed bank. For example, fires in consecutive years can cause the localized extinction of some shrub taxa (Zedler et al. 1983). As fire frequency increases, many native species that regenerate after fire, are replaced by non-native weedy annuals (Fig. 3.13). This fact is well known to land managers who, for many years, have used high fire frequency for type-conversion of chaparral to annual grassland (Sampson 1944; Arnold et al. 1951). Species that resprout after fire are generally less sensitive to high fire frequency than obligate seeding species, and depending upon the particular species, they may have disturbance-dependent recruitment or disturbance-free recruitment.

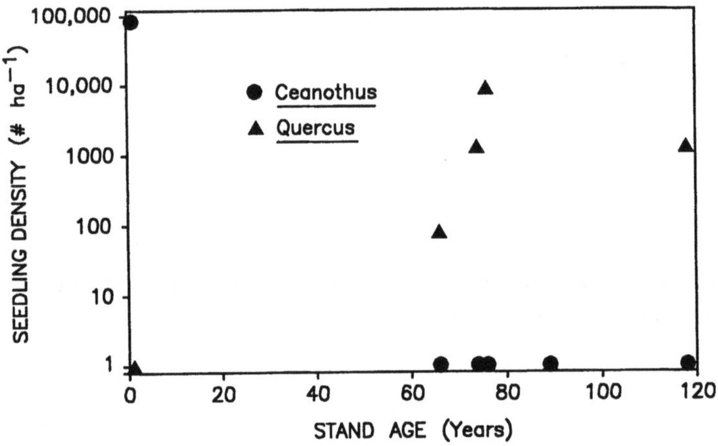

Fig. 3.12. Seedling recruitment and adult mortality as a function of stand age for *Ceanothus* and *Quercus*. (Keeley 1992a,b)

Extended fire-free periods are required for reproduction by species with disturbance-avoiding recruitment. Species in this functional group fail to recruit in postfire environments, and other gaps, because their seedlings are sensitive to drought stress (Davis 1989); thus recruitment is most successful under the canopy of older chaparral (Keeley 1992a,b).

Transition matrices provide a useful means of examining the effect of different fire frequencies on community composition of these two recruitment functional groups (Keeley, unpubl. data). By iteration we can model the demographic consequences of different recruitment modes in the disturbance-dependent mode of *Ceanothus* and the disturbance-avoiding mode of *Quercus*. Although *Ceanothus* suffers marked decline in older stands, their seed bank is capable of great persistence, making them reasonably resilient to long fire-free periods (Fig. 3.14A). The highly conservative resprouting mode of *Quercus* makes these shrubs relatively resistant to fire; however, seedling regeneration is markedly reduced by disturbance (Fig. 3.14B).

With this model we can predict how the demographic characteristics of these two functional groups will affect long-term community composition. If this transition matrix is iterated over repeated fire cycles of short and long fire return intervals we find substantial differences in the density of *Ceanothus* and *Quercus*. Over a period of repeated cycles of fires every 30 years, *Ceanothus* has the potential for population expansion (Fig. 3.15A). If fires are too frequent, *Ceanothus* declines due to insufficient time to accumulate a soil seed bank, and if fire frequencies are too infrequent, *Ceanothus* declines due to adult mortality and natural deterioration of the

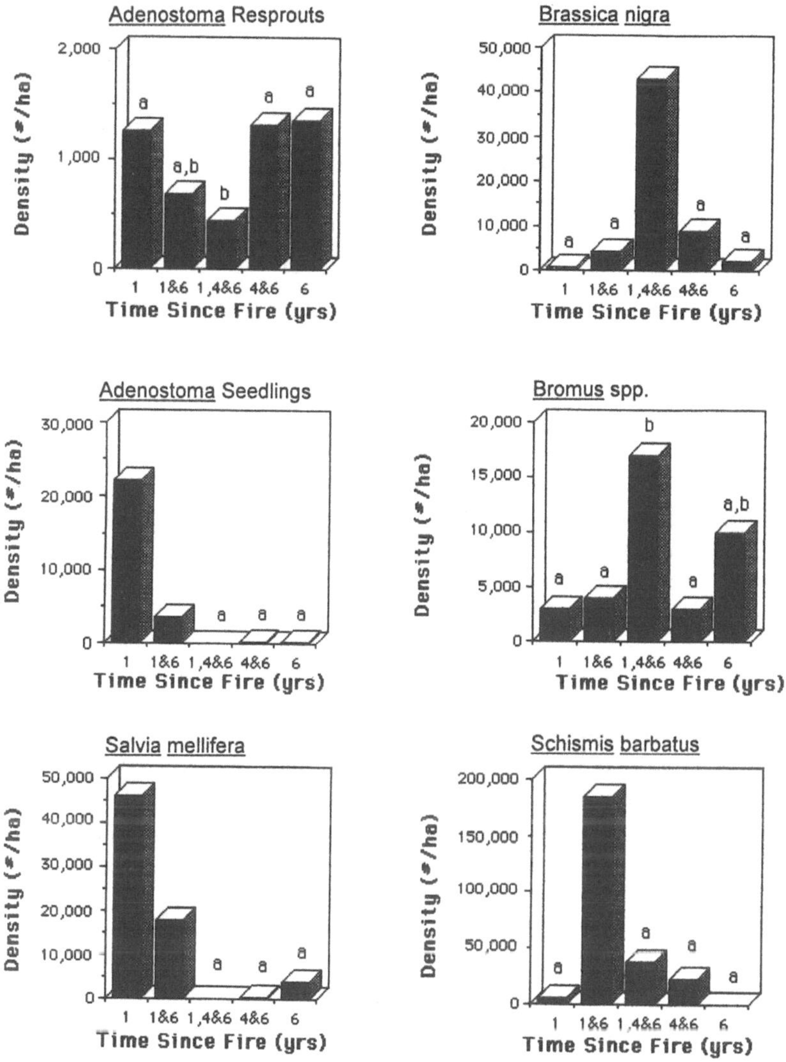

Fig. 3.13. Density of native shrubs and non-native annuals on sites subjected to different burning frequencies in southern California mixed chaparral. Each *bar* represents a different site subjected to burning during the indicated years of a 6-year cycle. Statistically significant differences are denoted by labels above the *bars*. (Haidinger and Keeley 1993)

soil seed bank (Fig. 3.15A). *Quercus* illustrates a more conservative pattern. It is largely unaffected by short or long fire-return intervals (Fig. 3.15B) and, in fact, will increase at fire return intervals of 150 years.

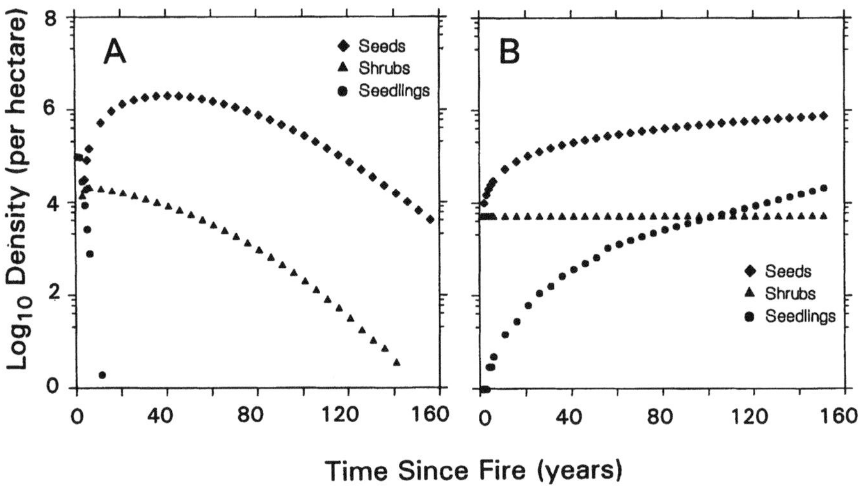

Fig. 3.14 A,B. Predicted sizes for state variables, shrubs, seeds, seedlings + juveniles for (**A**) *Ceanothus* and (**B**) *Quercus*. Transition probabilities and initial population sizes were based on values from the literature. (Keeley unpublished)

Under the conditions of these transition matrices, if fires occur at 30-year intervals, *Ceanothus* potentially could dominate, whereas at intervals of 100 years it declines and *Quercus* potentially dominates (Fig. 3.15A,B). Today, the shorter fire cycle is more typical of most chaparral sites due to the influence of humans on ignition of wildfires (Keeley 1982). Under natural conditions there are reasons to believe that, due to the stochastic nature of lightning ignitions, both short and long fire-return intervals were experienced by chaparral taxa. If we model such conditions (Fig. 3.15C) it is apparent that alternating long and short fire-return intervals result in an equilibrium in the density of *Ceanothus* and *Quercus*. In chaparral, co-existence of these functional groups is most likely enhanced by variable burning regimes, and thus biodiversity may be tightly linked to exploitation of the full range of fire regimes.

3.4.2.2 Diversity, Water Balance and Chaparral Functioning

In this summer-drought climate, water plays a critical role in the functioning of the chaparral ecosystem. Soil-water balance is intimately associated with the plant water balance and is a function of patterns of soil characteristics, precipitation, runoff, interception in the canopy, stemflow, throughfall, and infiltration. Plant water balance is intimately tied to potential plant biomass production and strongly affected by different plant characteristics (Fig. 3.16), and, not surprisingly, species differ in their sea-

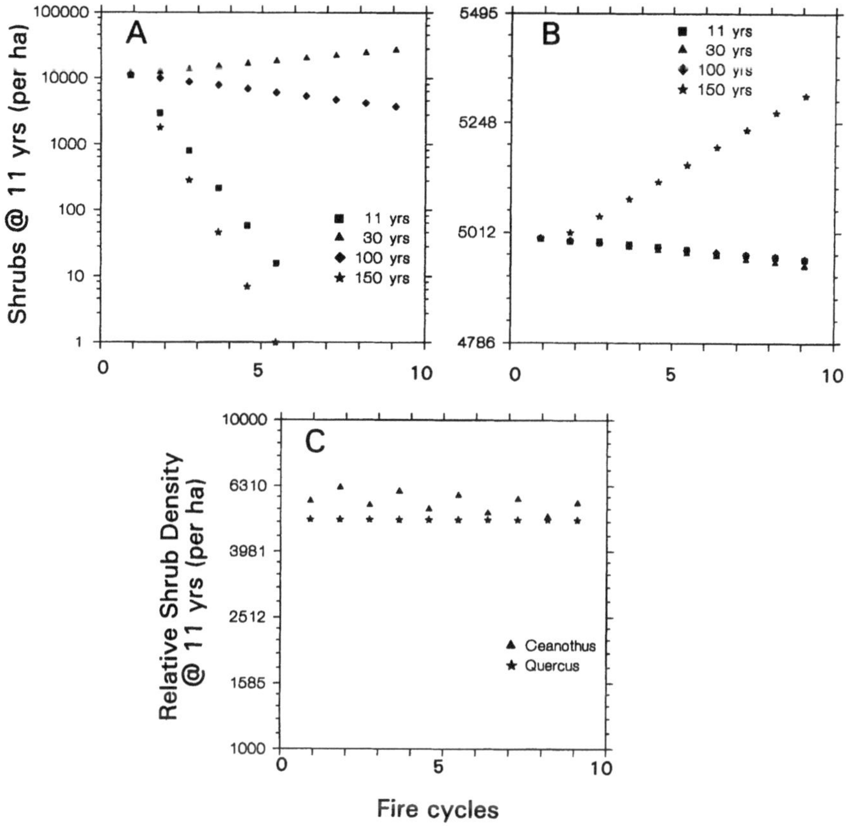

Fig. 3.15. A-C. Predicted shrub population density after 11 years at different fire-cycle return intervals for (**A**) *Ceanothus,* (**B**) *Quercus,* or (**C**) both taxa under an alternating fire regime of 30 and 100 years. Fire was introduced at the designated interval by multiplying the state vector of population size by a new "fire" transition matrix. (Keeley unpublished)

sonal patterns of water use (Poole et al. 1981). Water use efficiency (transpiration/precipitation) likewise varies between species and sites. Consequently, water capture efficiency for a site is strongly controlled by vegetation composition (Patric 1961) and different combinations of species may more fully utilize available soil moisture.

Chaparral can be categorized into different functional groups based on the mode of tolerating summer drought. Shrubs such as *Quercus, Rhamnus* and *Prunus* persist on arid chaparral sites, not because they are extremely drought-tolerant but rather due to their deep root systems, which penetrate fractured substrates and persist from one fire cycle to the next by resprouting; they survive summer drought because they are drought-

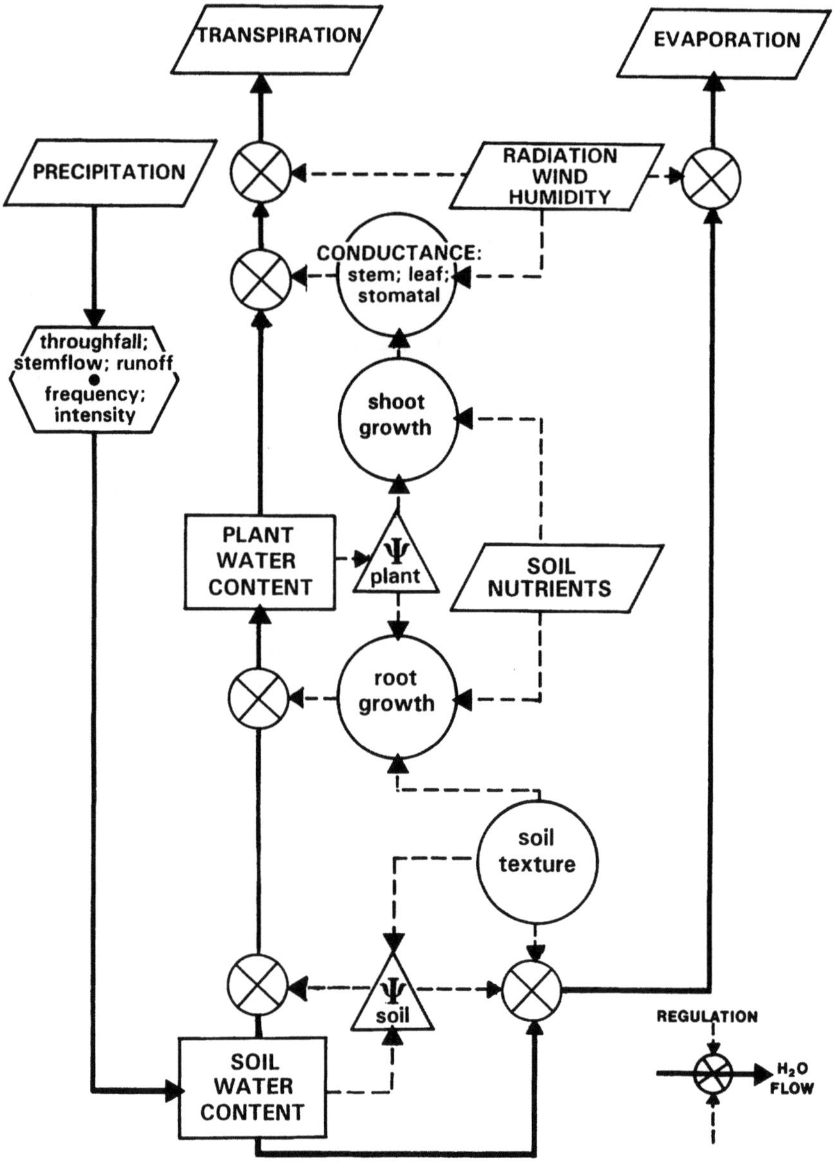

Fig. 3.16. Flow diagram of water through the soil-plant-atmosphere system, and controlling plant and environmental variables. *Heavy lines* flows of water; *broken lines* controls; *boxes* water storage; *circles* control variables; *psi* water potential. (Miller 1983)

avoiders. On the other, hand shrubs such as *Ceanothus* and *Arctostaphylos* are more drought-tolerant, but, because they regenerate after fire from seed, they have relatively shallow root systems. The vast majority of chaparral sites comprise mixed stands of deeply rooted and shallow rooted shrubs, which exploit water in very different ways. For example, on all but the most mesic sites, soil moisture is apparently insufficient to support pure stands of deep-rooted *Quercus*. The shallow-rooted *Ceanothus* fill in spaces between the *Quercus* and contribute shade and litter, both of which are required for *Quercus* seedling recruitment (Keeley 1992a).

The water balance for a site is markedly different after wildfires and is exploited differently by the diversity of growth forms normally present. Fire-following annuals, with rapid growth rates and fibrous roots exploit a very different part of the soil column than shrub seedlings or resprouts (Miller 1983). Seasonal patterns introduce a stochastic factor that makes some growth forms more efficient in years of light, evenly distributed rain, and other forms more efficient in years of heavy, infrequent storms. Also, subsurface topography, which may be quite heterogenous on a micro-topographic scale (Kummerow et al. 1981), means that sites most suitable for one species may be juxtaposed with sites more suitable for another. Very marked species-specific differences in growth rates of roots are just one way in which species may differ in their ability to exploit these microsite differences (Miller 1983). Although data are lacking, it is likely that the diversity of growth forms on a site will impact water yield following wildfires.

Other watershed processes tied to water yield include peak water flow, erosion and sedimentation. These likewise are linked to community composition through the diversity of root growth forms present on a site and are readily altered by disturbance such as fire or artificial type conversion. For example, in the San Dimas Experimental Forest of southern California, burned watersheds that were seeded and converted to annual grassland had an eightfold increase in landslides and sediment production (Knight et al 1984). Such impacts are temporary and in the long term, converted water sheds yield greater stream flow, although other site factors, such as evaporative losses, increase greatly (Rowe and Reimann 1961).

3.4.2.3 Diversity, Nutrient Cycling and Chaparral Functioning

For many species, leaf nitrogen content and photosynthetic capacity are highly correlated (Field and Mooney 1983). In chaparral, fertilizer application studies indicate some species such as *Adenostoma fasciculatum* are more nutrient-limited than others such as the nitrogen-fixing *Ceanothus* species (McMaster et al. 1982). The highest concentrations of soil nutrients tend to be in the upper soil layers (Christensen and Muller 1975), thus, shallow-rooted shrub species may have a competitive advantage. Rooting depth, of course, will impact water relations and thus different growth forms carry with them different trade-offs. This is most evident in the relationship between nitrogen use efficiency and water use efficiency (Fig. 3.17).

Fire acts to recycle nutrients tied up in organic matter but also generates substantial losses; up to two-thirds of the total nitrogen in litter and soil are lost during an intense fire (DeBano et al. 1979). Wet- and dry-fall inputs are so low that it may take >60 years to replenish these losses, although in polluted regions of southern California nitrogen input may be 10-20 times greater than elsewhere (Schlesinger and Gray 1982). Postfire successional species on many sites are nitrogen-fixers, e.g. annual *Lupinus* spp. or the subshrub *Lotus scoparius* (Rundel 1983), and of lesser importance is asymbiotic nitrogen-fixing microbes (Dunn and Poth 1979).

Also critical to the nitrogen balance in chaparral is the postfire annual flora that captures much of the mineralized nutrients that would otherwise be lost in runoff. The fire-specialist species are a highly diverse group and only occur on burns; disappearing after the first or second year. Other species in the postfire flora are less specialized and persist for much longer after fire. There is evidence that these two groups of species have very different nitrogen utilization strategies (Swift 1991). It appears that fire specialists such as *Phacelia brachyloba* and *P. minor* preferentially utilize ammonium and are apparently adapted to taking advantage of the high soil ammonium levels immediately after fire. These fire specialists, however, have much lower nitrogen-use efficiency than generalist species, that persist later in succession, such as *Cryptantha intermedia, Phacelia cicutaria* and *Brassica nigra.* It is unclear what the ecosystem impact would be if one or the other of these groups were eliminated from the chaparral, but it is possible that this diversity of postfire annuals plays an important role in the chaparral nutrient cycle. In light of the fine-scale pattern of soil nutrient distribution after fires (Rice 1993), a diversity of structural and functional growth forms may play a key role in the nutrient cycle of chaparral. Failure to retain or return nitrogen to a site potentially could result in site degradation to less diverse forms of chaparral such as *Adenostoma fasciculatum* dominated chaparral.

3.4.2.4 Energy Capture, Resource Use and Biodiversity

Photosynthetic capacity is one of the principal determinants of primary productivity. On a landscape scale, different plant attributes contribute to dramatic differences in predicted carbon gain along gradients of moisture

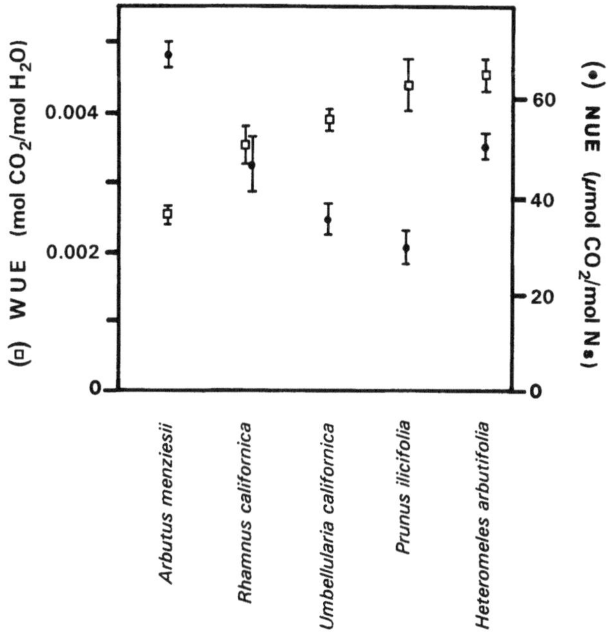

Fig. 3.17. Water-use efficiency (*WUE*) and nitrogen-use efficiency (*NUE*) in five Californian evergreen shrubs. Species range from those typical of moist sites on the *left* (viz. *Arbutus menziesii*), to those typical of dry sites on the *right*. (Field and Davis 1989)

and temperature (Oechel et al. 1981). There is a strong correlation between total seasonal carbon uptake and percentage cover of a species at a particular site, suggesting that no single species is best adapted to all parts of the landscape (Fig. 3.18). Increased growth form diversity results in a seasonal increase in duration of plant growth at a site (Fig. 3.19).

Chaparral taxa differ greatly in their ability to maintain net carbon gain under stress. In general, there are apparently physiological constraints that result in species capable of net photosynthesis under water stress being sensitive to light stress and vice versa (Oechel et al. 1981). In other words, shrubs seem to fall naturally into either functional groups of either drought-adapted or shade-adapted species. On a local scale one might predict that a mixed chaparral community would be more likely to optimize resources than monotypic stands of any single species.

In order to look at the role of biodiversity on resource utilization in chaparral, Miller et al. (1978) designed an ecosystem simulator called MEDECS. This multi-compartment model simulated seasonal patterns of resource use and demonstrated that shrub species had very different daily and seasonal patterns of uptake of water, solar energy and nitrogen. In ad-

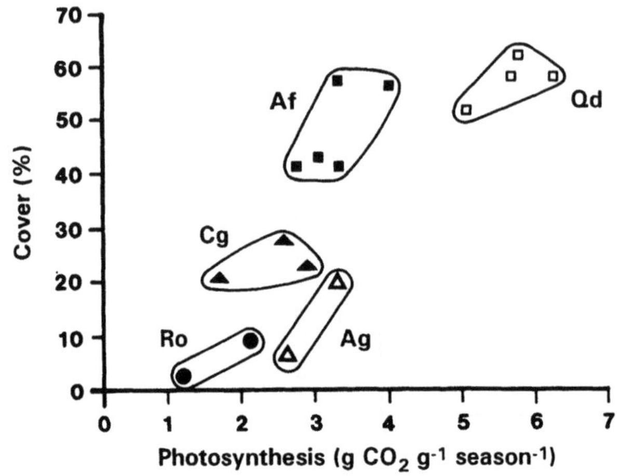

Fig. 3.18. Relationship between total seasonal photosynthesis and percentage cover of a species at a site from various elevations *Af Adenostoma fasciculatum*; *Qd Quercus dumosa*; *Cg Ceanothus greggii*; *Ag Arctostaphylos glauca*; *Ro Rhus ovata*). (Oechel et al. 1981)

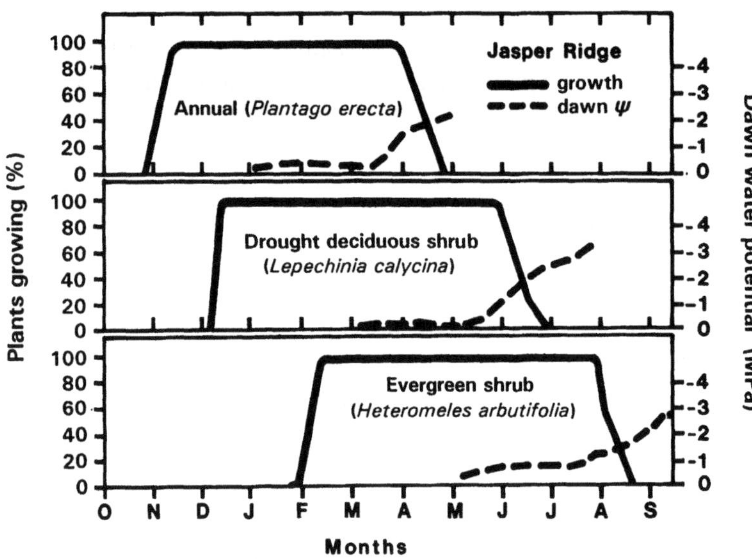

Fig. 3.19. Growth, and dawn water potentials of three different growth forms of mediterranean-climate plants from northern California. (Mooney 1983)

dition, simulations were run with various combinations of four species competing for light, water, nitrogen, and phosphorus. Unfortunately, Miller et al.'s (1978) simulations would predict quite different conclusions dependent upon assumptions made about root distribution and subsurface topology. Thus, an important factor that prevents accurate predictions about the ecosystem function of biodiversity in chaparral is the lack of information on underground conditions.

Despite this, their models are an important avenue for exploring the question of biodiversity and ecosystem function. Using different combinations of two species, *Adenostoma fasciculatum* and *Arctostaphylos glauca*, Miller et al. (1978) showed that mixed communities had greater net photosynthetic production than monotypic communities (Fig. 3.20A). However, for other aspects of resource use, such as nitrogen uptake, single species exhibited greater resource use than mixed communities (Fig. 3.20B). Thus, these simulation did not consistently predict that resource use by mixed species chaparral would be greater than by single-species communities. Rather, with respect to certain resources there may be a single optimum physiological type for any given site. However, even if this were true, landscape heterogeneity may select for greater biodiversity, dependent upon whether plants experience a coarse-grained or fine-grained environment.

3.5 Conclusions

In mediterranean-climate California, biodiversity is viewed as having an impact on ecosystem functioning in two fundamental ways. In grassland and chaparral there is reasonably good evidence that ecosystem functions, such as transfer of water, nutrients and carbon, are sensitive to the reductions in diversity of growth forms (above- and belowground), phenologies, physiologies and recruitment stratcgics. Species that are ecologically redundant in one character, e.g. recruitment mode, typically are not ecologically equivalent to the same taxa in other characters. Additionally, ecosystem processes, including trophic level linkages, are considered relevant ecosystem functions, since without them, ecosystems cannot persist. It is worth noting that some species may have limited control of ecosystem functions such as energy flow, but may be important controllers of ecosystem processes such as pollinator resources.

An impressive illustration of the key role played by certain species in ecosystem functioning is the massive extinction of late Pleistocene mammals, where loss of certain megaherbivores had a cascading effect, possibly causing the extinction of much of the mammalian fauna of California.

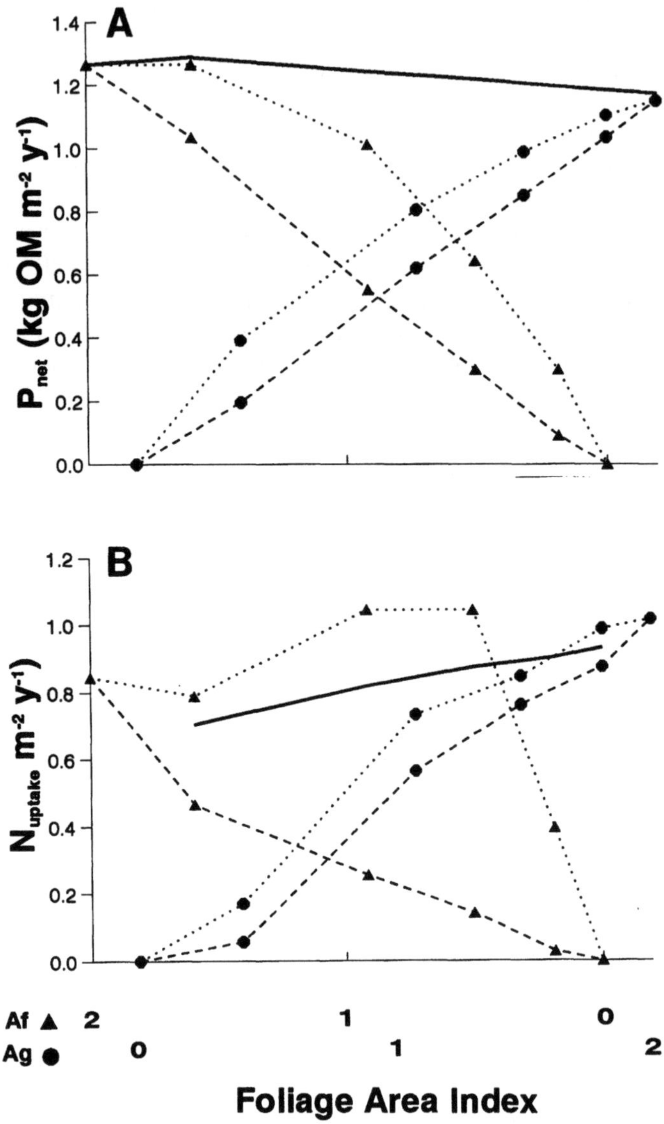

Fig. 3.20 A,B. Simulated patterns of **A** net photosynthesis and **B** total N uptake by *Arctostaphylos glauca* (*circles*) and *Adenostoma fasciculatum* (*triangles*) growing alone (*dotted line*) or with the other species (*dashed line*). The *solid line* is the sum of the two species growing together with a constant foliage area index of 2. The foliage area index of *A. glauca* increases from left to right (2 to 0). The differences between the *solid* and *dashed* lines of a species represent the effects of competition on the species. (Miller et al. 1978)

3.6 Future Research Needs

A definitive answer to the question of the role of biodiversity and ecosystem function requires experimental work specifically designed to address this question. Presently one such study is underway with the annual grassland ecosystem at Jasper Ridge (N. Chiariello, pers. comm.). Extending such studies to native ecosystems, such as chaparral and oak woodland would be important to the broader picture of biodiversity and ecosystem functioning in California. Due to problems inherent in large-scale manipulations of such communities, it may be logistically more feasible to utilize natural experiments that have manipulated biodiversity through grazing or type conversion projects and compare carbon, water and nutrient budgets for sites of differing diversity. While such an approach may tell us much about the functional significance of alpha diversity, other approaches may be necessary in order to evaluate other scales of diversity. Transplant studies that examine the degree of redundancy in species which appear to be ecological equivalents would be one way of examining the functional significance of gamma and beta diversity.

Acknowledgments. Many of the ideas presented here first took shape in a workshop attended by Martin Cody, Frank Davis, Jerry Powell, Mary Price, James Quinn, Philip Rundel and Paul Zedler. Their contribution is much appreciated.

References

Alberts AC, Richman AD, Tran D, Sauvajot R, McCalvin C, Bolger DT (1993) Effects of habitat fragmentation on populations of native and exotic plants in southern California coastal scrub. In: Keeley JE (ed) Interface between ecology and land development in California. Southern California Academy of Sciences, Spec Publ, Los Angeles, pp 103-110

Alten GR (1981) Post-fire avian community ecology. MS Thesis, California State Polytechnic University, Pomona

Armstrong WP, Platt JL (1993) The marriage between algae and fungi. Fremontia 22(2):3-12

Arnold K, Burcham LT, Fenner RL, Grah RF (1951) Use of fire in land clearing. Calif Agric 5(3):9-11; 5(4):4-5, 13

Atkins MD (1977) Insect biomass and diversity. In: Thrower NJW, Bradbury DE (eds) Chile-California mediterranean scrub atlas. Dowden, Hutchinson and Ross, Stroudsberg, pp 180-183

Atwood JL (1993) California gnatcatchers and coastal sage scrub: the biological basis for endangered species. In: Keeley JE (ed) Interface between ecology and land development in California. Southern California Academy of Sciences, Spec Publ, Los Angeles, pp 149-169

Axelrod DI (1958) Evolution of the Madro-Tertiary geoflora. Bot Rev 24:201-209

Axelrod DI (1973) History of the mediterranean ecosystem in California. In: di Castri F, Mooney HA (eds) Mediterranean ecosystems. Origin and structure. Springer, Berlin Heidelberg New York, pp 225-277

Axelrod DI (1975) Evolution and biogeography of Madrean-Tethyan sclerophyll vegetation. Ann Mo Bot Gard 62:280-334

Axelrod DI (1981) Holocene climate changes in relation to vegetation disjunction and speciation. Am Nat 117:847-870

Axelrod (1986) Cenozoic history of some western American pines. Ann Mo Bot Gard 70:629-657

Baker HG (1965) A timely note on the California oak moth. Four Seasons 1(3): 9-10

Ball CT, Keeley J, Mooney H, Seemann J, Winner W (1983) Relationship between form, function, and distribution of two *Arctostaphylos* species (Ericaceae) and their putative hybrids. Acta Oecol Oecol Plant 18(new ser 4):152-164

Barr WF, Linsley EG (1947) Distributional and biological notes on the species of the subgenus *Melanophila* occurring in western North America (Coleoptera: Buprestidae). Pan-Pac Entomol 23:162-166

Bartholomew B (1970) Bare zone between California shrub and grassland communities: the role of animals. Science 170:1210-1212

Bauder ET (1993) Genetic diversity: esoteric or essential? In: Keeley JE (ed) Interface between ecology and land development in California. Southern California Academy of Sciences, Spec Publ, Los Angeles, pp 35-41

Bazazz FA (1986) Life history of colonizing plants: some demographic, genetic, and physiological features. In: Mooney HA, Drake JA (eds) The ecology of biological invasions of North America and Hawaii. Springer, Berlin Heidelberg New York, pp 96-108

Beattie AJ (1989) The effects of ants on grasslands. In: Huenneke LF, Mooney HA (eds) Grassland structure and function: California annual grassland. Kluwer, Dordrecht, pp 105-116

Beetle AA (1947) Distribution of the native grasses of California. Hilgardia 17:309-347

Benson L, Walkington DL (1965) The southern California prickly pear invasion, adulteration and trial by fire. Ann Mo Bot Gard 52:262-273

Bock CE, Bock JH, Jepson KL, Ortega JC (1986) Ecological effects of planting African lovegrasses in Arizona. Nat Geogr Res 2:456-463

Bolger DT, Alberts AC, Soule ME (1991) Occurrence patterns of bird species in habitat fragments: sampling, extinction, and nested species subsets. Am Nat 137:155-166

Botkin DB, Mellilo JM, Wu LS-Y (1981) How ecosystem processes are linked to large mammal population dynamics. In: Fowler CW, Smith TD (eds) Dynamics of large mammal populations. John Wiley, New York, pp 373-388

Brown JW (1987) The peninsular effect in Baja California: an entomological assessment. J Biogeogr 14:359-365

Brown JH, Heske EJ (1990) Control of a desert-grassland transition by a keystone rodent guild. Science 250:1705-1707

Brown JH, Munger JC (1985) Experimental manipulation of a desert rodent community: food addition and species removal. Ecology 66:1545-1563

Brown LR, Eads CO (1965) A technical study of insects affecting the oak tree in southern California. Univ Calif Agric Exper Stn Bull 810

Brown RN, Lane RS (1992) Lyme disease in California: a novel enzootic transmission cycle of *Borrelia burgdorferi*. Science 256:1439-1442

Callaway RM, D'Antonio CM (1991) Shrub facilitation of coast live oak establishment in central California. Madrono 38:158-159

Callaway RM, Nakarni NM, Mahall BE (1991) Facilitation and interference of *Quercus douglasii* on understory productivity in central California. Ecology 72:1484-1499

Center DM, Vaughn CE, Jones MB (1989) Effects of management on plant production and nutrient cycling on two annual grassland sites. Hilgardia 57(1):1-40

Chiariello NR (1989) Phenology of California grasslands. In Huenneke LF, Mooney HA (eds) Grassland structure and function: California annual grassland. Kluwer, Dordrecht, pp 47-58

Chiariello N, Roughgarden J (1984) Storage allocation in seasonal races of an annual plant: optimal versus actual allocation. Ecology 65:1290-1301

Chiariello N, Hickman JC, Mooney HA (1982) Endomycorrhizal role for interspecific transfer of phosphorus in a community of annual plants. Science 217:941-943

Christensen NL, Muller CH (1975) Effects of fire on factors controlling plant growth in *Adenostoma* chaparral. Ecol Monogr 45:29-55

Cody ML (1973) parallel evolution and bird niches. In: di Castri F, Mooney HA (eds) Ecosystems of the world 11. Mediterranean-type shrublands. Elsevier, New York, pp 307-338

Cody ML (1974) Competition and structure of bird communities. Monographs in Population Biology no 7. Princeton University Press, Princeton

Cody ML (1986) Diversity, rarity, and conservation in mediterranean-climate regions. In Soule ME (ed) Conservation biology. Sinauer, Sunderland, pp 122-152

Cody ML, Fuentes ER, Glanz W, Hunt JH, Moldenke AR (1977) Convergent evolution in the consumer organisms of mediterranean Chile and California. In: Mooney HA (ed) Convergent evolution in Chile and California: mediterranean climate ecosystems. Dowden, Hutchinson & Ross, Stroudsberg, pp 144-192

Cornell HV (1979) Evolution of the richness-area correlation for cyniopid gall wasps on oak trees: a comparison of two geographic areas. Evolution 33:257-274

Cornell, HV (1985) local and regional richness of cynipine gall wasps on California oaks. Ecology 66:1247-1260

Cornell HV (1986) Oak species attributes and host size influence cynipine wasp species richness. Ecology 67:1582-1592

D'Antonio CM, Vitousek PM (1992) Biological invasions by exotic grasses, the grass/fire cycle, and global change. Annu Rev Ecol Syst 23:63-87

Davidson EA, Stark JM, Firestone MK (1990) Microbial production and consumption of nitrate in an annual grassland. Ecology 71:1968-1975

Davis SD (1989) Patterns in mixed chaparral stands: differential water status and seedling survival during summer drought. In: Keeley SC (ed) The California chaparral: paradigms reexamined. Nat Hist Mus Los Ang C Sci Ser 34:97-113

Davis SD, Mooney HA (1985) Comparative water relations of adjacent California shrub and grassland communities. Oecologia 66:522-529

DeBano LF, Eberlein GE, Dunn PH (1979) Effects of burning on chaparral soils: I. Soil nitrogen. Soil Sci Soc Am J 43:504-514

Diamond J (1992) The third chimpanzee. University of California Press, Los Angeles

Dobzhansky T (1953) Natural hybrids of two species of *Arctostaphylos* in the Yosemite region of California. Heredity 7:73-79

Dunn PH, Poth M (1979) Nitrogen replacement after fire in chaparral. In: Proceedings of the workshop on symbiotic nitrogen fixation in the management of temperature forests. Oregon State University, Corvallis, pp 287-293

Dunn PH, Barro SC, Poth M (1985) Soil moisture affects survival of microorganisms in heated chaparral soil. Soil Biol Biochem 17:143-148

Dutton D (1981) Resource partitioning and seasonal variation in an old, north slope chaparral avian community. MS Thesis, California State Polytechnic University, Pomona

Edwards S (1992) Observations on the prehistory and ecology of grazing in California. Fremontia 20(1):3-11

Ellstrand NC, Lee JM, Keeley JE, Keeley SC (1987) Ecological isolation and introgression: biochemical confirmation of introgression in an *Arctostaphylos* (Ericaceae) population. Acta Oecol Oecol Plant 8:299-308

Ewing AL (1981) Drought studies in California annual grassland communities. PhD Dissertation, University of California, Berkeley

Field CB, Davis SD (1989) Physiological ecology. In: Keeley SC (ed) The California Chaparral: paradigms reexamined. Nat Hist Mus Los Ang C Sci Ser 34:97-113

Field CB, Mooney HA (1983) Leaf age and seasonal effects on light, water, and nitrogen use efficiency in a California shrub. Oecologia 56:348-355

Force DC (1981) Postfire insect succession in southern California chaparral. Am Nat 117:575-582

Force DC (1982) postburn insect fauna in southern California chaparral. In: Conrad CE, Oechel WC (eds) Proceedings of the symposium on dynamics and management of mediterranean-type ecosystems. Gen Tech Report PSW.-58, Pacific Southwest Forest and Range Experiment Station, Forest Service, USDA, Berkeley, pp 234-240

Force DC (1990) Ecology of insects in California chaparral. USDA For Serv, Pacific Southwest Research Station, Research Paper PSW-201

Fordham DK (1977) Soil microfungi associated with vegetation zones of Sycamore Creek Canyon, Mount Diablo, California. MS Thesis, San Francisco State University, San Francisco

Freudenberger DO, Fish BE, Keeley JE (1987) Distribution and stability of grasslands in the Los Angeles Basin. Bull S Calif Acad Sci 86:13-26

Gilbert LE (1980) Food web organization and the conservation of neotropical diversity. In: Soule ME, Wilcox BA (eds) Conservation biology: An evolutionary-ecological perspective. Sinauer, Sunderland, pp 11-33

Gordon DR, Welker JM, Menke JW, Rice KJ (1989) Competition for soil water between annual plants and blue oak (*Quercus douglasii*) seedlings. Oecologia 79:533-541

Gottlieb LD (1968) Hybridization between *Arctostaphylos viscida* and *A. canescens* in Oregon. Brittonia 20:83-93

Graham RW (1992) Late Pleistocene faunal changes as a guide to understanding effects of greenhouse warming on the mammalian fauna of North America. In: Peters RL, Lovejoy TE (eds) Global warming and biological diversity. Yale University Press, New Haven, pp 76-87

Grinnell J (1917) The niche-relationships of the California thrasher. Auk 34:427-433

Grinnell J (1936) Up-hill planters. Condor 38:80-82

Groves RH, di Castri F (eds) (1991) Biogeography of mediterranean invasions. Cambridge University Press, Cambridge

Grubb PJ (1977) The maintenance of species richness in plant communities: the importance of the regeneration niche. Biol Rev 52:107-145

Gulmon SL, Chu CC (1981) The effects of light and nitrogen on photosynthesis, leaf characteristics, and dry matter allocation in the chaparral shrub *Diplacus aurantiacus*. Oecologia 49:207-212

Gulmon SL, Chiariello NR, Mooney HA, Chu CC (1983) Phenology and resource use in three co-occurring grassland annuals. Oecologia 58:33-42

Guthrie RD (1984) Mosaics, allelochemics and nutrients. An ecological theory of late Pleistocene megafaunal extinctions. In: Martin PS, Klein RG (eds) Quaternary extinctions. A prehistoric revolution. University of Arizona Press, Tucson, pp 259-298

Haidinger TL, Keeley JE (1993) Role of high fire frequency in destruction of mixed chaparral. Madrono 40:141-147

Hobbs RJ (1985) Harvester ant foraging and plant species distribution in annual grassland. Oecologia 67:519-523

Hobbs RJ, Hobbs VJ (1987) Gophers and grassland: a model of vegetation response to patchy soil disturbance. Vegetatio 69:141-146

Hobbs RJ, Mooney HA (1985) Community and population dynamics of serpentine grassland annuals in relation to gopher disturbance. Oecologia 67:342-351

Hobbs RJ, Mooney HA (1986) Community changes following shrub invasion of grassland. Oecologia 70:508-513

Hobbs RJ, Mooney HA (1991) Effects of rainfall variability and gopher disturbance on serpentine annual grassland dynamics. Ecology 72:59-68

Hobbs RJ, Gulmon SL, Hobbs VJ, Mooney HA (1988) Effects of fertilizer addition and subsequent gopher disturbance on a serpentine annual grassland community. Oecologia 75:291-295

Hopkins NA (1987) Mycorrhizae in a California serpentine grassland community. Can J Bot 65:484-487

Huenneke LF (1989) Distribution and regional patterns of Californian grasslands. In: Huenneke LF, Mooney HA (eds) Grassland structure and function: California annual grassland. Kluwer, Dordrecht, pp 1-12

Huenneke LF, Hamburg SP, Koide R, Mooney HA, Vitousek PM (1990) Effects of soil resources on plant invasion and community structure in California serpentine grassland. Ecology 71:478-491

Huffaker, CB (1951) The return of native perennial bunchgrass following the removal of Klamath weed (*Hypericum perforatum* L.) by imported beetles. Ecology 32:443-458

Hufstader RW (1976) Precipitation, temperature, and the standing crop of some southern California grassland species. J Range Manage 29:433-435

Hull JC, Mooney HA (1990) Effects of nitrogen on photosynthesis and growth rates of four California annual grasses. Acta Oecol 11:453-468

Hull JC, Muller CH (1976) Responses of California annual grassland species to variations in moisture and fertilization. J Range Manage 29:49-52

Hunt JH (1977) Ants. In: Thrower NJW, Bradbury DE (eds) Chile-California mediterranean scrub atlas. A comparative analysis. Dowden, Hutchinson and Ross, Stroudsberg, pp 195-198

Hunter JE (1991) Grazing and pocket gopher abundance in a California annual grassland. Southwest Nat 36:117-118

Jackson LE, Roy J (1986) Growth patterns of mediterranean annual and perennial grasses under simulated rainfall regimes of southern France and California. Oecol Plant 7:191-212

Jain SK (1982) Variation and adaptive role of seed dormancy in some annual grassland species. Bot Gaz 143:101-106

Jensen DB, Torn M, Harte J (1990) In our own hands: a strategy for conserving biological diversity in California. California Policy Seminar, University of California, Berkeley

Joern A (1989) Insect herbivory in the transition to California annual grasslands: did grasshoppers deliver the coup de grass? In: Huenneke LF, Mooney HA (eds) Grassland structure and function: California annual grassland. Kluwer, Dordrecht, pp 117-134

Joffre R (1990) Plant and soil nitrogen dynamics in mediterranean grasslands: a comparison of annual and perennial grasses. Oecologia 85:142-149

Johnson DL (1977a) The late Quaternary climate of coastal California: evidence for an ice age refugium. Quat Res 8:154-179

Johnson DL (1977b) The California ice-age refugium and the Rancholabrean extinction problem. Quat Res 8:149-153

Johnson WC, Thomas L, Adkisson CS (1993) Dietary circumvention of acorn tannins by blue jays. Implications for oak demography. Oecologia 94:159-164

Johnston GE, Lathrop EW (1978) Temporal segregation in southern California annual grassland. In: 1st Int Rangeland Congr. pp 234-238

Karron JD (1991) Patterns of genetic variation and breeding systems in rare plant species. In: Falk DA, Holsinger KE (eds) Genetics and conservation of rare plants. Oxford University Press, Oxford pp 87-98

Keeley JE (1976) Morphological evidence of hybridization between *Arctostaphylos glauca* and *A. pungens* (Ericaceae). Madrono 23: 427-434

Keeley JE (1982) Distribution of lightning and man-caused wildfires in California. In: Conrad CE, Oechel WC (eds) Proc symp dynamics and management of mediterranean-type ecosystems. USDA Forest Service, Pacific Southwest Forest and Range Experiment Station, General Technical Report PSW-58, pp 431-437

Keeley JE (1990) The California valley grassland. In: Schoenherr AA (ed) Endangered plant communities of southern California. Southern California Botanists, Fullerton, Spec Publ 3, pp 1-23

Keeley JE (1992a) Demographic structure of California chaparral in the long-term absence of fire. J Veg Sci 3:79-90

Keeley JE (1992b) Recruitment of seedlings and vegetative sprouts in unburned chaparral. Ecology 73:1194-1208

Keeley JE (ed) (1993) Interface between ecology and land development in California. Southern California Academy of Sciences, Spec Publ, Los Angeles. 297 pp

Keeley JE, Hays RL (1976) Differential seed predation on two species of *Arctostaphylos* (Ericaceae). Oecologia 24:71-81

Keeley JE, Keeley SC (1988) Chaparral. In: Barbour MG, Billings WD (eds) North American terrestrial vegetation. Cambridge University Press, Cambridge, pp 165-207

Keeley JE, Keeley MB (1993) Mitigation plan for *Pentachaeta lyonii* on the Lake Eleanor hills project site. Proposal for California Department of Fish and Game, Sacramento

Keeley JE, Sandquist DR (1991) Diurnal photosynthesis cycle in CAM and non-CAM seasonal pool aquatic macrophytes. Ecology 72:716-727

Keeley SC, Keeley JE, Hutchinson SM, Johnson AW (1981) Postfire succession of the herbaceous flora in southern California chaparral. Ecology 62:1608-1621

Kessel EL (1947) American smoke flies (Microsania: Clythiidae). Wasmann Coll 7:23-30

Kilgore BM, Taylor D (1979) Fire history of a sequoia-mixed conifer forest. Ecology 60:129-142

Knight A, Connor E, Rubenstein P (1984) Effects of shrubland management on streams. In: DeVries JJ (ed) Shrublands in California: literature review and research needed for management. Water Resources Center, University of California, Davis, Rep 161, pp 61-73

Kummerow J, Montenegro G, Krause D (1981) Biomass, phenology, and growth. In: Miller PC (ed) Resource use by chaparral and matorral: a comparison of vegetation function in two mediterranean type ecosystems. Springer, Berlin Heidelberg New York, pp 69-96

Langen TA, Bolger DT, Case TJ (1991) Predation on artificial bird nests in chaparral fragments. Oecologia 86:395-401

Larson JR, Duncan DA (1982) Annual grassland response to fire retardant and wildfire. J Range Manage 35:700-703

Laudenslayer WF Jr, Grenfell WE Jr, Zeiner DC (1991) A check-list of the amphibians, reptiles, birds, and mammals of California. Calif Fish Game 77:109-141

Ledig FT (1992) Human impacts on genetic diversity in forest ecosystems. Oikos 63:87-108

Lidicker WZ Jr (1989) Impacts of non-domesticated vertebrates on California grasslands. In: Huenneke LF, Mooney HA (eds) Grassland structure and function: California annual grassland. Kluwer, Dordrecht, pp 135-150

Lillywhite HB (1977) Effects of chaparral conversion on small vertebrates in southern California. Biol Conserv 11:171-184

Lundelius EL, Graham RW, Anderson E, Guilday J, Holman JA, Steadman DW, Webb SD (1983) Terrestrial vertebrate faunas. In: Porter SC (ed) Late-Quaternary environments of the United States. Vol 1. The late Pleistocene. University of Minnesota Press, Minneapolis, pp 311-353

MacArthur RH (1955) Fluctuations of animal populations and a measure of community stability. Ecology 36: 533-536

MacArthur RH (1972) Geographical ecology. Harper and Row, New York

Marshall DR, Jain SK (1969) Interference in pure and mixed populations of *Avena fatua* and *A. barbata*. Ecology 51:886-891

Martin PS (1984) Prehistoric overkill: the global model. In Martin PS, Klein RG (eds) Quaternary extinctions. University of Arizona Press, Tucson, pp 354-403

Martin MM, Harding J (1981) Evidence for the evolution of competition between two species of annual plants. Evolution 35:975-987

Mauffette Y, Oechel WC (1989) Seasonal variation in leaf chemistry of the coast live oak *Quercus agrifolia* and implications for the California oak moth *Phryganidia californica*. Oecologia 79:439-445

McClaran MP, Bartolome JW (1989) Effect of *Quercus douglasii* (Fagaceae) on herbaceous understory along a rainfall gradient. Madrono 36:141-153

McCloskey RT (1972) temporal changes in populations and species diversity in a California rodent community. J Mammal 53: 657-676

McCown RL, Williams WA (1968) Competition for nutrients and light between the annual grassland species *Bromus mollis* and *Erodium botrys*. Ecology 49:981-990

McMaster G, Jow W, Kummerow J (1982) Response of *Adenostoma fasiculatum* and *Ceanothus greggii* chaparral to nutrient additions. J Ecol 70: 745-756

Meserve PL (1976a) Food relationships of a rodent fauna in a California coastal sage scrub community. J Mammal 57: 300-319

Merserve PL (1976b) Habitat and resource utilization by rodents of a California coastal sage scrub community. J Animal Ecol 45:647-666

Miller PC (1983) Comparison of water balance characteristics of plant species in "natural" versus modified ecosystems. In: Mooney HA, Godron M (eds) Disturbance and ecosystems: components of response. Springer, Berlin Heidelberg New York, pp 188-212

Miller PC, Stoner WA, Richards SP (1978) MEDECS, a simulator for mediterranean ecosystems. Simulation 30:173-190

Mills JN (1983) Herbivory and seedling establishment in post-fire Southern California chaparral. Oecologia 60:267-270

Mills JN (1986) Herbivores and early postfire succession in southern California chaparral. Ecology 67:1637-1649

Moldenke AR (1975) Niche specialization and species diversity along a California transect. Oecologia 21:219-242

Moldenke AR (1976) California pollination ecology and vegetation types. Phytologia 34:305-361
Moldenke AR (1977) Insect-plant relations. In: Thrower NJW, Bradbury DE (eds) Chile-California mediterranean scrub atlas. Dowden, Hutchinson and Ross, Stroudsberg, pp 199-217
Moldenke AR (1979a) Pollination ecology within the Sierra Nevada. Phytologia 42:223-282
Moldenke AR (1979b) Pollination ecology as an assay for ecosystemic organization: convergent evolution in Chile and California. Phytologia 43:415-454
Mooney HA (1983) Carbon-gaining capacity and allocation patterns of mediterranean-climate plants. In: Kruger FJ, Mitchell DT, Jarvis JUM (eds) Mediterranean-type ecosystems: the role of nutrients. Springer, Berlin Heidelberg New York, pp 103-132
Mooney HA (1986) Lessons from mediterranean-climate regions. In: Wilson EO (ed) Biodiversity. National Academy Press, Washington DC, pp 157-165
Mooney HA, Drake JA (eds) (1986) Ecology of biological invasions of North America and Hawaii. Springer, Berlin Heidelberg New York
Mooney HA, Hamburg SP, Drake JA (1986a) The invasions of plants and animals into California. In: Mooney HA, Drake JA (eds) Ecology of biological invasions of North America and Hawaii. Springer, Berlin Heidelberg New York, pp 250-274
Mooney HA, Hobbs RJ, Gorham J, Williams K (1986b) Biomass accumulation and resource utilization in co-occurring grassland annuals. Oecologia 70:555-558
Moore JA, Taylor CE, Moore BC (1979) The *Drosophila* of southern California. I. Colonization after a fire. Evolution 33:156-171
Muller CH, Muller WH, Haines BL (1964) Volatile growth inhibitors produced by aromatic shrubs. Science 143:471-473
Mulroy TW, Rundel PW (1977) Annual plants: adaptation to desert environments. Bioscience 27:109-114
Nadkarni NM, Odion DC (1986) Effects of seeding an exotic grass *Lolium multiflorum* on native seedling regeneration following fire in a chaparral community. In: DeVries JJ (ed) Proceedings of the chaparral ecosystems research conference. California Water Resources Center, University of California, Davis, Rep 62, pp 115-121
Oechel WC Lawrence W, Mustafa J, Martinez J (1981) Energy and carbon acquisition. In: Miller PC (ed) Resource use by chaparral and matorral: a comparison of vegetation function in two mediterranean type ecosystems. Springer, Berlin Heidelberg New York, pp 151-183
Opler PA (1974) Biology, ecology, and host specificity of Microlepidoptera associated with *Quercus agrifolia* (Fagaceae). Univ Calif Publ Entomol 75:1-83 (+ plates)
Owen-Smith N (1987) Pleistocene extinctions. The pivotal role of megaherbivores. Paleobiology 13:351-362
Owen-Smith N (1988) Megaherbivores. The influence of very large body size on ecology. Cambridge University Press, Cambridge
Owen-Smith N (1989) Megafaunal extinctions: the conservation message from 11000 B.P. Conserv Biol 3:405-412
Parsons DJ (1991) Planning for climate change in national parks and other natural areas. Northwest Environ J 7:255-269
Parsons DJ, Stohlgren TJ (1989) Effects of varying fire regimes on annual grasslands in the southern Sierra Nevada of California. Madrono 36:154-168
Patric JH (1961) The San Dimas large lysimeters. J Soil Water Conserv 16:13-17
Pavlik BM, Johnson PMS, Popper M (1991) Oaks of California. Cachuma Press, Los Olivos
Peart DR (1989) Species interactions in a successional grassland. III. Effects of canopy gaps, gopher mounds and grazing on colonization. J Ecol 77:267-289

Pitt MD, Heady HF (1978) Responses of annual vegetation to temperature and rainfall patterns in northern California. Ecology 59:336-350

Poole DK, Roberts SW, Miller PC (1981) Water utilization. In: Miller PC (ed) Resource use by chaparral and matorral: a comparison of vegetation function in two mediterranean type ecosystems. Springer, Berlin Heidelberg New York, pp 123-149

Powell JA (1992a) An estimate of Lepidoptera biodiversity in central coastal California. unpublished

Powell JA (1992b) Interrelationships of Yuccas and yucca moths. Trends Ecol Evol 7:10-15

Powell JA, Hogue C (1979) California insects. University of California Press, Los Angeles

Price MV (1978) The role of microhabitat in structuring desert rodent communities. Ecology 59:910-921

Price MV, Waser NM (1984) On the relative abundance of species: postfire changes in a coastal sage scrub rodent community. Ecology 65:1161-1169

Quinn JF, Robinson GR (1987) The effects of experimental subdivision on flowering plant diversity in a California annual grassland. J Ecol 75:837-856

Quinn RD (1986) Mammalian herbivory and resilience in mediterranean-climate ecosystems. In: Dell B, Hopkins AJM, Lamont BB (eds) Resilience in mediterranean-type ecosystems. Junk, Dordrecht, pp 113-128

Quinn RD (1991) Herbivory after fire in Californian chaparral and other mediterranean-type ecosystems. In: Thanos CA (ed) MEDECOS VI Plant-Animal interactions in mediterranean-type ecosystems. Univ Athens, Greece, pp 28-38

Raven PH (1973) The evolution of mediterranean floras. In: di Castri F, Mooney HA (eds) Mediterranean type ecosystems: origin and structure. Chapman & Hall, London, pp 213-224

Raven PH, Axelrod DI (1978) Origin and relationships of the California flora. Univ Calif Publ Bot 72:1-134

Rice SK (1993) Vegetation establishment in postfire *Adenostoma* chaparral in relation to fine-scale pattern in fire intensity and soil nutrients. J Veg Sci 4:115-124

Richerson PJ, Lum K-L (1980) Patterns of plant species diversity in California: relation to weather and topography. Am Nat 116:504-536

Rowe PB, Reimann LF (1961) Water use by brush, grass, and grass-forb vegetation. J For 59:175-181

Rundel PW (1971) Community structure and stability in the giant sequoia ecosystem in the Sierra Nevada, California. Am Midl Nat 85: 478-492

Rundel PW (1983) Impact of fire on nutrient cycles in mediterranean-type ecosystems with reference to chaparral. In: Kruger FJ, Mitchell DT, Jarvis JUM (eds) Mediterranean-type ecosystems, the role of nutrients. Springer, Berlin Heidelberg New York pp 192-207

Saiz F (1977) Soil beetles. In: Thrower NJW, Bradbury DE (eds) Chile-California mediterranean scrub atlas. A comparative analysis. Dowden, Hutchinson and Ross, Stroudsberg, pp 184-194

Sampson AW (1944) Plant succession on burned chaparral lands in northern California. Agric Exp Stn Bull 685, Univ Calif, Berkeley

Samson DA, Philippi TE, Davidson DW (1992) Granivory and competition as determinants of annual plant diversity in the Chihuahuan desert. Oikos 65:61-80

Sauer JD (1988) Plant migration. The dynamics of geographic patterning in seed plant species. University of California Press, Los Angeles

Sauvajot RM, Buechner M (1993) Effects of urban encroachment on wildlife in the Santa Monica Mountains. In: Keeley JE (ed) Interface between ecology and land development in California. Southern California Academy of Sciences, Los Angeles, pp 171-180

Schimel JP, Jackson LE, Firestone MK (1989) Spatial and temporal effects on plant-microbial competition for inorganic nitrogen in a California annual grassland. Soil Biol Biochem 21:1059-1066

Schlesinger WH, Gray JT (1982) Atmospheric precipitation as a source of nutrients in chaparral ecosystems. In Conrad CE, Oechel WC (eds) Proceedings of the symposium on dynamics and management of mediterranean-type ecosystems. USDA Forest Service, Pacific Southwest Forest and Range Experiment Station, General Technical Report PSW-58, pp 279-284

Schulze E-D, Mooney HA (1993) Ecosystem function of biodiversity: a summary. In: Schulze E-D, Mooney HA (eds) Ecosystem function of biodiversity. Springer, Berlin Heidelberg New York pp 497-510

Shmida A, Whittaker RH (1981) Pattern and biological microsite effects in two shrub communities, southern California. Ecology 62:234-251

Sholes ODV, Beatty SW (1987) Influence of host phenology and vegetation on the abundance of *Tamalia coweni* galls (Homoptera: Aphididae) on *Arctostaphylos insularis* (Ericaceae). Am Midl Nat 118:198-204

Smith MF (1978) Relationships between genetic variability and niche dimensions among coexisting species of *Peromyscus*. PhD Dissertation, University of California, Berkeley

Soulé ME, Simberloff DS (1986) What do genetics and ecology tell us about the design of nature reserves. Biol Conserv 29:27-46

Soulé ME, Bolger DT, Alberts AC, Sauvajot RS, Wright J, Sorice M, Hill S (1988) Reconstructed dynamics of rapid extinctions of chaparral-requiring birds in urban habitat islands. Conserv Biol 2:75-92

Stamp NE (1990) Production and effect of seed size in a grassland annual (*Erodium brachycarpum*, Geraniaceae). Am J Bot 77:874-882

Stebbins GL (1976) Colonizing species of the native California flora. In: Baker HG, Stebbins GL (eds) The genetics of colonizing species. Academic Press, New York, pp 173-195

Stebbins GL, Major J (1965) Endemism and speciation in the Californian flora. Ecol Monogr 35:1-35

St John T (1993) Benefits of mycorrhizae in revegetation and restoration. In: Keeley JE (ed) Interface between ecology and land development in California. Southern California Academy of Sciences, Los Angeles, pp 269-272

Stock C (1930) Rancho La Brea: a record of Pleistocene life in California. Nat Hist Mus Los Ang C Mus Sci Ser 20

Swift C. (1991) Nitrogen utilization strategies in post-fire chaparral annual species. PhD Dissertation, University of California, Los Angeles

Swift C, Collins A, Gutierrez H, Lam H, Ratiner I (1993) Habitat linkages in an urban mountain chain. In: Keeley JE (ed) Interface between ecology and land development in California. Southern California Academy of Sciences, Los Angeles, pp 189-199

Swisher RG (1979) A survey of the insect fauna on *Eriogonum fasiculatum* in the San Gabriel Mountains, Southern California. MS Thesis, California State University, Los Angeles

Talbot MW, Biswell HH, Hormay AL (1939) Fluctuations in the annual vegetation of California. Ecology 20:394-402

Vale TR, Vale GR (1976) Suburban bird populations in west-central California. J Biogeogr 3:157-165

Van Vuren D, Coblentz BE (1987) Some ecological effects of feral sheep on Santa Cruz Island, California, USA. Biol Conserv 41:253-268

Vuilleumier F (1991) Invasions in the mediterranean avifaunas of California and Chile. In: Groves RH, di Castri F (eds) Biogeography of mediterranean invasions. Cambridge University Press, New York, pp 327-376

Warter JK (1976) Late Pleistocene plant communities - evidence from the Rancho la Brea Tar Pits. In: Latting J (ed) Symposia proceedings. Plant communities of southern California. California Native Plant Society, Spec Publ 2, pp 32-39

Welker JM, Gordon DR, Rice KJ (1991) Capture and allocation of nitrogen by *Quercus douglasii* seedlings in competition with annual and perennial grasses. Oecologia 87:459-466

Wells PV (1962) Vegetation in relation to geological substratum and fire in the San Luis Obispo quadrangle, California. Ecol Monogr 32:79-103

Wells PV (1969) The relation between mode of reproduction and extent of speciation in woody genera of the California chaparral. Evolution 23: 264-267

Westman WE (1987) Implications of ecological theory for rare plant conservation in coastal sage scrub. In: Elias TS (ed) Conservation and management of rare and endangered plants. California Native Plant Society, Sacramento, pp 130-140

Wilcove DS, McLellan CH, Dobson AP (1986) Habitat fragmentation in the temperate zone. In: Soulé ME (ed) Conservation biology. The science of scarcity and diversity. Sinauer, Sunderland, pp 237-256

Wirtz W (1979) Effects of fire on birds in chaparral. In: Proc 10th annu joint conf western section of the Wildlife Society and the California-Nevada chapter of the American Fisheries Society, pp 114-124

Yoder SE (1976) The relationship of avian species composition, density, diversity, and foraging behavior to habitat parameters of fire-disturbed and mature chaparral. MS Thesis, California State University, Los Angeles

Zedler PH (1987) The ecology of southern California vernal pools: a community profile. Biological Report 85(7.11), National Wetlands Research Center, Fish and Wildlife Service, US Department of the Interior, Washington DC

Zedler PH, Gautier CR, McMaster GS (1983) Vegetation change in response to extreme events: the effect of a short interval between fires in California chaparral and coastal scrub. Ecology 64:809-818

4 Functional Approaches to Biodiversity in the Mediterranean-Type Ecosystems of Central Chile

E.R. Fuentes, G. Montenegro, P.W. Rundel, M.T.K. Arroyo, R. Ginocchio and F.M. Jaksic

4.1 Introduction

The political boundaries of Chile coincide with topographic features that are also biogeographic boundaries. To the east lies the formidable barrier of the high Cordillera de los Andes. Although major uplifting of this range dates back to the middle Tertiary, the Andes remain tectonically active today, and only in the south are there mountain passes low enough to provide potential migration routes for plants and animals between Chile and Argentina. The severe climatic regime of Patagonia to the south allows only a relatively small number of plants and animals to survive. The hyper-arid Atacama Desert to the north effectively limits the interchange of species with northern subtropical regions. The Pacific Ocean to the west provides the final isolating barrier. Thus, although the biota of Chile has had many origins that date back to Gondwanaland, the region remains biogeographically isolated.

The extreme climatic diversity within Chile (di Castri and Hajek 1976) is due in part to the large latitudinal gradient (greater than any other country in the world), extending from 17°30'S in the north to 56°S at the islands around Tierra del Fuego. Mean annual temperatures range from 17°C in the coastal northern desert to less than 5°C in the high Andes and southern tip of Patagonia. Mean annual precipitation ranges from virtually zero at Arica in the north to more than 7000 mm in the austral islands of the Magellanic region (Romero 1985). The great topographic diversity adds further variation. The Andean Cordillera forms a continuous high wall along the eastern margin of northern and central Chile, with individual peaks reaching above 6000 m. The Andes are lower in the austral forest regions to the south. The geologically young coastal Cordillera de la Costa in Chile is also tall; Cerro Campana (1910 m), Campanita (1510 m) and El Roble (2220 m) are notable peaks west of Santiago and Sierra Vicuña MacKenna in northern Chile south of Antofagasta reaches 3000 m.

G.W. Davis and D.M. Richardson (Eds.)
Ecological Studies, Vol.109

4.2 Biodiversity in Chile

Modern accounts of patterns of plant and animal diversity in Chile are mainly limited to broad overviews. Individual groups such as trees, mammals, birds, reptiles and amphibians have been dealt with for the country as a whole, providing sufficient information to describe general patterns of diversity. There are very few detailed regional studies that have evaluated local patterns of β- and γ-diversity for any of these groups.

4.2.1 Vascular Plants

The vascular plant flora of Chile includes 4669 native species in 873 genera and 181 families (Table 4.1). This includes 154 ferns and fern relatives, 16 gymnosperms, 908 monocots and 3590 dicots. Gajardo (1987) estimated that 62% of this native flora was endemic to Chile. The 20 largest genera in the native Chilean flora, led by *Senecio* with 218 species, make up 30% of the flora (Table 4.2). *Adesmia* with 140 species, *Oxalis* with 111 species, *Calceolaria* with 86 species and *Calandrinia* with 70 species complete the group of the five largest genera. Rates of endemism within the largest 20 genera are mostly higher than for the entire flora, with 11 of these genera having 80% or more endemic species.

The age and geographic isolation of the Chilean flora is indicated by the relatively large number of families that are mainly endemic to Chile and adjacent areas of austral South America. The Thyrsopteridaceae (Pteridophyta), Aextoxicaceae, Gomortagaceae and Lactoridaceae (Juan Fernades Islands) are all confined to Chile. The Malesherbiaceae, Nolanaceae, Mizodendraceae, Vivianiaceae, Desfontaineaceae, Arachnitaceae, Francoaceae (a segregate of the Saxifragaceae), Tropaeolaceae, Alstroemeriaceae and Heterostylaceae are also largely Chilean. The majority of these endemic families are characteristic of the austral forests of southern Chile. Only Malesherbiaceae, Nolanaceae, Tropaeolaceae and Alstroemeriaceae have distributions centred in the mediterranean-climate and arid regions of Chile.

Endemism is also relatively high at the generic level in the Chilean flora. From data presented by Gajardo (1987), we estimate that about 16% of the native genera are strict Chilean endemics. Another 17% of genera are endemic to austral South America, while 22% have broader South American distributions, and 10% are of Gondwanaland origin with extant representatives in Australia and/or New Zealand.

Table 4.1. The native vascular plant flora of Chile. Data taken from Marticorena and Quezada (1985) by subtracting exotic species.

	Native families	Native genera	Native species
Pteridophyta	25	50	154
Gymnospermae	4	9	16
Angiospermae			
Dicotyledoneae	126	644	3 590
Monocotyledoneae	25	169	908
TOTAL	181	873	4 669

Table 4.2. The 20 largest genera of vascular plants in Chile. Totals of native and introduced species are based on Marticorena and Quezada (1985). Level of endemism and origin for each genus is based on Gajardo (1987).

	No. of Species			
Genus	Native	Introduced	Endemism (%)	Origin[a]
Senecio (Asteraceae)	218	4	55	COS
Adesmia (Papilionaceae)	140	-	89	SA-A
Oxalis (Oxalidaceae)	111	4	86	COS
Calceolaria (Scrophulariaceae)	86	-	89	SA
Calandrinia (Portulacaceae)	70	-	95	GON
Viola (Violaceae)	69	3	84	COS
Cristaria (Malvaceae)	62	-	100	SA-A
Haplopappus (Asteraceae)	61	-	88	SA-A
Poa (Poaceae)	60	5	72	DIS
Solanum (Solanaceae)	60	4	72	COS
Neochilenia (Cactaceae)	57	-	100	CH
Carex (Cyperaceae)	51	-	35	DIS
Berberis (Berberidaceae)	50	-	80	COS
Astragalus (Papilionaceae)	46	-	78	COS
Stipa (Poaceae)	44	-	37	COS
Loasa (Loasaceae)	45	-	88	SA
Leucheria (Asteraceae)	43	-	60	SA-A
Valeriana (Valerianaceae)	43	-	71	COS
Dioscorea (Dioscoriaceae)	42	-	100	TROP
Baccharis (Asteraceae)	42	-	50	SA

[a] COS = cosmopolitan, SA-A = austral South America, SA = South America, GON = Gondwanaland, DIS = disjunct between north and south temperate zones, TROP = tropical.

Table 4.3. Species richness of plant communities in central Chile (Montenegro and Tellier in Westman 1988). The number of woody perennials, herbaceous perennials, and annuals, respectively, are shown in parentheses.

Vegetation type	Site	Quadrat size			
		1 m²	10 m²	100 m²	1 000 m²
Coastal matorral	Papudo	9 (1,6,2)	25 (8,11,6)	43 (14,18,11)	85 (32,33,20)
Matorral	Fundo Santa Laura	8 (1,3,4)	44 (6,17,21)	94 (22,31,41)	114 (29,39,46)
Montane matorral	Paso Marchant	6 (2,3,1)	17 (3,6,8)	46 (9,21,16)	102 (26,44,32)
Espinal	Polpaico	7 (0,3,4)	12 (1,3,8)	20 (2,3,15)	25 (5,5,15)

Although there is a checklist of the Chilean flora (Marticorena and Quezada 1985), there is no manual of the flora of the mediterranean-type ecosystems (MTEs) in central Chile. There are about 2100 vascular plant species between latitudes 32° and 36°S (M.T.K. Arroyo et al., unpubl. data). Data on species richness in plant communities of central Chile are available for four sites (Westman 1988). The matorral site at Fundo Santa Laura was the richest with 114 species, 29 of them woody perennials, in 0.1 ha (Table 4.3). This site contained 94 species, 22 of them woody species, in 100 m^2. The greatest richness of woody species in 0.1 ha was found in a coastal matorral site at Papudo, where 32 woody species were counted among the 85 species present.

4.2.2 Mammals

Mammal diversity in Chilean MTEs, like that of temperate South America overall, is relatively low. Whereas major feeding niches in temperate South America are occupied by mammals of similar size to those in North America, the number of species within each niche is generally smaller in South America. Eisenberg and Redford (1982) suggest that glacial cycles in the Andes during the Pleistocene may have had a strong impact on South American mammal faunas. They note that temperate South America currently has many ecological niches that appear to be only partially occupied by mammals in comparison to North America. Rodents dominate in many herbivore niches in Chile, and have adapted to insectivore and omnivore niches usually filled by other taxa. Community studies of small mammals carried out at three sites in central Chile indicate that the fauna is quite small. These sites are a matorral shrubland at Fundo Santa Laura, a mesic coastal matorral at Zapallar, and a dry coastal scrub at Fray Jorge (Glanz 1977a; Meserve 1981; Glanz and Meserve 1982). Ten species were trapped on these three sites - nine rodents and one small marsupial, *Marmosa elegans*. A sampling of similar sites in California yielded 17 species.

Specialization on seed resources is surprisingly rare among Chilean rodents. Only two species, *Oryzomys longicaudatus* and *Phyllotis darwini*, are primarily granivorous, and even these are relatively concentrated around existing cover rather than open habitats. There is no Chilean mammal that is ecologically equivalent to a hetermyid rodent specialized for a granivorous diet in open habitats. In contrast, Chilean small mammals show higher than expected utilization of insects in their diet. The mouse opposum, *Marmosa elegans*, is almost entirely insectivorous, and two cricetid rodents, *Akodon longipilis* and *A. olivaceus*, also feed heavily on insects. Chilean MTEs lack a morphological equivalent of the chip-munksand ground squirrels that play an important ecological role in California. *Octodon degus*, a diurnally active hystricognath rodent, is a rough equivalent of the ground squirrel but is smaller in size and more similar to North American wood rats in body proportions (Glanz and Meserve 1982). At lower latitudes, the small mammal fauna of southern Chile and Patagonia becomes increasingly dominated by rodents in the New World subfamily Sigmodontinae. A notable characteristic of the *Nothofagus* forest communities, in addition to this sigmodontine dominance, is the complete absence of squirrels and other arboreal frugivores and folivores (Pearson and Pearson 1982).

4.2.3 Birds

The community structure and diversity of bird faunas in central Chile were described in some detail by Cody (1970, 1973, 1974). Nine Chilean sites were sampled during the International Biological Program (IBP); these ranged from simple communities of coastal and valley grasslands through sclerophyll shrublands to montane forests. Parallel studies were carried out in California. Excluding aerial feeders, most parasites, raptors and nocturnal species with specialized niches, the Chilean gradient comprised only 39 bird species in 37 genera, whereas the California gradient exhibited much more α-diversity, with 69 species in 51 genera. Chilean sites were poorer in diversity in all of the structurally complex sclerophyll shrubland, woodland and forest matching sites. Evergreen sclerophyll woodlands of *Cryptocarya* in the coastal range of central Chile supported only 18 bird species, compared with 31 in the structurally similar oak woodlands of southern California. Changes in β-diversity over these pairs of habitat gradients in Chile and California showed a similar pattern, with Chilean sites having only about half the β-diversity of the California sites with woody vegetation. As with mammal populations, however, γ-diversity of birds in Chile is higher than for California, and species replacements occur at a higher rate in the former area than in the latter at the same scale (Cody et al. 1977). The resulting totals for bird species are remarkably similar: 230 for Chile and 235 for California.

4.2.4 Reptils and Amphibians

Broad descriptions of the reptiles and amphibians (frogs) of Chile are given by Donoso-Barros (1966) and Cei (1962) respectively. The reptile fauna of Chile is notable for the relative paucity of snakes, with only five terrestrial species in three genera. There are 13 genera of lizards, but the large genus *Liolaemus* (Iguanidae) is overwhelmingly dominant. This remarkable genus is ecologically quite plastic and fills a wide variety of ecological niches (Fuentes 1976; Fuentes and Jaksic 1980). Lizard studies during the IBP matched sites in coastal scrub, sclerophyll shrubland and montane woodland in California and Chile. In contrast to bird and small mammal studies, α-diversity at these sites was virtually identical, with nine species present in Chile (three, five and four in the three habitats respectively). Point diversity was somewhat higher in Chile, however, as this region had more generalists whereas many California lizards show microsite specialization. Beta-diversity was also greater in Chile with 1.5 times higher turnover of species between habitats (Fuentes 1976; Cody et al. 1977). Chile, with 58 species (34 of them in the genus *Liolaemus*; Donoso-Barros 1966), has a richer lizard fauna than California, which has 35 species.

The frog fauna of Chile includes 35 species, 5 in the Bufonidae and 30 in the Leptodactylidae. Species diversity is highest in the Valdivian forest region, but individual species occur over practically the full geographical range of Chile, from Arica in the north to Magallanes in the south, and from the coast to the high Andes.

4.2.5 Invertebrates

Two groups of invertebrates, ants and pollinating insects, have been studied in some detail in central Chile. The ant fauna of Chile comprises about 65 species (Snelling and Hunt 1975), about one-third that of California. Hunt (1973) surveyed ant populations in four habitat types in Central Chile –

coastal succulent scrub, coastal scrub, sclerophyll scrub, and montane forest. He found 6, 9, 15 and 10 ant species in these habitats respectively. These numbers overrate α-diversity, however, since these ants were highly specific to particular microhabitats. The 14 common ant species at the primary IBP site occupied evergreen woodland (5 species), matorral (9 species), and open, disturbed sites (six species). These α-diversities were 40-75% of α-diversities at similar microhabitats in California (Cody et al. 1977).

As with ants, the bee fauna of central Chile is lower in diversity than that of southern California. The four study sites used during the IBP in Chile contained 65 bee species, compared with 206 in California along a similar gradient (Cody et al. 1977). Overall, however, the bee fauna of arid and mediterranean-climate Chile is large, with about 800 species (Moldenke in Cody et al. 1977).

4.3 Mediterranean-Type Ecosystems in Central Chile

The MTEs of central Chile have been studied intensively for the last 20 years, and are probably the best studied of all natural ecosystems in Chile. The initial thrust was provided by the IBP (Mooney 1977; Cody and Mooney 1978), but later individual and group efforts pursued many of the unresolved questions posed by those initial studies (di Castri et al. 1981; Miller 1981).

Eu-mediterranean ecosystems in Chile (sensu di Castri 1973) occur on lowlands (below 1500 m elevation), approximately between Combarbalá (31°11'S) and Linares (35°51'S). Total annual rainfall is too low north of Combarbalá, and desert conditions prevail. Above 1500 m elevation and south of Linares winter temperatures are frequently too low and rainfall is too high. Within these bounds, rainfall occurs mostly in winter, temperatures very infrequently reach freezing point, and mean total annual rainfall is between 100 and 200 mm.

Topography and human action are the two major factors other than climate that explain the distribution of vegetation and ecosystems in the Chilean area with mediterranean-type climate. Briefly, most of the Chilean population (ca. 14 million) lives in the low flatlands near the coast or in the Valle Central, a tectonic depression between the parallel high Andes and the lower Coastal Ranges. Human impacts and transformations have been largest on the low flatlands of the Central Valley, intermediate on the Coastal Ranges to the west of the Central Valley, and lowest on the Andes Ranges. As a consequence, the better conserved sclerophyllous vegetation occurs on two mid-altitude belts on the two mountain ranges. There are latitudinal, altitudinal, edaphic and exposure differences within these belts that, when added to human disturbance, explain current landscape patterns (Mooney 1977; Thrower and Bradbury 1977; Armesto and Martínez 1978; Miller 1981; Hoffmann and Hoffmann 1982). This chapter focusses on the matorral component of evergreen sclerophyll shrubland of central Chile, and on associated coastal matorral and sclerophyll woodlands (Rundel 1981a). It should, however, be noted that the mediterranean-climate region of Chile extends southward into much more mesic forest communities and northward into arid desert communities.

A diverse assemblage of forest communities form the dominant vegetation cover in south-central and southern Chile. The Valdivian forests, including evergreen and deciduous communities, occur in a region with relatively high winter rainfall and summer drought, and are thus MTEs in the broad sense (see Hobbs et al., Chap. 1, this Vol.). Deciduous *Nothofagus* forests predominate in the most environmentally stressed areas of the Chilean forest region - low latitude with moderate rainfall, high elevation, and southern Magellanic forests. These deciduous forests comprise relict populations of *Nothofagus obliqua* in the Coastal Range of central Chile (Rundel and Weisser 1975). Temperate evergreen forests dominated by *Nothofagus*, other broad leaved evergreen trees and conifers predominate in higher rainfall areas without extremes of cold stress (Veblen et al. 1983).

The northern coast of Chile forms the hyper-arid Atacama Desert region, a mediterranean-type desert with low and irregular rainfall in the winter months and predictable summer drought (Rundel et al. 1991). With its moderate temperatures, low mean annual rainfall, and characteristic winter fogs brought on by the cold Humboldt Current, this region is biogeographically parallel with the strandveld of South Africa's west coast, and the Pacific coast of Baja California in the Sonoran Desert of Mexico.

4.4 The Role of Diversity in Ecosystem Function

Ecological research in central Chile, as in most other MTEs, has not been directed towards determining the role of diversity in ecosystem function (Hobbs et al. Chap. 1, this Vol.). This question must therefore be addressed by focussing on information already collected with other purposes in mind. In this chapter we present some case studies where a relationship between these variables seems apparent. First, we will briefly discuss some of the general aspects of these relationships (see also Hobbs et al., Chap. 1, this Vol.).

One of the extreme forms in which the question of the relationships between biodiversity and ecosystem function can be posed is by asking whether diversity plays any role at all in the functioning of nature (Solbrig 1991). The null hypothesis is thus that ecosystem properties, other than biodiversity, would not change if most within- and between-species diversity is removed. At the broadest level it is obvious that species diversity (e.g. that of producers, consumers and decomposers) plays a role in ecosystems. The answer is not so easy at finer levels of resolution. This is because most researchers have tended to study "representative species" selected a priori, usually on the basis of physiognomy, from broad species groups, rather than to study the similarity of species in a group, and hence how representative the chosen species is. Implicit in the approach of selecting "representative species" is the notion that there is some degree of redundancy. In other words, it is taken for granted that: (1) there are several species that are similar enough to replace each other; and (2) that it is only when one compares differences between representatives of *these groups* that real (significant) differences will be obtained. The central question posed in this Volume is thus assumed to have been answered without the experiments or observations having been made.

Some progress has been made in studying density compensation effects (sensu MacArthur 1972) in the Chilean fauna. There is density compensation when, in the absence of one or more species in a group (e.g. birds, lizards, mammals, ants), the remaining ones exhibit higher densities, thus suggesting that the same overall performance at the ecosystem level can be realized with fewer species at higher densities. There is evidence that in the Chilean matorral there is density compensation among birds (Cody 1974), lizards (Fuentes 1976), and perhaps among vertebrate predators, such as foxes and raptorial birds (Jaksic, pers. observ.). It is not known whether all taxonomic groups show density compensation, and to what extent there could also be between-taxonomic group compensation, for example when mammals and ants compensate for each other's absences (Brown et al. 1986). To the extent that there are broad density compensation effects in nature, there could be a high degree of species redundancy not only within, but also between, taxonomic groups of organisms.

Ecosystem function means different things to different authors, and ecosystems do not have a single "natural" function (or purpose). Investigators attribute functions to ecosystems or their parts. Consequently, all ecosystem functions and classifications of organisms into functional groups are biased by the questions that are posed. For example, a matorral classification by an ecologist interested in energy flow may consider that litre (*Lithrea caustica*), formerly a common tree in central Chile and now mostly a common shrub, is a key primary producer. An ecologist interested in plant-soil relationships may argue that it is a most important type of soil cover. Another ecologist might say that because this species fruits during summer, when food is otherwise scarce, it serves to save foxes from starvation during long summers. The ecologist exploring the consequences of global warming might say that litre belongs to the functional type that, as a seedling, needs shade and nurse-plants, and is therefore likely to decrease or even go extinct with drying conditions. This exercise could continue with other litre attributes and with other species. In different models, we would focus on different species' attributes and define different functions. These attributed functions include the fact that, under some circumstances and within some constraints, species diversity may play a role in enhancing species diversity. That is, species diversity and some keystone species may function to increase the species richness that an area may sustain. All matorral shrubs could be classified by any or all of the above criteria. In most such classifications the various species would segregate differently, ultimately reminding us of the role of opportunism in the evolution of the taxa. Which one of these classifications would represent true functional groups? All and none of them, because they all have meaning only within the model in which they have been defined. Hence, when exploring redundancy in a group of species, the functional significance of the biodiversity of matorral, or assessing whether removing a particular functional group has an impact on ecosystem function, the criteria invoked must be explicitly stated. Another general bias is that, just as the selection of "representative species" might prevent us from perceiving the importance of species diversity, we must also face the problem of the few representative sites that have been investigated and the relative absence of long-term studies and of studies carried out in several nearby localities. To the extent that the different species may play a keystone role at different times or in different places, species diversity may have a very important role that is not obvious by studying, as most researchers do, a single site at a particular time (season, year). Species may, as do some enzymes and isozymes, play complementary roles from the perspective of some general ecosystem features.

In the next section we offer a few examples where ecosystem functions appear to be related to local species diversity.

4.5 Plant Species Diversity and Growth-Form Diversity

It would be interesting to test correlations of changes in species diversity along gradients of productivity and nutrient availability in central Chile. The necessary data bases are now being developed jointly by the Universities of Chile and Concepción, and the Museo de Historia Natural. About

2100 vascular plant species occur between 32°00S and 35°59S (M.T.K. Arroyo et al., unpubl. data). In order to test hypotheses of resource availability in relation to changes in γ-diversity in matorral plant communities, it would be necessary to have detailed information on local floras. In comparison to other mediterranean-climate regions, there are relatively few such regional flora available. Modern regional floras include those for the Santiago Basin (Navas 1977-1979), the Zapallar area along the central coast (Johow 1948), and Fray Jorge (Muñoz and Pizarro 1947; Skottsberg 1950). Other checklists of local floras are available, but without annotations on ecological patterns of occurrence. A project to produce a modern Flora of Chile was only recently initiated as a collaborative international effort. A checklist of the Chilean flora has been assembled by Marticorena and Quezada (1985). A specimen-based data base of the flora of mediterranean-climate Chile will soon be available for generating local species lists for any sized area between 30° and 40° S, and a checklist for the region with mediterranean-type climate is in preparation (M.T.K. Arroyo, unpubl. data). This data base will also identify areas that are poorly collected in central Chile.

An appropriate area for future studies of α- and γ-diversity of plant communities in relation to environmental gradients of moisture and nutrient availability in central Chile is the Parque Nacional La Campana in the Cordillera de la Costa between Valparaíso and Santiago (Rundel and Weisser 1975). This park, less disturbed than many other areas of the Cordillera de la Costa, is remarkably high in γ-diversity because of the topographic and climatic diversity represented in a relatively small area. Matorral, dry sclerophyll forest and hygrophilous forest are all present (Villaseñor 1980; Villaseñor and Serey 1981).

Some indirect indications of the role of nutrient availability and vascular plant diversity can be seen in a comparison of matorral communities from the Cordillera de la Costa with montane matorral from similar climatic zones on the west slopes of the Andes (Rundel 1982). The montane matorral in the Andes occurs on geologically young and unstable soils where nutrient availability appears to be lower than in the coast ranges. There have been no detailed floristic studies in the montane matorral, but species diversity in this zone is somewhat lower than in the more typical matorral communities. Dominant shrubs in the montane matorral include matorral species, sibling species from matorral genera, and a relatively small number of endemics. Sclerophyllous *Kageneckia angustifolia* woodland is rich in dioecious and genetically self-incompatible species (Arroyo and Uslar 1993). In contrast, coastal matorral communities appear to be relatively rich in species (Johow 1948) despite more arid conditions for growth than in the matorral of the Cordillera de la Costa. These coastal matorral communities contain a remarkable mixture of growth forms

(Mooney and Schlegel 1967; Weisser and Rundel 1980), and this may be a factor in their high diversity. Species diversity and growth-form diversity have been found to be positively related in coastal desert communities at Paposo in northern Chile (Rundel and Mahu 1976).

Patterns of vegetation structure and composition in central Chile have been heavily impacted by human activities, particularly grazing and charcoal collection, for four centuries (Bahre 1979). Changes in the structure of matorral communities continue today around Santiago (Fuentes et al. 1984). This effect complicates interpretations of diversity in relation to community structure and function. The only data on net primary production of matorral are from studies carried out during the IBP project on convergent ecosystems (Mooney et al. 1977). These data suggest that the net primary production of matorral at Fundo Santa Laura in central Chile is comparable to that of mediterranean-scrub communities from California and the Mediterranean Basin which have similar levels of species diversity. Net primary production of matorral is higher than that of South African fynbos and Western Australian kwongan (Richardson and Cowling 1992), both of which have much higher levels of species diversity.

Do keystone plant species exist in the matorral zone of central Chile? As with other MTEs, it is difficult to identify species whose ecological significance in the stability and function of ecosystem processes outweighs its relative biomass role. The relative importance of nitrogen-fixing shrubs, as described below, suggests that facilitation of soil nutrient availability may be important, particularly in successional sequences on disturbed sites. Such a relationship has, however, not been directly demonstrated. Because of the lower frequency and extent of natural fires in comparison to other mediterranean-type ecosystems (Kruger 1983), post-fire specialists are not well represented in the flora. There is no evidence that matorral herbs act as "nutrient dams" following fires, as has been shown in California chaparral (Rundel and Parsons 1984).

Plant growth form (above- and belowground), and related aspects of physiological function provide important aspects of interspecific competition for limited resources and thus community diversity. Plants of very similar morphological form and architecture and physiological traits hypothetically compete more intensely for water and nutrient resources than do unlike plant species. The relationship between plant diversity and competition for water and nutrients has been poorly studied in Chilean MTEs. It is, nonetheless, useful to consider the availability of water and nutrients in relation to patterns of biogeography and community structure.

4.6 Environmental Resources as a Factor in Diversity

4.6.1 Soil Moisture Resources

Variability in the spatial and temporal availability of soil moisture provides important dimensions for the niche diversification in woody matorral species. Horizontal gradients in soil moisture are obvious components of community structure away from streams in central Chile. Such gradients are evident at Quebrada de Tiltil, Fundo Santa Laura (Fig. 4.1; Rafols and Kummerow 1977). At this site, located at 750 m elevation in the Cordillera de la Costa, many woody species were restricted to streamside sites which experience little or no soil moisture stress. These species included *Aristotelia chilensis, Drimys winteri, Myrceugenella chequen* and *Salix chilensis*, all of which (except the phreatophytic *Salix*) are more typical of mesic Valdivian rain forests to the south.

Relict forests with Valdivian rainforest species occur in a number of sites along the coast of central Chile from 30° to 35°S. Three types of such forests have been identified: (1) hilltop with almost entirely relict forest species; (2) hillsides with a mixture relict forest species and sclerophyllous woody species from the mediterranean zone of Chile; and (3) the base of pole-facing slopes in mesic east-west ravines. The remarkable forests at Fray Jorge are the most well-known example of the first category where fog produces moisture conditions comparable to mesic Valdivian rainforest to the south. Relict species of trees, shrubs, herbs and epiphytes are all present (Muñoz and Pisano 1947; Skottsberg 1950). *Aextoxicon punctatum, Griselinia scandens, Myrceugenia corraefolia* and *Rhaphithamnus spinosus*, and two species of epiphytic *Peperomia* are notable relictual species. Mixed relict/sclerophyll forests have been described on mountain slopes with fog moisture inputs at Cerro La Campana and El Roble in the Cordillera de la Costa, with *Nothofagus obliqua* (Rundel and Weisser 1975; Villaseñor and Serey 1981).

Vertical profiles of water availability in matorral soils have led to divergent strategies of rooting architecture and physiological response to drought, much as in other MTEs (Rundel 1993). The classic paradigm for ecological adaptation in an evergreen sclerophyll shrub of a MTE is a dual root system to provide shallow roots to take up surface precipitation and deep tap roots to utilize semi-permanent groundwater pools (Hellmers et al. 1955). Although this model of adaptation is still widely cited, root systems in Chilean and other MTEs exhibit a great diversity in the architecture, morphology and depth of roots (Specht and Rayson 1957; Kummerow et al. 1972; Hoffmann and Kummerow 1978; Dodd et al. 1984; Higgins et al. 1987). Such complexity in rooting structure is not unexpected, given the differential selective factors related to spatial and temporal patterns of

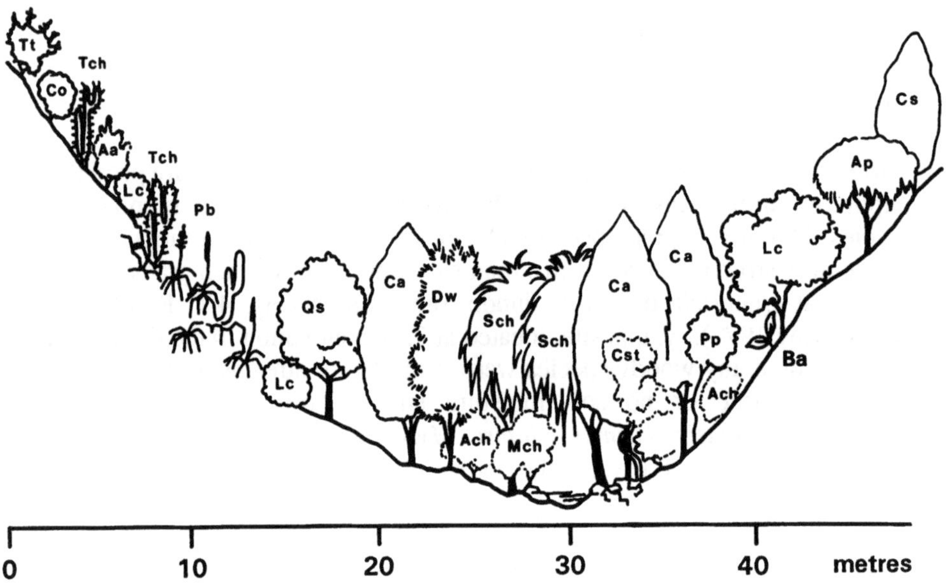

Fig. 4.1. Vegetation transect through the Quebrada de Tiltil at 750 m a.s.l. in the Cordillera de la Costa. (Rafols and Kummerow 1977). Species abbreviations are as follows: *Aa: Adesmia arborea; Ach: Aristotelia chilensis; Ap: Azara petiolaris; Ba: Blechnum auriculatum; Ca: Cryptocarya alba; Co: Colliguaja odorifera; Cst: Cissus striata; Dw: Drimys winteri, Ei: Escallonia illinita; Lc: Lithraea caustica; Mch: Myrcengenella chequen; Qs: Quillaja saponaria; Pb: Puya berteroniana; Pp: Proustia pyrifolia; Sch: Salix chilensis; Tch: Trichocereus chilensis; Tt: Trevoa trinervis. Vertical* and *horizontal scales* are the same

availability for water and nutrients and for below ground storage of carbohydrates to facilitate growth after fire and drought (Rundel 1993).

Although data on the water relations of Chilean matorral shrubs are limited, it is clear that physiological mechanisms to avoid or tolerate seasonal drought stress are much less obvious than rooting architecture; both groups of traits seem to be important. Divergent patterns of seasonal changes in pre-dawn water potentials of matorral shrubs show the physiological significance of rooting depth on water availability. Water potentials for the deep-rooted *Lithrea caustica* and *Quillaja saponaria* dropped to below -2.2 and -2.0 MPa respectively during summer (Fig. 4.2). Pre-dawn potentials of -6.0 and -5.6 MPa were recorded for *Colliguaja odorifera* and *Cryptocarya alba*, which have shallower root systems (Martínez in Poole et al. 1981). Similar differences in the seasonal progression of midday water potentials between co-ocurring matorral shrubs was shown by Giliberto and Estay (1978). They also found little seasonal

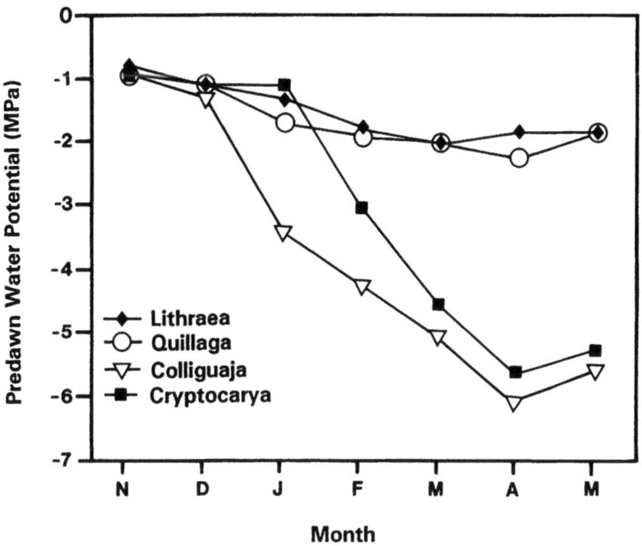

Fig. 4.2. Seasonal change in predawn water potential for matorral shrubs at Fundo Santa Laura. Data are shown for *Lithraea caustica, Quillaja saponaria, Colliguaja odorifera* and *Cryptocarya alba.* (Poole et al. 1981)

change in *L. caustica* and *Q. saponaria* compared to *C. alba* and *C. odorifera. Proustia cinerea,* a shallow-rooted deciduous shrub tolerant of drought, showed a third pattern, with low water potentials throughout the summer and early autumn.

Analyses of the components of tissue water potential in dominant Chilean matorral species have shown parallel differences in the relationships of water potential, osmotic potential and turgor potential to relative water contents of leaf tissues. Osmotic potentials at full turgor were about -2.5 MPa in *L. caustica* and *Q. saponaria,* the two more deeply rooted species, compared with -3.0 to -3.5 MPa in the more drought-adapted *C. odorifera* and *C. alba* (Fig. 4.3). Although all four species had similar relative water contents of about 70% at zero turgor, the osmotic potentials at zero turgor demonstrated two distinctive groups with values of around -3.5 MPa for *L. caustica* and *Q. saponaria* and -4.0 to -4.5 MPa in *C. alba* and *C. odorifera.*

Drought-deciduous shrubs, more abundant in the Chilean matorral than in old-growth chaparral in California, appear to have a pattern of seasonal water relations distinctive from that of the two groups of evergreen matorral plants. Seasonal progressions of pre-dawn and midday water potentials in *Satureja gilliesii* and *Trevoa trinervis* are most similar to those seen in shallow-rooted evergreen shrubs. The drought-deciduous phenology,

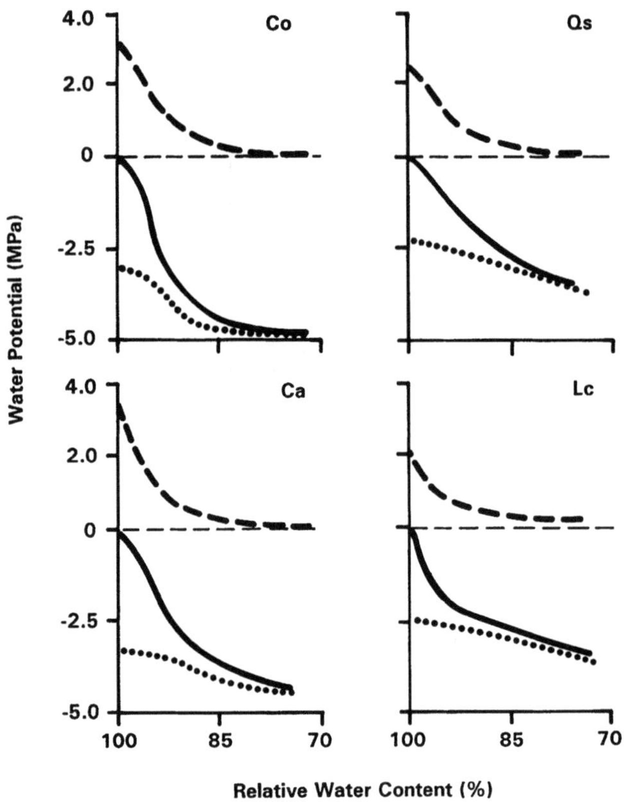

Fig. 4.3. Relationship of relative water content to water potential (leaf, osmotic and turgor) for matorral shrub species: *Co: Colliguaja odorifera; Qs: Quillaja saponaria, Ca: Cryptocarya alba; Lc: Lithraea caustica.* (Poole et al. 1981)

however, moderates the impact of summer drought as leaf tissues are shed (Martínez in Poole et al. 1981). It is interesting to note that *C. odorifera*, one of the most xerophytic matorral shrubs, is semi-deciduous to the extent that up to 50% of its leaf area may be lost in summer when vegetative growth ceases (Hoffmann and Hoffmann 1976). Its water relations appear to be intermediate between *C. alba*, the shallow-rooted and less xerophytic and the drought-deciduous species.

The importance of seasonal availability of water to growth-form diversity in matorral shrubs can be seen in the increased dominance of drought-deciduous shrubs in the coastal matorral (Weisser and Rundel 1980; Rundel 1981a). Unlike California, where coastal sage scrub often shows distinct boundaries with chaparral, coastal matorral represents the end point in a gradient of change in community dominance from evergreen sclerophyllous shrubs inland to drought-deciduous, or malacophyllous semi-evergreen shrubs along the coast. Semi-woody members of the Asteraceae and Labiatae are an important component of this community. *Baccharis concava*, *B. linearis*, *Bahia ambrosioides*, *Eupatorium gleachonophyllum*, *E. salvia*, *Lepechinea salviae* and *Podanthus mitique* are all common. *Flourensia thurifera*, a semi-evergreen shrub, is abundant on low-elevation xeric sites away from the coast. Some of the most diverse groups among the perennial flora of central Chile are drought-deciduous subshrubs. The genera *Senecio* (222 spp), *Calceolaria* (70 spp. in Chile), *Haplopappus* (61 spp.) and *Baccharis* (42 spp) all fall into this category (Marticorena and Quezada 1985). Although woody evergreen shrubs occur all along the coast of Chile only three species (*Lithrea caustica*, *S. latifolius* and *Schinus polygamus*) are widespread and ecologically important. Several large woody and drought-deciduous shrubs are important. *Fuchsia lycioides* is endemic to this habitat (Atsatt and Rundel 1982), whereas *Adesmia arborea* and *Proustia pungens* occur along the coast and on xeric inland slopes.

4.6.2 Nutrient Availability

Competition for limiting nutrients has long been recognized as an important factor in defining plant niches and structuring communities (Tilman 1988). The role of nutrient availability is of particular interest in MTEs since they represent a wide range of soil nutrient conditions. The highest diversity in South African and Western Australian MTEs is associated with nutrient-poor soils, although the causal factors have not been fully explained (Hobbs et al., Chap. 5, this Vol.; Richardson et al. Chap. 6, this Vol.). Soils in the MTEs of Chile have formed from a heterogeneous mosaic of parent materials. Concentrations of total phosphorous in the soil are higher than those in California chaparral soils. Soil nitrogen concentrations range from 0.02-0.33% by weight (Fig. 4.4). This pattern of soil fertility suggests that nutrients, particularly phosphorous, may be less limiting for growth in Chile than in other MTEs (see also Fig. 1.5 in Chap. 1).

Fertilization studies on matorral shrubs have shown highly species-specific responses in fine-root density and shoot growth (Kummerow et al. 1982). Phosphorous treatment alone did not significantly affect leaf production or shoot growth in any of the four species studied. Addition of nitrogen, and nitrogen + phosphorous significantly increased growth in *Colliguaja odorifera*, *Cryptocarya alba* and *L. caustica*. Nitrogen fertilization enhanced fine-root production in these three species, but decreased production in *Trevoa trinervis*, a nitrogen-fixing shrub. Response

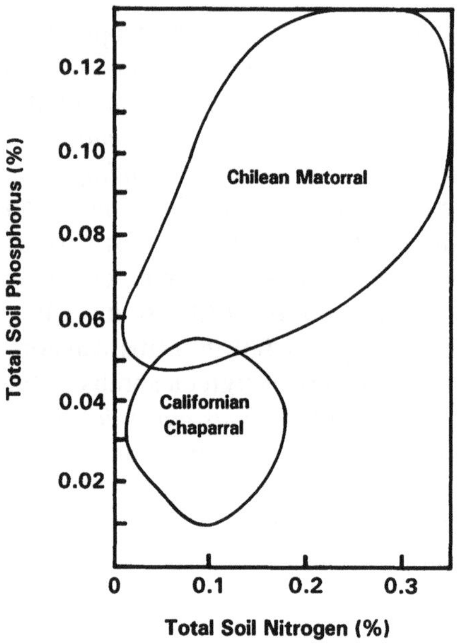

Fig. 4.4. Relationship of total soil nitrogen and phophorous in surface soils from Chilean matorral and California chaparral communities. (Rundel 1988)

to phosphorous fertilization was less clear; fine root production was stimulated in *L. caustica* and *T. trinervis* but effects on other shrubs were inconclusive. Clearly nitrogen fixation is an important attribute of many matorral shrubs, particularly those in disturbed or xeric sites. Nitrogen-fixing nodules are present in woody legumes (e.g. *Adesmia arborea* and *Sophora macrocarpa*), and in several genera of Rhamnaceae (*Colletia, Talguenea, Trevoa* and probably in *Retanilla*; Rundel and Neel 1978).

Comparative nitrogen use efficiencies (NUE, in g annual production per mg of leaf nitrogen expended; g mg^{-1} N) of matorral shrubs have been calculated to assess the significance of nitrogen use to net primary production (Rundel 1982). Values of NUE for seven matorral shrubs ranged from 0.43 to 2.00 g mg^{-1} N. As expected, the two drought-deciduous shrubs, *Satureja gilliesii* and *Trevoa trinervis*, had low values of NUE. Surprisingly, however, their NUEs were not significantly lower than those of several evergreen shrubs, suggesting that selection for high NUE has not been a significant aspect of the evolution of sclerophyllous evergreen leaves in matorral shrubs. The relationship between NUE and net production was weak for matorral species, in contrast to shrubs in California chaparral (Rundel 1982).

Leaf phosphorous levels of Chilean matorral and coastal matorral shrubs are highly variable (0.8 - 0.26%), while leaf nitrogen shows less of a range (1.0 - 1.7%; Fig. 4.5). California chaparral shrubs were lower in leaf nitrogen but no different from matorral shrubs in mean leaf phosphorous. California maritime chaparral species from near Monterey, however, are extremely low in phosphorous (Rundel 1988).

Statistical analyses of leaf chemical attributes for shrubs in MTEs have been carried out to compare intra- and interregional patterns of leaf characteristics (Rundel 1988). A PCA ordination was used to first compare traits of five leaf nutrients (N, P, K, Ca, Mg) and four chemical and structural traits (leaf-specific weight, lignin, cellulose and ether extractives). Chilean matorral and California chaparral leaves showed notable clustering compared to two groups of leaves from South African fynbos (Fig. 4.6). Higher nutrient levels and lower lignin and cellulose concentrations were the significant variable separating the Chilean and California leaves from the proteoid and restioid species in South African fynbos. Within Chilean communities, however, a similar PCA ordination was unable to separate matorral, coastal matorral, montane matorral and hydrophilous forest species (Fig. 4.7). Thus, the functional guilds of leaf structure and nutrition evident in South Africa do not seem to be present in these Chilean species (Rundel 1988).

4.6.3 Fire Frequency and Biodiversity

Natural fires, a significant ecological factor in all MTEs of the world, seem much less common in Chile than in the other regions. Convective thunderstorms and associated lightning are uncommon in central Chile, thereby providing few ignition sources under natural circumstances (Rundel 1981b). Lower natural fire frequencies may well have influenced the biodiversity of the Chilean flora in several ways. Natural fires provide a critically important pathway for decomposition and cycling of nutrients in MTEs with low nutrient availability. There is evidence to suggest that natural selection has acted to maintain relatively flammable shrub structure and chemistries which would favour efficient use of such fire cycling. This hypothesis suggests that shrubs in Chilean matorral with relatively moderate fertility should show less adaptations for flammability than California shrubs growing in more nutrient-poor soils. Rundel (1981c) showed that this expected pattern was present for the heat of combustion of volatile fuels, leaf energy content, leaf ether extractives, and leaf ash content. Although there have been no definitive studies, we suggest that the relative abundance of succulent species and paucity of geophytes may be related to low fire frequency.

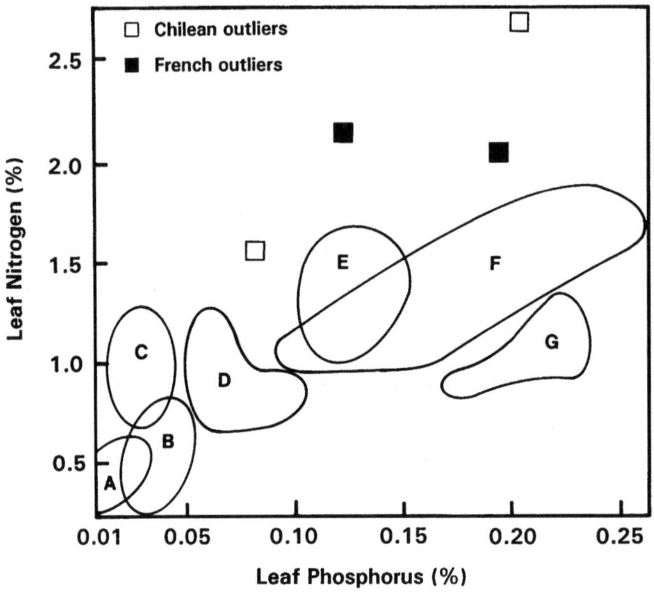

Fig. 4.5. Relationship of leaf phosphorous and leaf nitrogen concentration in shrub species from Chilean matorral (**F**), Californian chaparral (**G**), Californian maritime chaparral (**D**), French maquis and garrique (**E**), and South African fynbos (**A** Restiod; **B** Proteoid; **C**: Ericoid. Rundel 1988). The Chilean data include species from true matorral, coastal matorral and montane matorral which are not nitrogen-fixing

A unique feature of the mediterranean-climate region of Chile is the diversity and ecological significance of succulent plants, most notably members of the Cactaceae. Large arborescent or semi-arborescent cacti in the genera *Trichocereus* and *Eulychnia* (now often lumped with *Echinopsis*) are often abundant on xeric north-facing slopes of matorral or coastal bluffs. The most widespread of these species is *T. chilensis* which occurs from near sea level to more than 2000 a.m.s.l. in the Andes (Rundel 1974, 1977). Smaller cacti are also diverse in central Chile, including species of *Austrocactus, Eriocyse, Maihuenia, Neoporteria* and *Opuntia* (Hoffmann 1989a). The chaparral ecosystems of California, in contrast, have very low diversities of Cactaceae, and arborescent cacti drop out rapidly in the transition from the Sonora Desert to mediterranean-type scrub communities in northwestern Baja California. Cold tolerance cannot explain this pattern of occurrence. The northern limit of many arborescent Sonora Desert cacti, such as *Carnegia gigantea*, is certainly influenced by extreme frosts (Nobel 1980), but *T. chilensis* is more frost-tolerant. The relatively low fire frequency in central Chile compared to other regions with MTEs may well

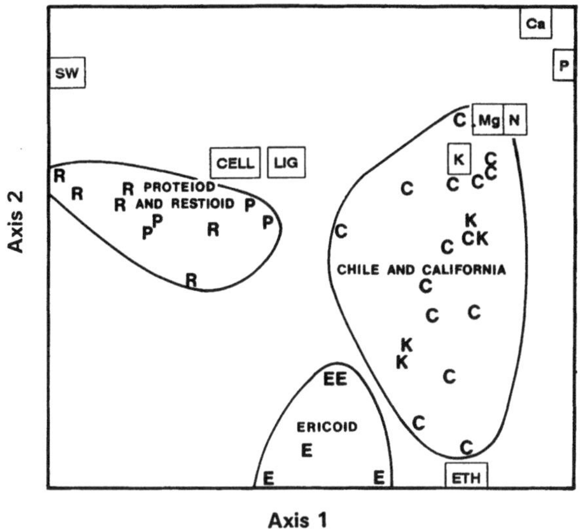

Fig. 4.6. PCA ordination of leaf nutrient and chemical data for species from California chaparral (**K**), Chilean matorral (**C**), and South African fynbos representatives of proteoid (**P**), restioid (**R**) and ericoid (**E**) elements. (Rundel 1988). Data on leaf nitrogen, phosphorous, calcium, potassium, magnesium, ether extractive, lignin, cellulose and specific weight were utilized for each species

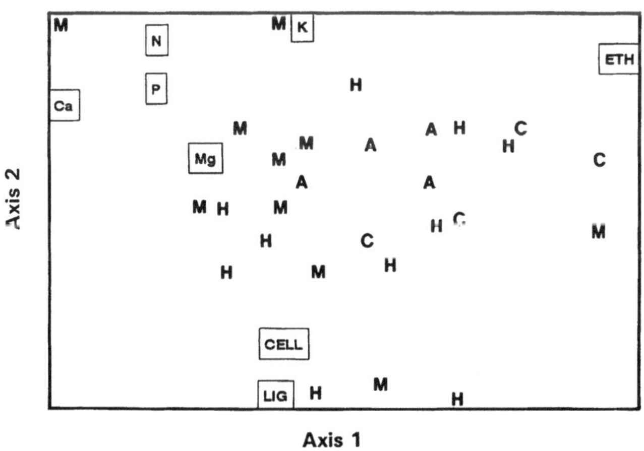

Fig. 4.7. PCA ordination of leaf nutrient and chemical data for Chilean species from matorral (**M**), coastal matorral (**C**), montane matorral (**A**), and hygrophilous forest species (**H**). Data as in Fig. 4.6

be the critical element promoting the survival of large cacti. Fire in grasslands and shrublands commonly kills cacti and other succulents (Humphrey 1958; Niering and Lowe 1984). *Bergerocactus emoryi*, the only shrub-form cactus that is abundant in mediterranean-type scrub communities of northwestern Baja California, has the adaptive trait of resprouting from root tissues after fire. The relatively high diversity and abundance terrestrial bromeliads in central Chile, particularly species of *Puya*, may also related to low fire frequency. Like the cacti, *Puya* species, of which there are 11 in Chile, are common on xeric north-facing slopes of matorral and coastal bluffs in Chile. There are also terrestrial species of *Fascicularia*, *Greigia* and *Ochagavia* (Hoffmann 1989b).

In contrast, the overall diversity of geophytes is surprisingly small in Chile (Hoffmann 1989c; Fig. 4.8) compared to many other MTEs (see Le Maitre and Brown 1992 for a summary). Perhaps this is also related to fire frequency. Despite the presence of a number of notable geophytes, including species of *Alstroemeria*, *Hippeastrum* and *Tecophilea*, only some 6% of the flora of central Chile are geophytes. This is a slightly lower value than that found in California chaparral and far lower than their relative frequency of this growth form in maquis communities in the Mediterranean Basin (Rundel 1993). Overall, the diversity of geophytes in California is far greater than in Chile. Although little is known about the adaptive strategies utilized by geophytes, these species are clearly well adapted to fire and other types of disturbance. Less frequent fires and/or other disturbance impacts may also be a factor promoting the greater diversification of herbaceous perennials relative to geophytes. Annuals are

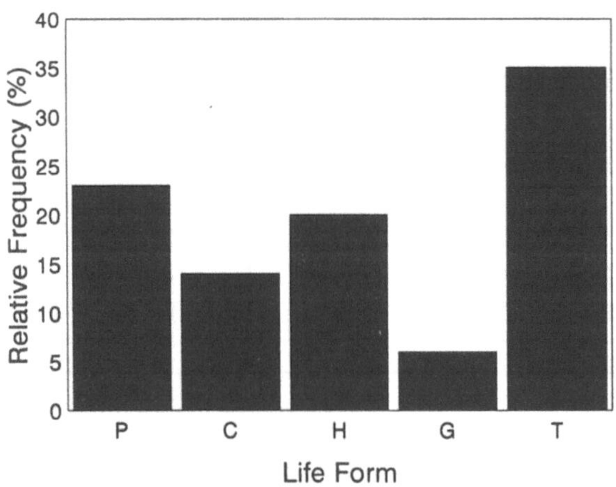

Fig. 4.8. Relative life from distribution of the matorral flora. (Data from Parsons and Moldenke 1975). **P** Phanerophytes; **C** Chamaephytes; **H** Hemicryptophytes; **G** Geophytes; **T** Therophytes

also diverse, and make up about 17% of the vascular flora between 32° and 36°S in central Chile (M.T.K. Arroyo, unpubl. data), a higher figure than for South Africa, but much lower than for California. Differences in fire frequency may also play a determining role in this phenomenon.

4.7 Storage of Species Diversity

Species diversity could have a positive feed-back function on species diversity if the presence of some components is contingent upon the presence of others. We describe five cases of this phenomenon.

Firstly, in central Chile, as in other regions with MTEs, human action has favoured species diversity at the local level by increasing habitat diversity. Only in the most extreme cases do human impacts decrease landscape and species diversity. This local increase in diversity, however, usually implies reduced population size of resident species. The final outcome of the latter is not easily measured. Secondly, herbivores are known to increase the diversity of the shrub and herbaceous layers (Fuentes and Etchegaray 1983). Thirdly, animal prey species diversity seems to drive the diversity of predators feeding on them (Jaksic et al. 1992). For example, in the matorral areas near Santiago it has been shown that the rodent *Octodon degus* plays a keystone role in sustaining several predators, including three species of raptors and two species of foxes (Jaksic 1986; Fuentes and Jaksic 1988). Fourthly, among pollinator-plant relationships there is also a diversity-dependent feedback loop. Specifically, different plant species in a community seem necessary for the maintenance of the pollinator pool of the component species. For montane sclerophyllous forest, a very large proportion of the species pairs analyzed have mutually exclusive pollinators (Table 4.4). There are, thus, many plant species whose presence is not vital for maintaining the pollinator pool of others. However 5.6% of the species pairs tested share pollinators, but flower at different times. Elimination of some plant species in this group could negatively influence the availability of pollinators for other plant species of the group, particularly those that flower latter in the season. In subalpine scrub the percentage of species comparisons in this last category is higher (9.2%). Clearly, the longer the activity period of the pollinator species in question, the greater the importance of maintaining the total plant species pool. Further analyses of this kind for dispersal mechanisms and other attributes might reveal similar patterns, giving a range of plant species from those functionally redundant at all levels, to those redundant for certain features, to keystone species. Fifthly, a relatively well studied example of the im-

Table 4.4. Flowering and pollinator overlap in bee-pollinated species in some mediterranean-type climate vegetation types in central Chile. The data are for pairwise comparisons between the number of species indicated. F/P: species pairs flowering simultaneously and sharing pollinators; F/-: species pairs flowering simultaneously, but not sharing any pollinators; -/P: species pairs sharing pollinators but with mutually exclusive flowering periods; -/-: species pairs with mutually exclusive flowering periods which have no pollinators in common. N = number of plant species studied; n = number of pairwise combinations of those species (based on phenology data of Arroyo et al. (1981) and Uslar (1982), and pollination data reported in Arroyo et al. (1982, 1985) and Uslar (1982)).

		Overlap categories			
	N (n)	F/P	F/-	-/P	-/-
Sclerophyllous forest 1 800 - 1 900 m; 33°S	22 (231)	7.8	62.8	5.6	23.8
Subandean shrub 2 100 - 2 300 m; 33°S	16 (120)	26.7	45.6	9.2	18.3

portance of particular species in maintaining diversity concerns rabbits, European annual grasses, and native birds, shrubs and trees. The clumpy structure of the matorral on valley bottoms and north-facing slopes is dependent on the contingency of the following phenomena: the strong competition between shrub seedlings and European grasses with grasses inhibiting shrub seedlings (Anabalon et al. pers. comm.), the ability of European rabbits and hares in central Chile to freely use the open spaces between shrubs and sparsely covered habitats (Jaksic and Fuentes 1991), and eat shrub seedlings there (Fuentes et al. 1983), and the role of native birds in depositing and concentrating seeds and fruits under already established shrubs (Fuentes et al. 1984, 1986). Without rabbits and hares, the grasses or the birds, the system would have a different configuration.

In the five examples given above, there are positive relationships between species in different trophic levels indicating a diversity-storage function of biodiversity. In other words, some species are present only if some other species are also present. These examples provide average figures or data for particular points in time and space. Relationships between species in different trophic levels are, however, not static. This is particularly true for MTEs, which are well known for their between year variability in rainfall (Fuentes et al. 1988). In the next section we review a study of a predator-prey community in the more northern and arid part of the Chilean area with a mediterranean-type climate. This study was conducted over 5 consecutive years and shows how biodiversity and its components change over time.

4.8 Biodiversity in Vertebrate Predator/Prey Systems

The question of whether prey diversity ensures predator diversity or vice versa is generally considered to be settled in the sense of Paine's (1966) keystone predator concept. Strong preference by a dominant predator on a competitively dominant prey species enhances local diversity by allowing "inferior" competitors to thrive on resources released by the competitive "superior". Although intuitively appealing, the reverse phenomenon - that of prey species diversity driving predator species diversity - is less well documented (Jaksic 1986; Jaksic and Simonetti 1987). Indeed, for that to be the case, theoretical considerations would require that predators be fairly specialized on certain prey, and that they select prey species not consumed by sympatric predators. Perhaps this situation is common among predator/prey and parasitoid/host systems involving insects, but it should not be expected among more generalized predators such as carnivorous vertebrates. Here, we document a puzzling situation in which the diversity of generalist predators is positively correlated with prey diversity over time.

In this section we focus on a suite of predatory vertebrates belonging to two classes and several orders, for which it is reasonable to assume that interactions are stronger among these members than with other members of the local community. They interact with a rather uniform suite of mammalian prey in only two orders. Focus on such a predator/prey system, in a markedly seasonal environment, and over the span of several years during which at least one major climatic disturbance occurred, enables us to address the following issues: (1) is biodiversity a stable attribute for a given site, or is it a "fuzzy" feature?; (2) if the latter, then what ranges of variation are involved?; (3) do those ranges differ between trophic levels (predators versus prey species)?; and (4) are there "redundant" species in the system (species whose presence/absence does not ostensibly affect community dynamics)?

The study area was Las Chinchillas National Reserve, near Aucó, Chile (31°30 S; 71°06 W), which encompasses 4229 ha of rugged terrain with semi-desert vegetation. Annual rainfall averages 206 mm, with 85% in winter (May-September). There is considerable variation in rainfall (see also Fuentes et al. 1988); during the study it was: 513 mm in 1987; 58 mm in 1988; 104 mm in 1989; 64 mm in 1990, and 200 mm in 1991. Precipitation in 1986, before the start of the study, was below average (158 mm). This series shows that 1987 was an extremely wet year, when precipitation exceeded the mean by roughly 150%. Primary productivity in this area is closely associated with precipitation (Fuentes et al. 1988; Jaksic, pers. observ.); this was reflected by a population explosion of small mammals that reached its peak in 1987 and declined steadily thereafter

(Jiménez et al. 1992). We assessed the results of this ongoing study up to February 1992, thus covering 5 full years. Two biologically meaningful seasons were recognized: a breeding season spanning September through February (spring and summer), and a non-breeding season spanning March through August (autumn and winter).

The target group of species in this study is a predator/prey system comprising small mammals and both avian and mammalian carnivores (Jaksic et al. 1993a). There were eight small mammal species at the site: the rodents *Abrocoma bennetti, Akodon longipilis, Akodon olivaceus, Chinchilla lanigera, Octodon degus, Oryzomys longicaudatus, Phyllotis darwini*, and the marsupial *Marmosa elegans*. These species were preyed upon by four falconiform hawks (*Buteo polyosoma, Falco sparverius, Geranoaetus melanoleucus, Parabuteo unicinctus*), four owls (*Athene cunicularia, Bubo virginiatus, Glaucidium nanum, Tyto alba*) and two mammalian carnivores (*Pseudalopex culpaeus, Pseudalopex griseus*).

The diversity of mammalian prey remained constant at eight species until the non-breeding season of 1989, then declined steeply over the next 2 years to only three species. A dramatic recovery started in the breeding season of 1991 (Table 4.5). Overall, mammalian diversity ranged from three to eight species, i.e., almost a 2.7-fold difference. Three species, namely *A. olivaceus* (an omnivore), *M. elegans* (an insectivore) and *P. darwini* (a granivore), were present throughout the entire (or most) of the study period. One granivore (*O. longicaudatus*), one insectivore (*A. longipilis*) and three folivores (*A. bennetti, O. degus* and *C. lanigera*) disappeared from the study site. Predator diversity peaked at ten species during the early phases of the study and crashed to four, with that number increasing by only one species in the breeding season of 1991 (Table 4.6) when the system apparently started recovering. There was thus a 2.5-fold difference between peak and crash predator diversity. The four species that were resident at the site were the foxes *P. culpaeus* and *P. griseus* (which were in fact omnivorous), and the three owls, *G. nanum* (omnivorous) *Bubo virginianus* and *T. alba* (both carnivorous), which were present for most of the time and may be considered as residents. Three of the four falconiforms at the site (*G. melanoleucus, B. polyosoma* and *P. unicinctus*, all large and carnivorous) were present only during the early stages of the study, and only the omnivorous species (*F. sparverius*) persisted for a prolonged period. These species are all migrants.

Although there was a close match between prey and predator diversity at any given time, causation was not that simple: predator species do not consume prey species – they consume individuals. What is clear, though, is that when small mammal abundance decreased below 100 individuals ha^{-1} (in the 1988 breeding season), predator species began to disappear (Tables 4.5 and 4.6). Why falcons should disappear before owls is unclear.

Table 4.5. Density (animals/ha) of small mammals at Aucó, by season. B = breeding season; NB = non-breeding season

		1987		1988		1989		1990		1991	
Mammal species	wt(g)	NB	B	NB	B	NB	B	NB	B	NB	B
Abrocoma bennetti	182	-	1.1	0.9	0.5	0.4	0.2	0.7	0.0	0.0	0.0
Akodon longipilis	61	-	1.4	0.5	0.9	0.4	0.0	0.0	0.0	0.0	0.4
Akodon olivaceus	28	-	30.0	37.7	14.5	3.4	2.5	1.4	0.5	0.0	1.6
Chinchilla lanigera	306	-	5.9	5.3	4.6	2.5	1.8	1.1	0.0	0.0	0.0
Marmosa elegans	26	-	6.0	8.8	5.0	2.3	1.6	2.0	2.5	1.8	6.8
Octodon degus	123	-	11.5	9.3	4.8	2.5	1.1	0.2	0.0	0.0	1.4
Oryzomys longicaudatus	27	-	1.4	5.9	2.7	1.1	0.0	0.2	0.0	0.7	4.1
Phyllotis darwini	47	-	164.9	140.3	65.2	17.6	17.3	10.9	19.3	13.4	59.1
Number of resident mammals			8	8	8	8	6	7	3	3	6
Density (no./ha)			222.2	208.7	98.2	30.2	24.5	16.5	22.3	15.9	73.4
No. resident predators			9	10	9	7	5	6	6	4	5

Table 4.6. Numbers of pellets and faeces collected at Aucó, by season. B = breeding season; NB = non-breeding season; # = species did not reside at the site during that season

Predator species		1987		1988		1989		1990		1991		Total
	wt (g)	NB	B	NB	B	NB	B	NB	B	NB	B	
Geranoaetus melanoleucus	2378	15	8	9	#0	#0	#0	#0	#2	#0	19	53
Buteo polyosoma	975	0	40	19	69	#0	#0	#0	#0	#0	#0	128
Parabuteo unicinctus	876	#4	#0	28	27	11	#0	#0	#0	#0	#0	70
Falco sparverius	116	15	38	4	5	#0	#2	#0	31	#0	#0	95
Tyto alba	307	51	104	181	71	21	#0	56	#0	#3	88	575
Glaucidium nanum	81	100	76	129	118	81	68	73	23	#1	#0	669
Athene cunicularia	247	124	214	25	101	59	103	217	154	63	#0	1060
Bubo virginianus	1227	184	194	260	433	62	64	71	10	10	309	1597
Pseudalopex culpaeus	4317	42	88	402	189	242	158	143	124	163	270	1821
Pseudalopex griseus	2495	42	88	402	189	242	157	143	125	163	271	1882
No. resident predators		8	9	10	9	7	5	6	6	4	5	4–10
No. pellets/faeces		577	850	1459	1202	718	552	703	469	403	957	7890

Every falcon species can be matched with every owl species on the basis of diet (Table 4.7). Although these matches are presented here at very general levels of prey identification, they are based on a species-level resolution (Jaksic et al. 1993a). A major difference between the three larger falcon species and their owl counterparts is that the former preyed more strongly on *O. degus* (the only strictly diurnal prey species at the site); their disappearance might, therefore, be connected with the decline of that prey. However, the omnivorous *F. sparverious* also disappeared before its natural counterpart, the continuously active owl *A. cunicularia* (Table 4.6).

Curiously, the four owl and two fox species that resided at the site did not concentrate on the most abundant mammal prey, *P. darwini* (Table 4.5). Indeed, they consistently took this prey either less than expected from its abundance in the field, or just at the expected rate (Table 4.8). The resident predators showed strong preferences for species that were never too abundant, and that eventually disappeared from the site, such as *A. bennetti, O. degus* and *O. longicaudatus* (Tables 4.5 and 4.8). Clearly, biodiversity is not a fixed attribute through time at a given site. Perhaps more interesting is the fact that at both predator and prey trophic levels, the ranges of variation in biodiversity were very similar (2.5 to 2.7-fold). One of the most puzzling aspects of this research is that prey species in the same trophic guild (or functional group) did not react similarly (population-wise) to the factors that triggered the outbreak of small mammals. Interestingly, no two members in the same guild increased their populations simultaneously (Jiménez et al. 1992). Although one granivore at the site (*P. darwini*) irrupted, another (*O. longicaudatus*) did not. The more grani- vorous of the two omnivores, *A. olivaceus*, irrupted but its congener *A. longipilis* (the more insectivorous) did not, even though the insectivorous *M. elegans* also irrupted. It is not clear whether species in the same prey trophic guild compete for food resources. If this is the case, perhaps com- petition from the first species to irrupt impedes its guild-mate from doing so. At least *O. longicaudatus* is known to irrupt more strongly than its granivore counterpart *P. darwini* under some circumstances (Péfaur et al. 1979; Fuentes and Campusano 1985; Meserve and Le Boulengé 1987; Iriarte et al. 1989). In this sense, there may not be species redundancy at the prey trophic guilds. If guild-mates were redundant, they should show the same population trends when faced with changing environmental con- ditions. The other interesting finding is that there seems to be strong redundancy at the predator level, as demonstrated by some tight trophic guilds being present throughout the 5-year study at Aucó (Jaksic et al. 1993a). *Bubo virginianus* and *T. alba* preyed on essentially the same prey species at roughly the same frequencies, suggesting redundancy at this trophic level. The same may be said of the two *Pseudalopex* foxes, and of the owls *A. cunicularia* and *G. nanum*, with which the falcon *F. sparverius*

Table 4.7. Overall diets of predators at Aucó from the non-breeding season of 1987 until the breeding season of 1991. For full binomials of predators see Table 4.6

	Migrant predators				
Prey	*P. unicinctus*	*B. polyosoma*	*F. sparverius*	*G. melanoleucus*	
Small mammals (%)	62.5 - 100	41.1 - 52.4	0 - 16.7	43.3 - 60.0	
Other vertebrates (%)	0 - 37.5	22.0 - 30.9	11.8 - 100	22.0 - 48.0	
Invertebrates (%)	0	19.0 - 36.9	0 - 82.3	0 - 34.1	
No. of prey	92	348	420	121	
	Resident predators				
Prey	*T. alba*	*G. nanum*	*A. cunicularia*	*B. virginianus*	*P. culpaeus/P. griseus*
Small mammals (%)	92.0 - 100	9.1 - 52.1	0.3 - 9.2	69.1 - 100	6.2 - 33.9
Other vertebrates (%)	0 - 8.0	5.4 - 30.4	1.3 - 7.7	0 - 5.7	1.7 - 14.6
Invertebrates (%)	0 - 0.9	41.9 - 90.9	86.4 - 98.4	0 - 30.9	51.5 - 90.6
No. of prey	1 061	1 721	9 958	3 127	22 986
% Faeces with plant materials					29.8 - 75.0

Table 4.8. Summary of prey preferences for resident predators at Auc6. + = taken more frequently than expected from trapping data; - = taken less frequently; 0 = taken about as expected; +/0 = apparent preference shift. See Tables 4.5 and 4.6 for full binomials of prey and predator species, respectively

Mammal prey			Predators					
				A. cunicularia	*B. virginianus*	*G. nanum*	*T. alba*	*P. culpaeus* & *P. griseus*
			Wt(g)	247	1227	81	307	4317 & 2495
	Wt(g)	Activity[a]	Activity[a]	CN	N	DCN	N	CN
A. bennetti	182	N		-	+	0	+	+
A. longipilis	61	N		-	0	-	0	-
A. olivaceus	28	DCN		0	0	+/0	+	0
C. lanigera	306	N		-	-	-	-	-
M. elegans	26	N		+	+/0	+/0	+/0	-
O. degus	123	DC		-	-	+	-	+
O. longicaudatus	27	N		0	+	0	+	0
P. darwini	47	N		-	0	-	0	-

[a] D = diurnal; C = crepuscular; N = nocturnal

was frequently associated. The three large falcons (*B. polyosoma, G. melanoleucus* and *P. unicinctus*) had similar diets as well.

A partly simultaneous study conducted over 2 years in Fray Jorge (Jaksic et al. 1993b) found high redundancy between *B. virginianus* and *T. alba*, and somewhat less between *A. cunicularia* and *Pseudalopex culpaeus*. An earlier guild-level analysis conducted in central Chile on short-term data (Jaksic et al. 1981; see also Simonetti 1988) detected the same redundancy, essentially among the same species as in Aucó. In the case of the predators, however, there is no indication that early population increase of a guild-mate affects its trophic counterparts. They all appear to increase in parallel, at least during good prey-years. Unfortunately, the "acid test" for predator redundancy, i.e. density compensation by resident predators when guild-mates migrate away, cannot be performed, as no tallies of absolute numbers of predators were kept.

The above cases illustrate a possible storage function of species diversity. The only other documented case of a link between ecosystem function and species diversity in Chilean MTEs is a human-related function: the commercial production of honey using the introduced bee, *Apis mellifera*, and the matorral as the agents of production agents. In the next section we explore whether plant species diversity plays a role in the functioning of this system.

4.9 Production of Honey at the Community Level

Most commercial beekeepers in Chile exploit flower resources in the southern forest dominated by *Nothofagus dombeyii* and *N. obliqua*, or in woodlands dominated by *Aextoxicum punctatum, Drimys winteri* and *Luma apiculata* (Sempe et al. 1989). However, the shrublands of the MTEs in the semi-arid part of central Chile also provide pollen and nectar during late autumn-winter and early spring, a critically important time for the industry (Montenegro et al. 1989a; Varela et al. 1991; Montenegro et al. 1992; Poblete and Montenegro 1992). *Apis mellifera* is known to be highly selective for flowers as a source of pollen and/or nectar (Free 1963; Corbert 1978; Eisikowitch and Loper 1984). That is, they do not visit all the available vegetation in flower in a given community. How does plant diversity in the matorral, in terms of species and growth forms, relate to this "service" provided to the honey-producing industry?

4.9.1 Species Representation in Pollen Load Samples

Microscopical analyses of pollen loads gathered by the honeybees, and of the nectar accumulated in the hive cells, can be used to determine the floristic origin of pollen or nectar in a determined territory. Two beehives established in a region of shrubland in the semi-arid mediterranean region of Chile (15 km from the city of Paine, 33°49S; 70°49W) were sampled for pollen using pollen traps attached to the entrance of the hives. A list was made of all species flowering within a 500 m radius of the hives, and relative cover of shrub species was also assessed (Armesto and Gutiérrez 1980). The number of species utilized by bees as pollen sources was always significantly lower ($p < 0.05$) than the species in flower (Fig. 4.9). The average number of pollen types in each hive was about six, with a range of three to 11. Honeybees used only about 35 of the 130 species available at the site, confirming that *Apis mellifera* is a selective species (van der Moezel et al. 1987). *Apis mellifera* also does not use a constant fraction of the species in flower at any one time, but rather shows what seems to be a saturation curve (Fig. 4.9). The explanation of this behaviour might lie in optimal foraging theory, considering aspects as such the reward obtained by the bees (Hobbs 1962), the protein content of the pollen grains, the differential cover of the species (Severson and Parry 1981), the colour and form of the flowers (Heinrich 1975) and the ease with which bees handle the various flowers. Our analyses have shown that only 14.3% of the species utilized by honeybees have tubular corolla (Montenegro et al. 1990), suggesting that species with actinomorphic corollas are favoured. In addition, bees exhibit within-year variability in their harvesting behaviour. Analyzing the composition of pollen traps in hives located at the Paine study site between October 1988 and January 1989, species contributing more than 10% of the pollen grains at any one sampling time were determined (Fig. 4.10). Statistical analysis (ANOVA-Tukey test) revealed two levels of pollen gathering by the bees (Varela et al. 1991). One level comprised the species whose contribution was above 50% at any one sampling time. Species such as *Galega officinalis* and *Lithrea caustica* and three Brassicaceae species, *Hirschfeldia incana, Raphanus sativus* and *Rapistrum rugosum,* were in this category. A seasonal replacement occurs among these species (Fig. 4.10). When the Brassicaceae species show an abrupt decrease, a concomitant increase of *L. caustica* pollen utilization was observed, and when this latter species decreases, *G. officinalis* increases. The second level of pollen utilization, significantly different from the previously described one ($p < 0.05$), is shown by the rest of the species that in general contribute less than 20% of the corbicular pollen samples. The highest contribution in this category is shown by the native species *Colliguaja odorifera, Equinopsis chilensis, Kageneckia oblonga, Podanthus*

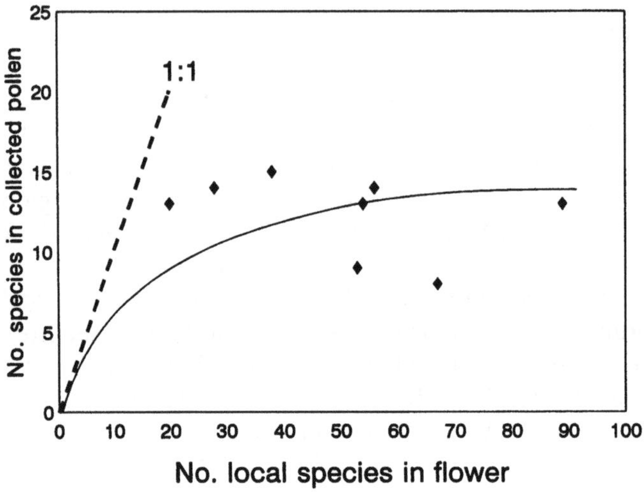

Fig. 4.9. Bee visits and flower diversity. Species appearing in collected pollen samples as a function of the number of local flowering species. Notice that there is a saturation in the number of species visited, suggesting that only some species are important

mitique and *Trevoa trinervis*. No significant differences were found between hives with respect to the types of pollen observed (chi-square similarity test - Shanon Index, $\underline{p} < 0.01$). These results suggest that Brassicaceae species (*Hirschfeldia incana, Raphanus sativus* and *Rapistrum rugosum*), as well as *Galega officinalis, Lithrea caustica* and *Trevoa trinervis*, all of them used as pollen sources when flowering, could be considered to be "key pollen species". The quantity of collected pollen per hive will depend on the presence of these key species more than on the total plant species diversity at any given site.

Interestingly, more than 50% of the key pollen species were introduced herbs, indicating that the presence of these species actually increases the value of the services provided by the matorral to the honey producing industry. As might have been expected, a correspondence exists between the distribution pattern of the human population and that of introduced herbs (Fig. 4.11). Also, there has been a constant increase in the number of introduced herbs since 1854 (Fig. 4.12). These findings indicate that, although human activities modify the landscape, generating a mosaic where native matorral mixes with areas covered by introduced herbs, the current net result and its projections into a more human-populated future, could be an increase in the diversity of plants that are useful to the honey industry.

Using a somewhat different technique, namely identification of pollen recovered from honey samples from the hives (Maurizio 1975), two

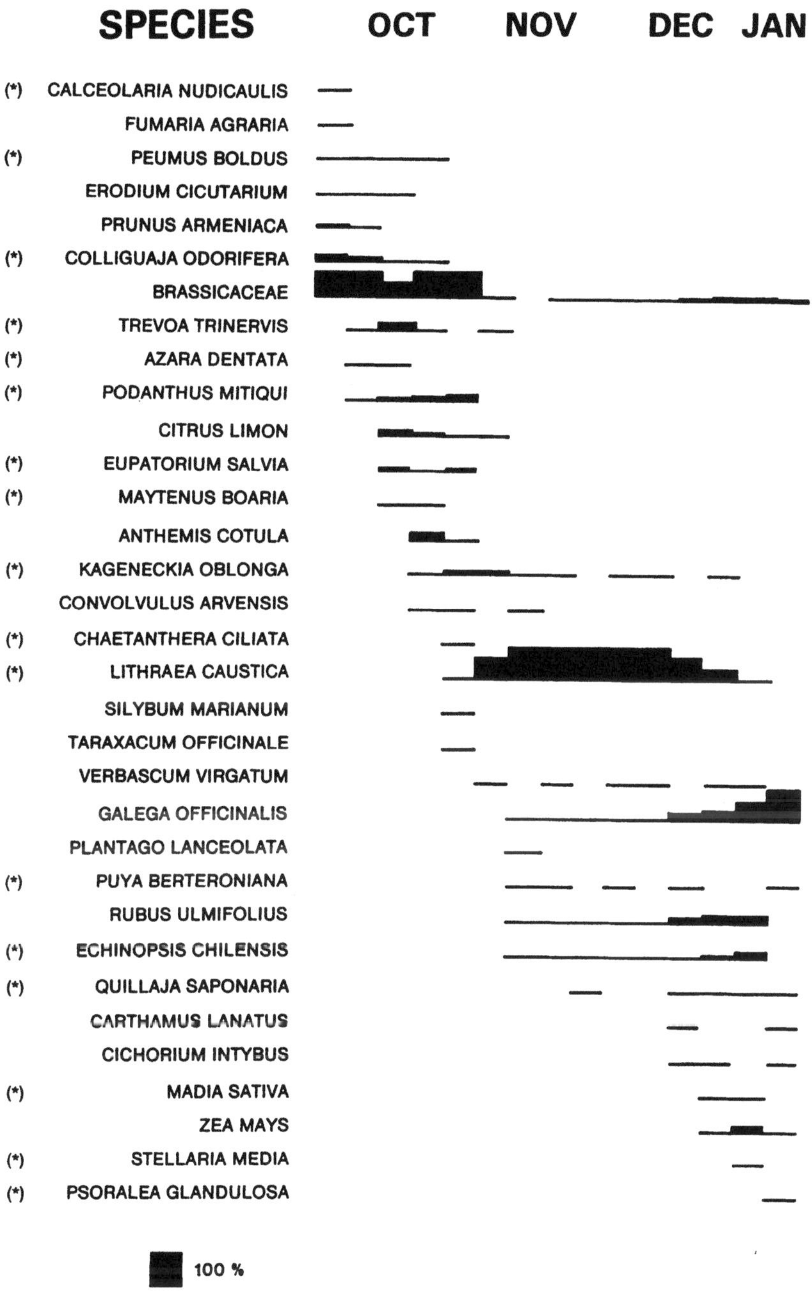

Fig. 4.10. Seasonal succession of bee visits. Main sources of corbicular pollen in the evergreen semi-arid matorral. Monthly percentage of contribution by species from October 1989 to January 1990 is shown. *Asterisks* indicate native species. Notice the seasonal replacement of species

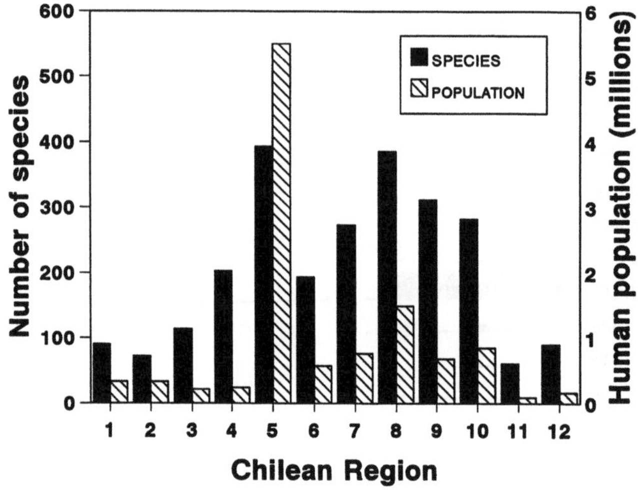

Fig. 4.11. Introduced herbaceous species versus human density. Number of introduced herb species present per Chilean Region and the human population density tend to be associated with each other. Here Chilean Region number **5** includes the Santiago Metropolitan Region (see below), which is where most of the population lives. Extreme Regions such as **1**, **2** and **11** and **12** have very low population densities. The regions are: **1** Tarapacá; **2** Antofagasta; **3** Atacama; **4** Coquimbo; **5** Valparaíso + Santiago Metropolitan; **6** Gen. Bernado O'Higgins; **7** Maule; **8** BíoBío; **9** Araucanía; **10** Lagos; **11** Gen. Carlos Ibañez del Campo; **12** Magallanes y de la Antártica Chilena. (O. Matthei, publ. data)

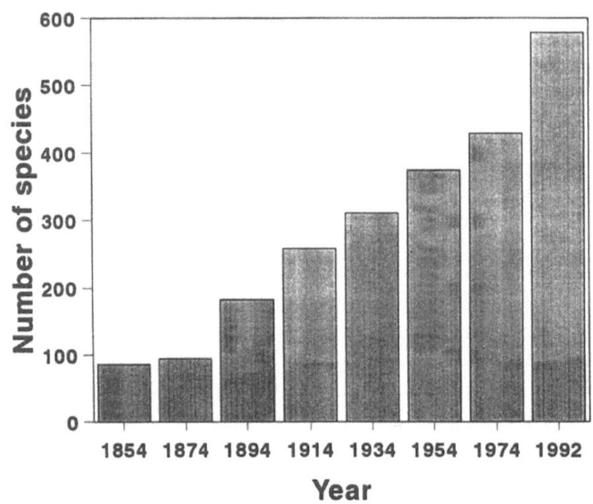

Fig. 4.12. Number of introduced herb species for 20 year intervals over the period 1854 to 1992

different geographical areas in the matorral were compared. The more northern site, near Pichidangui (see Thrower and Bradbury 1977), is dominated by drought-deciduous matorral, whereas the southern site is evergreen shrubland at the Paine site discussed above. Pollen analyses were made of samples of nectar taken from combs at frequent intervals throughout the season (Montenegro et al. 1990; Avila et al. 1992; Montenegro et al. 1992). Pollen grains from each taxon contained in nectar samples showed clear seasonal patterns at both locations (Fig. 4.13). Species such as *Colliguaja odorifera, Lithrea caustica, Puya berteroniana, Quillaja saponaria* and *Trevoa trinervis* appear to be keystone nectar sources in the semi-arid region throughout the year (Fig. 4.13, upper part). *Colliguaja odorifera* and *T. trinervis* supply nectar during the winter when hives are mostly inactive. *Quillaja saponaria* is used in the middle of the season and *P. berteroniana* and *L. caustica* are used in summer. In the arid region (Fig. 4.13, lower part), *C. odorifera* is again a very important nectar source for the honeybees, but the nectar from *Bahia ambrosioides* was a main item during 4 months, and the only species visited during 3 months. Again, keystone species for honey production can be distinguished, and can explain more than the total plant diversity of the area. In fact, the seasonal changes in the importance of the keystone plant species at the northern as well as the southern matorral site suggest that these species could be defined as functional groups from the perspective of commercial honey production.

In general, summer-deciduous shrubs flower in winter, whereas evergreen shrubs flower mainly in spring or summer (Montenegro et al. 1989b). This temporal succession in the flowering period means that these two growth forms differ in relative importance as food sources to honeybees throughout the year. For example, *L. caustica* (evergreen) and *T. trinervis* (drought-deciduous) are the species more used by honeybees as pollen (Fig. 4.10) and nectar (Fig. 4.13) sources. They have similar latitudinal distributions from 30 to 36°S. Whereas *L. caustica* shows morphological adaptations which allow it to live throughout the aridity gradient from equator-facing slopes to pole-facing slopes (Montenegro et al. 1980a,b; Kummerow et al. 1981), *T. trinervis* occupies predominantly equator-facing slopes (Donoso 1978). *Trevoa trinervis*, because of its winter-flowering period, accounts for the first weight increment of the hives after the winter (August to the end of September), representing a nectar recollection of about 6 kg day^{-1} (D. Varela and G. Montenegro, unpubl. data). *Lithrea caustica*, which flowers abundantly every 3 or 4 years, starts flowering in spring, before other evergreen species, and continues flowering until the middle of summer (Hoffmann and Alliende 1984). This species accounts for the second weight increment of the hives during spring

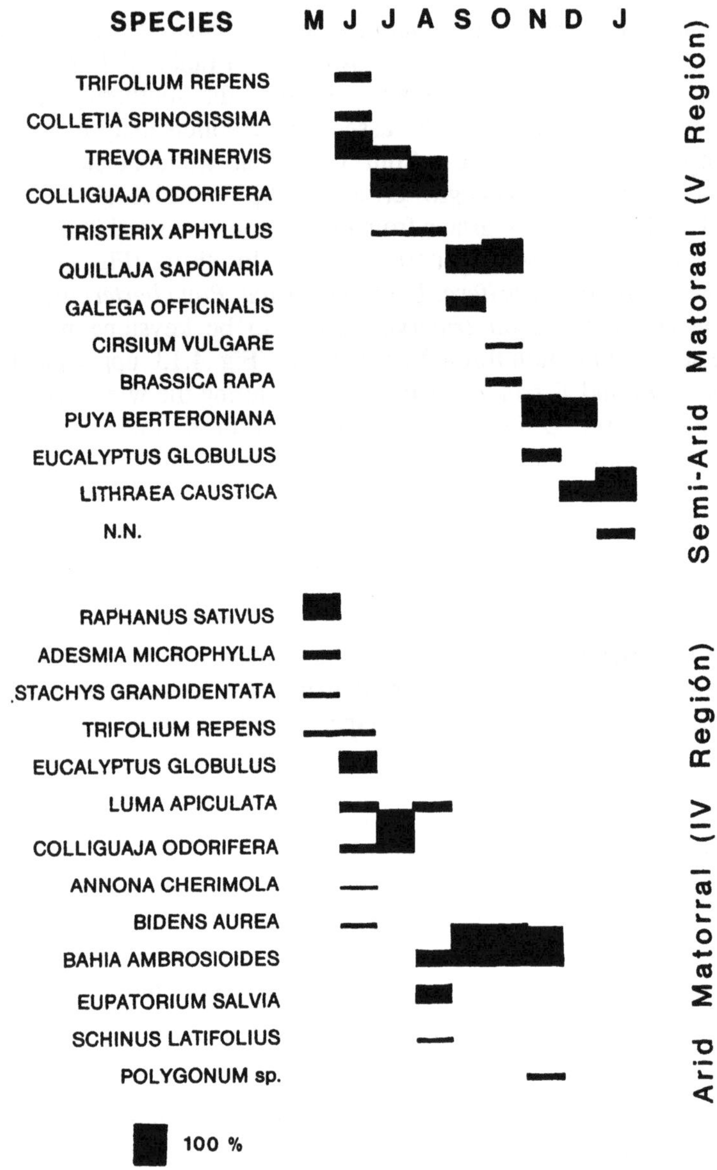

Fig. 4.13. Nectar sources at two matorral sites. Principal nectar sources in the arid (Region 4) and semi-arid (Region 5) matorral (see Fig. 4.11). Monthly percentage of contribution by species is from May 1990 to January 1991

(November), and can contribute up to 2 kg of nectar per day (D. Varela and G. Montenegro, unpubl. data).

In summary, the evidence suggests that, from the perspective of honey production, matorral species differ in importance. Some species at any site are keystones for the honey industry. On the other hand, there is a seasonal succession of the keystone species for *A. mellifera*, suggesting that they have, again from the perspective of the honey industry, an additional role as functional groups in protracting the honey production season. Native and introduced species combine to produce these effects. Little information is currently available on the breeding systems and pollination behaviour of plant communities in Chilean MTEs. Consequently, little can be said about the impact that *A. mellifera* will have on the shrubland communities of central Chile.

4.10 Impact of Changes in Climate and Land-Use on Biodiversity

There are no reliable means of predicting the consequences of climate change in Chilean ecosystems. Several authors have recently speculated on potential ecosystem shifts and plant groups most likely to be affected (Fuenzalida et al. 1989; Fuentes et al. 1991; Arroyo et al. 1993). Scenarios incorporating an increase in temperature and a decrease in rainfall suggest that MTEs should move polewards and uphill. Areas with mediterranean-type conditions may deteriorate rapidly under these conditions, due to the combined effects of increased aridity and increased human extraction of resources from already stressed systems (Fuentes and Muñoz 1994). The combination of longevity, genetic self-incompatibility, biotic pollination and fruit dispersal in many matorral dominants could make the flora of MTEs particularly vulnerable to extinction in the face of global climate change (Arroyo et al. 1993).

A recent model of climate change (Stouffer et al. 1990) suggests that areas in the southern hemisphere, at the latitudes where MTEs are now found, should experience very slow and moderate warming compared to areas at equivalent latitude in the northern hemisphere. If predictions by Stouffer et al. (1990) on expected temperature increases when the CO_2 concentration doubles are corrected by already observable shifts, it is conceivable that central Chile might experience an increase of about 1°C. Equivalent latitudes in the northern hemisphere could have increases of more than 3°C over the same period.

The other major component of global change, namely human population increase and its effects on the landscape, is likely to have a major effect in Chile. Two extreme scenarios of future human use can be envisioned. In the first one, tendencies toward land degradation increase, both in intensity and geographical extent. That is, human impact would increase in the lowlands and upland areas. In the second extreme scenario, industrialization and urbanization increase, population pressure on the land decreases and (as seen in many developed countries during the last century) some ecosystem recovery is observed. There is evidence that following industrialization in the area near Santiago, the matorral surrounding it has recovered in the period between 1955 and 1980 (Fuentes et al. 1991). It is premature to anticipate that this pattern will hold for the whole matorral area, or that a mosaic of situations will develop, resulting in a mixture of partially recovered and more intensively and degraded patches.

4.11 Conclusions

We found no strong evidence that ecosystem functions such as productivity, nutrient flows, or stability are related to biodiversity in Chilean MTEs. We did, however, find that diversity varies over space and time, and that the presence of some species is sometimes contingent upon the presence of other species. Species diversity can, therefore, explain species diversity and even the spatial configuration of matorral ecosystems. In other words, there may be postive feedbacks between high species diversities in interacting guilds of organisms. The results of the study of commercial honey production presented in this chapter clearly demonstrate the functional attributes of biodiversity in relation to this human activity. Another functional attribute that may, at least in principle, be related to species diversity is landscape aesthetics and attractiveness. However, in central Chile, the evidence obtained in interviews with various people suggests that physiognomy is more important than species diversity in attributing aesthetic value to landscape (Fuentes et al. 1984) (Fig. 4.14).

The current state-of-the-art is a direct and logical consequence of research questions in the past not having been directed toward determining the relationships between diversity and function. To address this question experimentally, long-term and multiple site studies with similar as well as different species will have to be planned and carried out. Functional attributes of interest (productivity, nutrient flows, stability, etc) will have to be measured and correlated with overall diversity or the presence of some keystone species. Components of α-, β- and γ-diversity must also be considered.

Fig.4.14. Vegetation structure in the mediterranean-climate region of Chile: *Trichocereus litoralis* in the costal matorral at Los Molles (*top*); open matoı ·l at Fundo Santa Laura in the Cordillera de la Costa with domesticated grazing llama (*bottom*); and degraded espinal savanna in the central valley north of Santiago (see page 226.). (Photos: P.W. Rundel)

Fig. 4.14. *continued*

Acknowledgements. The work in this chapter was financed in part by the following funding: FONDECYT Grants 1940655 to G Montenegro and 1930639 to FM Jaksic; a Fundación Andes doctoral fellowship to R Ginocchio; grants from the University of California Pacific Rim Research Program, the Ecological Research Division of the Office of Health and Environmental Research of the U.S. Department of Energy, and the National Science Foundation to RW Rundel; an Andrew Mellon Foundation grant to Facultad de Ciencias Biológicas of P. Universidad Catolica de Chile; and on NIH Grant N 1 U01 TW-CA 00316-01 awarded to B Timmermann. We also thank Dave Richardson and George Davis for their help in preparing the chapter for publication.

References

Armesto J, Gutiérrez J (1980) Aplicación de algunas técnicas de muestreo en el análisis de la vegetación en Chile central. Arch Biol Med Exp 13: 403-412

Armesto J, Martínez JA (1978) Relations between vegetation structure and slope aspect in the mediterranean region of Chile. J Ecol 66: 881-889

Arroyo MTK, Armesto JJ, Squeo F, Gutiérrez J (1993) Global change: the flora and vegetation of Chile. In: Mooney HA, Fuentes ER, Kronberg E, Fuenzalida H (eds) Northern and Southern Hemisphere responses to global change. Academic Press, London (in press)

Arroyo MTK, Armesto J, Villagrán C (1981) Plant phenological patterns in the high Cordillera de los Andes in central Chile. J Ecol 61: 205-233

Arroyo MTK, Primarck R, Armesto J (1982) Community studies in pollination ecology in the high temperate Andes of small central Chile. I. Pollination mechanisms and altitudinal variation. Am J Bot 69: 82-97
Arroyo MTK, Uslar P (1993) Breeding systems in a temperate mediterranean-type climate montane sclerophyllous forest in central Chile. Bot J Linn Soc 111: 83-102
Atsatt PR, Rundel PW (1982) Pollinator maintenance versus fruit production: a partitioned and facultative reproductive effort in subdeciduous *Fuchsia lycioides*. Ann Mo Bot Gard 69: 199-208
Avila G, Gómez M, Mujica AM, Montenegro G (1993) La importancia de la flora nativa como sustentadora de colmenas en Pichidangui IV Región de Chile. Rev Cien Invest Agrar (in press)
Bahre C (1979) Destruction of the natural vegetation of north-central Chile. Univ Calif Publ Geogr 23: 1-117
Brown J, Davidson DW, Munger JC, Inouye RS (1986) Experimental community ecology: the desert granivore system. In: Diamond J, Case TJ (eds) Community ecology. Harper and Row, New York, pp 41-62
Cei JM (1962) Batracios de Chile. Edic Univ Chile, Santiago
Cody ML (1970) Chilean bird distribution. Ecology 51: 455-464
Cody ML (1973) Parallel evolution in bird niches. In: Di Castri F, Mooney HA (eds) Mediterranean-type ecosystems. Springer, Berlin Heidelberg New York, pp 307-338
Cody ML (1974) Competition and the structure of bird communities. Princeton Univ Press, Princeton
Cody ML, Mooney HA (1978) Convergence versus non convergence in mediterranean climatic systems. Annu Rev Ecol Syst 9: 265-321
Cody ML, Fuentes ER, Galanz W, Hunt JH, Moldenke AR (1977) Convergent evolution in consumer organisms of Mediterranean Chile and California. In: Mooney HA (ed) Convergent Evolution in Chile and California. Dowden, Hutchinson and Ross, Stroudsburg, PA, pp 144-192
Corbert SA (1978) A bee's view of nectar. Bee World 59: 25-32
di Castri F (1973) Climatographical comparison between Chile and western coast of North America. In: di Castri F, Mooney HA (eds) Mediterranean type ecosystems. Origin and structure. Ecological studies 7. Springer, Berlin Heidelber New York, pp 21-36
di Castri F, Hajek ER (1976) Bioclimatologia de Chile. Univ Catolica, Santiago
di Castri F, Goodall DW, Specht RL (eds) (1981) Ecosystems of the World II: mediterranean-type sclerphyllous shrublands. Elsevier, Amsterdam
Dodd J, Heddle EM, Pate JS, Dixson KW (1984) Rooting patterns of sandplain plants and their functional significance. In: Pate JS, Beard JS (eds) Kwongan: plant life of the sandplain. University of Western Australia Press, Nedlands, pp 146-177
Donoso C (1978) Dendrología de árboles y arbustos Chilenos. Manual N°2, Universidad de Chile
Donoso-Barros R (1966) Reptiles de Chile. Edic Univ Chile, Santiago
Eisenberg JF, Redford KH (1982) Comparative niches structure and evolution of mammals of the Nearctic and southern South America. In: Mares MA, Genoways HH (eds) Mammalian biology in South America. Univ Pittsburgh, pp 77-84
Eisikowitch D, Loper GM (1984) Some aspects of flower biology and bee activity on hybrid cotton in Arizona, USA. J Apic Res 23: 243-248
Free JB (1963) The flower constancy of honeybees. J Anim Ecol 32: 119-131
Fuentes ER (1976) Ecological convergence of lizard communities in Chile and California. Ecology 57: 3-17

Fuentes ER, Campusano C (1985) Pest outbreaks and rainfall in the semi-arid region of Chile. J Arid Environ 8: 67-72

Fuentes ER, Etchegaray J (1983) Defoliation patterns in matorral ecosystems. In: Kruger FJ, Mitchell DT, Jarvis JV (eds) Mediterranean-type ecosystems. Ecological studies 43. Springer, Berlin Heidelberg New York, pp 525-542

Fuentes ER, Jaksic FM (1980) Ecological replacement of *Liolaemus* lizards along a habitat gradient. Oecologia 46: 45-48

Fuentes ER, Jaksic FM (1988) The hump-backed species diversity curve: why has it not been found among land animals? Oikos 53: 136-138

Fuentes ER, Muñoz M (1994) The role of humans in changing the landscapes of central Chile. In: Arroyo MKT, Zedler PH, Fox MD (eds) Ecology and biogeography of Mediterranean ecosystems in Chile, California, and Australia. Springer, Berlin Heidelberg New York (in press)

Fuentes ER, Jaksic FM, Simonetti J (1983) European rabbits versus native rodents in central Chile: effects on shrub seedlings. Oecologia 58: 411-414

Fuentes ER, Espinoza G, Fuenzalida I (1984) Cambios vegetacionales recientes y percepción ambiental. El caso de Santiago de Chile. Rev Geogr Norte Grande 11: 45-53

Fuentes ER, Hoffmann AJ, Poiani A, Alliende MC (1986) Vegetation change in large clearings: patterns in the Chilean matorral. Oecologia 68: 358-366

Fuentes ER, Hajek ER, Espinoza G (1988) Some consequences of rainfall variability for mediterranean type ecosystems in Chile. In: di Castri F, Floret C, Rambal S, Roy J (eds) Time scales and water stress. Proc 5th Int Conf Mediterranean Ecosystems. International Union of Biological Sciences, Paris, pp 347-360

Fuentes ER, Kronberg B, Mooney HA (1991) The west coasts of the Americas as indicators of global change. Trends Ecol Evol 6: 203-204

Fuenzalida HA, Bernal PA, Villagrán C, Fuentes ER, Montecino V, Santibáñez F, Peña H. Hajek E, Rutlant J (1989) El cambio global del clima y sus eventuales efectos en Chile. Comité Nacional del Programa Internacional de la Geosfera-Biósfera (IGBP). ICSU

Gajardo R (1987) La Vegetation Naturelle du Chile. PhD Dissertation, Univ Droite d'Economie et des Sciences d'Aix-Marseille, Marseille

Gauch H (1982) Multivariate analysis in community ecology. Cambridge University Press, Cambridge

Giliberto J, Estay H (1978) Seasonal water stress in some Chilean matorral shrubs. Bot Gaz 139: 236-240

Glanz WE (1977a) Comparative ecology of small mammal communities in California and Chile. PhD Thesis, University of California, Berkeley

Glanz WE (1977b) Small mammals. In: Thrower NSW, Bradbury DE (eds) Chile-California Mediterranean scrub atlas: a comparative analysis. Dowden, Hutchinson and Ross, Stroudsburg, PA, pp 232-237

Glanz, WE, Meserve PL (1982) An ecological comparison of small mammal communities in California and Chile. In: Conrad CE, Oechel WC (eds) Dynamics and management of mediterranean-type ecosystems. USDA For Serv Gen Tech PSW-58, pp 220-226

Heinrich B (1975) Bee flowers: a hypothesis on flower variety and blooming times. Evolution 29: 325-334

Hellmers H, Horton JS, Juhren G, O'Keefe J (1955) Root systems of some chaparral plants in southern California. Ecology 36: 667-678

Higgins KB, Lamb AJ, van Wilgen BW (1987) Root systems of selected plant species in mesic fynbos in the Jonkershoek Valley, South Western Cape Province. S Afr J Bot 53: 249-257

Hobbs GA (1962) Further studies on food-gathering behaviour of bumblebees (Hymenoptera: Apidae). Can Entomol 94: 538-541
Hoffmann AE (1989a) Cactáceas en la flora silvestre de Chile. Fundación Claudio Gay, Santiago, 272 pp
Hoffmann AE (1989b) The conservation status of Chilean succulent plants: a preliminary assessment. In: Benoit I (ed) Red book on Chilean terrestrial flora. Chilean Forest Service (CONAF), Santiago, pp 107-120
Hoffmann AE (1989c) Chilean geophyte monocotyledons: taxonomic synopsis and conservation status. In: Benoit I (ed) Red book on Chilean terrestrial conservation status. Chilean Forest Service (CONAF), Santiago, pp 140-151
Hoffmann AJ, Hoffmann AE (1976) Growth pattern and seasonal behaviour of buds of *Colliguaja odorifera*, a shrub from the Chilean mediterranean vegetation. Can J Bot 54: 1767-1774
Hoffmann AJ, Hoffmann AE (1982) Altitudinal ranges of phanerophytes and chamaephytes in central Chile. Vegetatio 48: 151-163
Hoffmann A, Kummerow J (1978) Root studies in the Chilean matorral. Oecologia 32: 57-69
Hoffmann AJ, Alliende MC (1984) Interactions in the patterns of vegetative growth and reproduction in woody dioecious plants. Oecologia 61: 109-114
Humphrey RR (1958) The desert grassland. A history of vegetational change and an analysis of causes. Bot Rev 24: 193-252
Hunt J (1973) Comparative ecology of ant communities of California and Chile. PhD Thesis, University of California, Berkeley
Iriarte JA, Contreras LC, Jaksic FM (1989) A long-term study of a small-mammal assemblage in the central Chilean matorral. J Mammal 70: 79-87
Jaksic FM (1986) Predation upon small mammals in shrublands and grasslands of southern South America: ecological correlates and presumable consequences. Rev Chil Hist Nat 59: 209-221
Jaksic FM, Fuentes ER (1991) Ecology of a successful invader: the European rabbit in central Chile. In: Groves RH, Di Castri F (eds) Biogeography of mediterranean invasions. Cambridge University Press, Cambridge, pp 73-83
Jaksic FM, Simonetti JA (1987) Predator/prey relationships among terrestrial vertebrates: an exhaustive review of studies conducted in southern South America. Rev Chil Hist Nat 60: 221-244
Jaksic FM, Greene HW, Yáñez JL (1981) The guild structure of a community of predatory vertebrates in central Chile. Oecologia 49: 21-28
Jaksic FM, Jiménez JE, Castro SA, Feinsinger P (1992) Numerical and functional response of predators to a long-term decline in mammalian prey at semi-arid neotropical site. Oecologia 89: 90-101
Jaksic FM, Feinsinger P, Jiménez JE (1993a) A long-term study on the dynamics of guild structure among predatory vertebrates at a semi-arid neotropical site. Oikos: 67: 87-96
Jaksic FM, Meserve PL, Gutiérrez JR, Tabilo E (1993b) The components of predation on small mammals in north-central Chile: preliminary results. Rev Chil Hist Nat 66: 305-321
Jiménez JE, Feinsinger P, Jaksic FM (1992) Spatiotemporal patterns of an irruption and decline of small mammals in northcentral Chile. J Mammal 73: 356-364
Johow F (1948) Flora de Zapallar. Rev Chil Hist Nat 49:1-566
Kruger FJ (1983) Plant community diversity and dynamics in relation to fire. In: Kruger FJ, Mitchell DT, Jarvis JV (eds) Mediterranean-type ecosystems. Ecological studies 43, Springer, Berlin Heidelberg New York, pp 447-472

Kummerow J, Krause D, Jow W (1972) Root systems of chaparral shrubs. Oecologia 29: 163-177

Kummerow J, Montengro G, Krause D (1981) Biomass, phenology and growth. In: Miller P (ed) Resource use by chaparral and matorral. Springer, Berlin Heidelberg New York, pp 69-96

Kummerow J, Avila G, Aljaro ME, Araya S, Montenegro G (1982) Effect of fertilizer on fine root density and shoot growth in Chilean matorral. Bot Gaz 143: 498-504

Le Maitre DC, Brown PJ (1992) Life cycles and fire-stimulated flowering in geophytes. In: van Wilgen BW, Richardson DM, Kruger FJ, van Hensbergen HJ (eds) Fire in South African mountain fynbos. Springer, Berlin Heidelberg New York, pp 145-160

MacArthur RH (1972) Geographical ecology. Harper and Row, New York

Marticorena C, Quezada M (1985) Catálogo de la flora vascular de Chile. Gayana 42: 1-157

Maurizio A (1975) Microscopy of honey. In: Crane E (ed) Honey: a comprehensive survey, London, pp 240-257

Meserve PL (1981) Resource partitioning in a Chilean semi-arid small mammal community. J Anim Ecol 40: 747-757

Meserve PL, Le Boulengé E (1987) Population dynamics and ecology of small mammals in the northern Chilean semiarid region. In: Patterson BD, Timm RM (eds) Studies in neotropical mammalogy: essays in honor of Philip Hershkovitz. Fieldiana: Zoology, New Ser 39: 413-431

Miller P (1981) Resource use by chaparral and matorral. A comparison of vegetation function in two mediterranean-type ecosystems. Ecological studies 39, Springer, Berlin Heidelberg New York

Montenegro G, Martínez J, Aljaro ME (1980a) Estructura de comunidades y estacionalidad de especies dominantes en un gradiente altitudinal de Chile central. Proc VIII Reunión Argentina de Ecología, 27 pp

Montenegro G, Riveros F, Alcalde C (1980b) Morphological structure and water balance of Chilean shrub species. Flora 170: 554-564

Montenegro G, Schuck M, Mujica AM, Teillier S (1989a) Flora utilizada por abejas melíferas (*Apis mellifera*) como fuente de polen en Paine, Región Metropolitana Chile. Rev Cien Invest Agrar 16(1-2): 47-54

Montenegro G, Avila G, Aljaro ME, Osorio R, Gómez M (1989b) Chile. In: Orshan G (ed) Plant pheno-morphological studies in mediterranean type ecosystems. Kluwer, Dordrecht, pp 347-387

Montenegro G, Teillier S, Arce P, Poblete V (1990) Introduction of plants into the mediterranean type climate of Chile. In: Groves RH, di Castri F (eds) Biogeography of mediterranean invasions. Cambridge University Press, Cambridge, pp 102-115

Montenegro G, Cardalda E, Avila G, Ginocchio R, Gómez M, Mujica AM (1992) Especies nativas de Chile y su importancia económica en la productividad de colmenas de *Apis mellifera*. Proc symp Etnobotánica, Córdoba, Spain (in press)

Mooney HA (1977) Convergent evolution in Chile and California. Dowden, Hutchinson and Ross, Stroudsburg

Mooney, HA, Schlegel F (1967) La vegetación costera del Cabo "Los Molles" en la provincia de Aconcagua. Bol Univ Chile 75: 27-32

Mooney HA, Kummerow J, Johnson AW, Parsons DJ, Keeley S, Hoffmann A, Hays RI, Giliberto J, Chu C (1977) The producers, their resources and adaptative responses. In: Mooney HA (ed) Convergent evolution in Chile and California. Dowden, Hutchinson and Ross, Stroudsburg, pp 85-143

Muñoz C, Pisano E (1947) Estudio de la vegetación y flora de los parques nacionales de Fray Jorge y Talinay. Agricultura Técnica, Santiago 7: 71-190

Navas LE (1977-1979) Flora de la cuenca de Santiago de Chile, vols 1-3. Univ de Chile, Santiago

Niering WA, Lowe CH (1984) Vegetation of the Santa Catalina Mountains: community types and dynamics. Vegetatio 58: 3-28

Nobel PS (1980) Influences of minimum stem temperatures on ranges of cacti in southwestern United States and central Chile. Oecologia 48: 194-198

Paine RT (1966) Food web complexity and species diversity. Am Nat 100: 65-75

Parsons DJ, Moldenke AR (1975) Convergence in vegetation structure along analogous climatic gradients in California and Chile. Ecology 56: 950-957

Pearson OP, Pearson AK (1982) Ecology and biogeography of the southern rainforests of Argentina. In: Mares MA, Genoways HH (eds) Mammalian biology in South America. Pymatuning Symposia in Ecology 6. University of Pittsburgh, Pittsburgh, pp 129-142

Péfaur JE, Yáñez JL, Jaksic FM (1979) Biological and environmental aspects af a mouse outbreak in the semi-arid region of Chile. Mammalia 43: 313-322

Poblete V, Montenegro G (1992) Diagnóstico de polen corbicular en colmenas ubicadas en la provincia seco estival nubosa de Chile Central. Rev Cien Invest Agrar 19: 23-30

Poole DK, Roberts SW, Miller PC (1981) Water utilization. In: Miller PC (ed) Resource use by chaparral and matorral. Springer, Berlin Heidelberg New York, pp 123-150

Rafols ES, Kummerow J (1977) Vegetations transects. In: Thrower NJW, Bradbury DE (eds) Chile-Califronia mediterranean scrub atlas: a comparative approach. Dowden, Hutchinson and Ross, Stroudsburg, pp 83-91

Richardson DM, Cowling RM (1992) Why is mountain fynbos invasible and which species invade? In: van Wilgen BW, Richardson DM, Kruger FJ, van Hensbergen HJ (eds) Fire in South African mountain fynbos, Springer, Berlin Heidelberg New York, pp 161-181

Romero H (1985) Geografía de Chile, Tomo XI: Geografía de los Climas Instit Geogr Militar, Santiago

Rundel PW (1974) *Trichocereus* in the mediterranean zone of central Chile. Catus Succ J 46: 86-88

Rundel PW (1977) Population variability in the genus *Trichocereus* (Cactaceae) in central Chile. Plant Syst Evol 127: 1-9

Rundel PW (1981a) Fire as an environmental factor. In: Lange OL, Nobel PS, Osmond CB, Ziegler H (eds) Physiological ecology, vol A. Interaction with the physical environment. Springer, Berlin Heidelberg New York, pp 501-559

Rundel PW (1981b) The matorral zone of central Chile. In: di Castri F, Goodall D, Specht RL (eds) Mediterranean-type shrublands. Elsevier, The Hague, pp 175-201

Rundel PW (1981c) Structural and chemical components of flammability. In: Mooney HA, Bonnicksen TM, Christianson NL, Lotan JE, Reiners WA (eds) Fire regimes and ecosystem properties. USDA Forest Serv Gen Tech Rep WO-26, pp 183-207, Washington DC

Rundel PW (1982) Nitrogen utilization efficiencies in mediterranean-climate shrubs of California and Chile. Oecologia 55: 409-413

Rundel PW (1988) Leaf structure and nutrition in mediterranean climate sclerophylls. In: Specht RL (ed) Mediterranean-type ecosystems: a data source book. Kluwer, Dordrecht, pp 157-167

Rundel PW (1993) Adaptive significance of some morphological and physiological characteristics in mediterranean plants: facts and fallacies. In: di Castri F, Roy J (eds) Time-scales of water stress response of mediterranean biota. Springer, Berlin Heidelberg New York (in press)

Rundel PW, Mahu M (1976) Community structure and diversity in a coastal fog desert in northern Chile. Flora 165: 493-505

Rundel PW, Neel J (1978) Nitrogen fixation in root nodules of *Trevoa trinervis* (Rhamnaceae). Flora 167: 127-132

Rundel PW, Parsons DJ (1984) Post-fire uptake of nutrients by diverse ephemeral herbs in chamise chaparral. Oecologia 61: 285-288

Rundel PW, Weisser P (1975) La Campana, a new National Park in central Chile. Biol Conserv 8: 35-46

Rundel PW, Dillon MO, Palma B, HA Mooney, Gulman SL, Ehleringer JR (1991) The phytogeography and ecology of the coastal Atacama and Peruvian deserts. Aliso 13: 1-49

Sempe J, Ramírez C, Montenegro G (1989) Flora utilizada como fuente de polen por *Apis mellifera* en la Provincia de Valdivia: análisis cuantitativo del polen corbicular. Rev Cien Invest Agrar 16(1): 55-65

Severson DW, Parry JE (1981) A chronology of pollen collection by honeybees. J Apicult Res 20(2): 97-103

Simonetti JA (1988) The carnivorous predatory guild of central Chile: a human-induced community trait? Rev Chil Hist Nat 61: 23-25

Skottsberg C (1950) Apuntes sobre la flora y vegetación de Fray Jorge (Coquimbo, Chile). Medd Göteborgs Bot Träj 18: 84-91

Snelling RR, Hunt JH (1975) The ants of Chile. Rev Chil Entomol 89:63-129.

Solbrig OT (ed) (1991) From genes to ecosystems: a research agenda for biodiversity. IUBS, Cambridge, MA

Specht RL, Rayson P (1957) Dark Island Heath (ninety-mile plain, South Australia). III. Root systems. Aust J Bot 5: 103-114

Stouffer R, Manabe S, Bryan K (1989) Interhemispheric asymmetry in climate response to a gradual increase of atmospheric CO_2. Nature 342: 660-662

Thrower NJW, Bradbury DE (eds) (1977) Chile-California mediterranean scrub atlas: a comparative approach. Dowden, Hutchinson and Ross, Stroudsburg

Tilman D (1988) Plant strategies and the dynamics and structure of plant communities. Princeton University Press, Princeton

Uslar P (1982) Sistemas de reproducción de plantas: zona ecotonal entre la zona andina y el matorral esclerófilo de Chile central. Licenciatura Thesis, Univ de Chile, Santiago

van der Moezel PG, Delfs JC, Pate JS, Loneragan WA, Bell DT (1987) Pollen selection by honeybees in shrublands of the Northern sandplains of Western Australia. J Apicult Res 26: 224-232

Varela D, Schuck M, Montenegro G (1991) Selectividad de *Apis mellifera* en su recolección de polen en la vegetación de Chile central (Región Metropolitana). Rev Cien Invest Agrar 18(12): 73-78

Veblen TT, Schlegel FM, Oltremari JV (1983) Temperate broad-leaved evergreen forests of South America, In: Ovington JD (ed) Temperate broad leaved evergreen forests. Elsevier, Amsterdam, pp 5-31

Villaseñor C (1980) Unidades fisionómicas y florísticas del Parque Nacional La Campana. Anal Mus Hist Nat Val 13: 65-70

Villaseñor C, Serey I (1981) Estudio fitosociológico de la vegetación del Cerro La Campana (Parque Nacional La Campana), en Chile central. Atti Ist Bot Lab Critt Univ Pavia 14: 69-91

Weisser P, Rundel P (1980) Estudion cuantitative de un matorral costero en Pichidangni (prov. Coquimbo, Chile). Anal Mus Hist Nat Valp 13: 47-57

Westman WE (1988) Vegetation, nutrition and climate - data-tables (3) Species richness. In: Specht RL (ed) Mediterranean-type ecosystems : A data source book. Kluwer, Dordrecht, pp 81-91

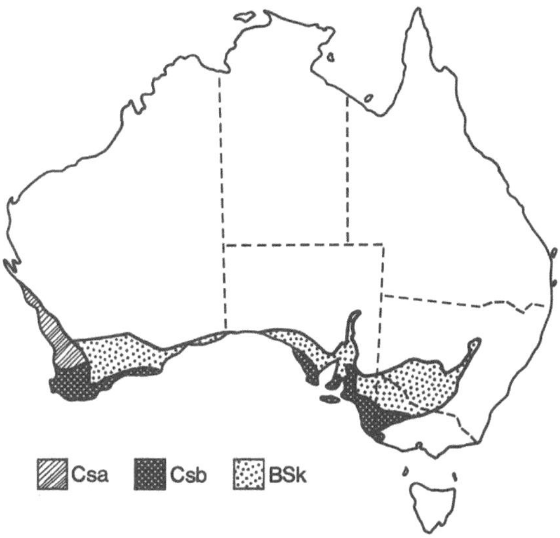

Fig. 5.1. Köppen climatic zones which define the limits of mediterranean-type climates (*Csa* and *CSb*) in southern Australia, with adjoining semi-arid regions (*Bsk*). (Dick 1975)

Mediterranean-type climates vary in the length of summer drought and merge into the arid types at one extreme and the humid types at the other (di Castri 1981). As such, the mediterranean climate is transitional between temperate and dry tropical climates. Strong rainfall gradients are experienced over the MTEs of Australia. In the southwestern region, rainfall declines from about 1500 mm per year in the extreme southwest (Gentilli 1989) to about 250 mm at the eastern edge of the area which could be classed as mediterranean, and the length of the summer drought increases correspondingly.

Hopper (1979) recognised three major rainfall zones (Fig. 5.2b). These are the High Rainfall Zone (800-1200 mm annual rainfall), the Transitional Rainfall Zone (300-800 mm), and the Arid Zone (less than 300 mm). The High Rainfall Zone encompasses forests and woodlands dominated by jarrah (*Eucalyptus marginata*), marri (*E. calophylla*), karri (*E. diversicolor*) and tuart (*E. gomphocephala*), with substantial coastal heath and banksia low woodlands (Fig. 5.2d).

The Transitional Rainfall Zone is roughly equivalent to the major agricultural wheatbelt of the southwest. It contains the most species-rich areas of wildflowers, including the Mt Lesueur-Kalbarri region to the north of Perth, the lateritic uplands of the western edge of the wheatbelt, and the Stirling Range, Fitzgerald River and Cape Arid National Parks of the southern heathlands and mallee (see Fig. 5.2a for locations). Vegetation is

5 Function of Biodiversity in the Mediterranean-Type Ecosystems of Southwestern Australia

R.J. Hobbs, R.H. Groves, S.D. Hopper, R.J. Lambeck, B.B. Lamont, S. Lavorel, A.R. Main, J.D. Majer and D.A. Saunders

5.1 Introduction

A workshop on the ecosystem function of biodiversity in the mediterranean-type ecosystems (MTEs) of Australia was held in Perth in December 1991 (Hobbs 1992a). In this chapter we elaborate on the workshop findings. We first provide an outline of characteristics of Australian MTEs. We mostly concentrate on southwestern Australia since the workshop focussed exclusively on this region. We examine the patterns of biodiversity by considering vegetation, floristic and landscape patterns, and discuss the vegetation response to disturbance and environmental constraints. We also examine how human activities have affected these patterns. We discuss the ecosystem role of diversity in southwest Australian MTEs by presenting a number of topics and case studies which highlight some of the difficulties involved in interpreting the available information. Finally, further areas of research on the topic are discussed.

5.2 Mediterranean-Type Ecosystems in Australia

5.2.1 Physical Characteristics

Köppen's climatic zones *Csa* and *Csb* define the areas classed as mediterranean-type in Australia, but allied ecosystem types extend into the semi-arid region, classed as *BSk* (Dick 1975; Specht 1981; Fig. 5.1). The distinction between *Csa* and *Csb* is in mean summer temperatures, that in *Csa* being over 22°C, while *Csb* is under 22°C. Here we restrict our discussions to the area of true mediterranean-type climate indicated as *Csa* and *Csb* in southwestern Australia. Details of MTEs in South Australia can be found in Specht (1972), and Noble and Bradstock (1989) discuss the mallee vegetation found in the *BSk* region.

G.W. Davis and D.M. Richardson (Eds.)
Ecological Studies, Vol.109

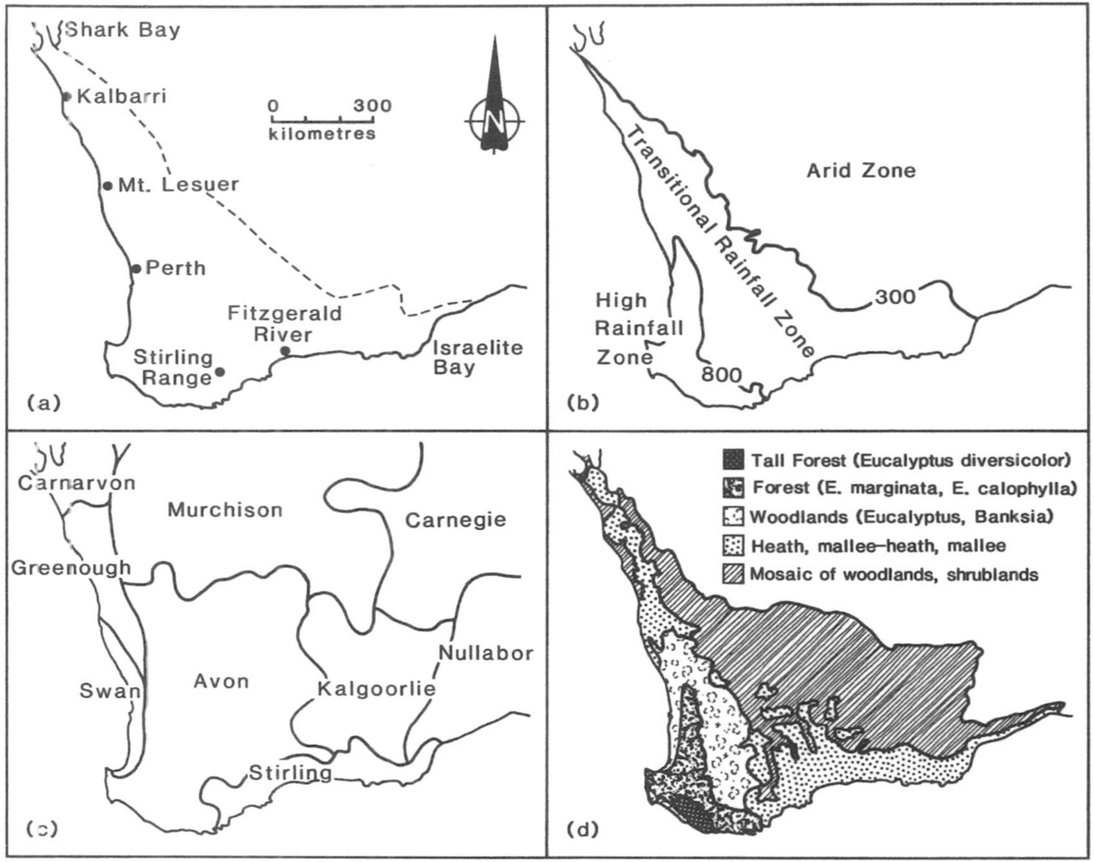

Fig. 5.2 a-d. Southwestern Australia **a** Locations mentioned in text. The *dotted line* shows the eastern extent of the Southwest Botanical Province (after Hopper 1992). **b** Rainfall isohyets showing the rainfall zones of Hopper (1979). **c** Landform regions; see Table 5.1 for details (after McArthur and Bettenay 1979). **d** Major vegetation zones, as defined by Beard in Hopper (1979)

complex, but dominated by kwongan (shrublands), mallee and woodlands. The Transitional Rainfall Zone also extends inland to encompass some of the Goldfields woodlands and pastoral country beyond the present wheatbelt.

Australian mediterranean-type climates, along with those of South Africa, differ from those of the other mediterranean regions (i.e. the Mediterranean Basin, California and Chile) because summer rainfall is a common, if unpredictable, occurrence in these regions (di Castri 1981). The current winter rainfall regime has been in effect for only the last 2.5 million years, during which period there have been numerous warm wet/cold dry fluctuations (Churchill 1968; Bowler 1982; Lamont et al. 1984).

The soils of southwestern Australia and South Africa are, with a few exceptions, very old, weathered and relatively poor in nutrients, and this again sets them apart from the other regions (di Castri 1981; Kruger et al. 1983). Southwestern Australia comprises four distinct landform regions (Table 5.1, Fig. 5.2c). The Swan Coastal Plain contains sections of piedmont and riverine alluvium, coastal marine and estuarine deposits, and coastal dunes, while the other regions are dominated by an extensive lateritised plateau dissected into broad valleys, with deep in situ weathering to form mottled and pallid zones, with alluvium in the broad valleys and sands and gravels elsewhere (McArthur and Bettenay 1979). Details of major soil types and vegetation associated with these are given in Table 5.1.

5.2.2 Vegetation Patterns

Because of the range of mean annual rainfalls covered by the mediterranean-type climate, coupled with variations in soils and landforms, the MTEs of southwestern Australia encompass a wide array of ecosystem types, ranging from closed forests to open savannah woodlands and shrublands (Table 5.2; Beard 1983,1990; Beard and Sprenger 1984). At a broad scale in southwestern Australia, patterns of vegetation composition and structure can be related to rainfall gradients, and to changes in soil and landform type (Table 5.1; Fig. 5.3; Beard 1983; Hobbs 1992b; Hopper 1992). Generally, forest vegetation is restricted to the higher rainfall areas (>750 mm), and at lower rainfalls the distribution of woodland, mallee, shrubland and heath vegetation types is determined mainly by substrate type (Beard 1983). Thus within any given region, soil type and landform generally determine the broad vegetation patterning (Fig. 5.4; Beard 1983; McArthur 1993). The main vegetation types are also largely characterised by their dominant growth forms (i.e. trees, mallees, shrubs). In addition to the major vegetation types listed are a number of important types which

Table 5.1. Landforms and soils of southwestern Australia. (McArthur and Bettenay 1979)

Region	Geology	Landform	Soils	Vegetation
Swan	Quaternary, Tertiary, Mesozoic and Paleozoic sediments	Coastal plain	Red or yellow earths or duplex soils	*Eucalyptus marginata* – *E. calophylla* – *E. wandoo* forest
			Shallow podzolized sands	*Eucalyptus gomphocephala* forest
			Podzols; calcareous sands	*Banksia/Eucalyptus* woodland; heath
Avon	Precambrian granite and gneisses with dolerite dykes	Lateritized plateau, dissected at fringes; saline drainage lines inland	Red earths on fringe dissections	*Eucalyptus marginata* – *E. diversicolor* forest
			Gravelly sands and duricrust on Darling range	*Eucalyptus marginata* – *E. calophylla* – *E. wandoo* forest
			Yellow sand on inland residuals	*Allocasuarina/Acacia* shrublands
			Red or yellow duplex soils on valley fill	*Eucalyptus salubris/* *E. salmonphloia* woodland
			Shallow red duplex soils on dissections	*Eucalyptus loxophleba/* *Acacia accuminata* woodland
Stirling	Tertiary sediments: Proterozoic gneisses, magmatites and quartzites	Lateritized plateau, dissected at fringes; emergent quartzite ranges	Yellow gravelly duplex soils on plains	Heath and shrubland
			Yellow duplex soils in dissections	*Eucalyptus/Banksia* scrub
			Shallow sandy soils on ranges	Heath
Greenough	Jurassic and Permian sediments; Proterozoic granites	Lateritized plateau, dissected at fringes	Yellow sand on laterite residuals	Heath and shrubland
			Red earths or duplex soils in dissections and on granites	*Eucalyptus loxophleba/* *Acacia accuminata* woodland
			Yellow gravelly duplex soils on Dandaragan plateau	Heath and scrub

Table 5.2. List of physiognomic units in the southwest botanical province showing their total original area, their percentage contribution to the area of the province, and the percentage of each unit which has been cleared. (After Beard and Sprenger 1984)

Unit no.	Physiognomic unit	Floristic markers	Area (km^2)	% of total	% cleared
1	Tall forest	*Eucalyptus diversicolor, E. marginata, E. calophylla*	4004	1.29	14
2	Tall woodland	*Eucalyptus gomphocephala*	69	0.02	94
3	Forest	*Eucalyptus marginata*	31 802	10.26	44
4	Woodland	*Eucalyptus calophylla–E. wandoo*	16697	5.39	94
5	Woodland	*Eucalyptus astringens–E. accedens*	958	0.31	72
6	Woodland	*Eucalyptus gomphocephala±E. marginata*	1039	0.34	90
7	Woodland	*Eucalyptus loxophleba, E. salmonophloia, E. wandoo*	41 126	13.27	97
8	Woodland	*Eucalyptus samonophloia–E. salubris*	17580	5.67	78
9	Woodland	*Eucalyptus lesouefii–E. transcontinentalis–E. oleosa*	950	0.31	0
10	Woodland	*Eucalyptus oleosa–E. flocktoniae*	1575	0.51	0
13	Open woodland	Mainly *Eucalyptus wandoo*	221	0.07	94
14	Low forest	*Eucalyptus marginata*	964	0.31	82
15	Low forest	*Callitris preissii*	29	0.01	100
16	Low forest	*Eucalyptus platypus* or *E. cornuta–E. lehmannii*	113	0.04	2
17	Low forest	*Acacia rostellifera*	327	0.10	98
22	Low woodland	*Agonis flexuosa*	242	0.08	74
23	Low woodland	*Banksia* spp.,±*Eucalyptus marginata*	6229	2.01	61
25	Low woodland	*Allocasuarina huegeliana–Eucalyptus loxophleba*	88	0.03	88
26	Low woodland	*Eucalyptus marginata–Allocasuarina fraserana*	901	0.29	41
27	Low woodland	*Melaleuca* spp.	926	0.30	49
30	Low woodland with scattered emergent trees	*Banksia* spp.+*Eucalyptus marginata*	680	0.22	100
31	Thicket with same	*Allocasuarina campestris+Eucalyptus wandoo* or *Melaleuca uncinata+E. loxophleba*	441	0.14	98
35	Scrub with scattered trees	*Acacia spp.+Eucalyptus* and/or *Callitris*	5354	1.73	83
36	Thicket	*Acacia–Allocasuarina–Melaleuca*	35496	11.46	80
37	Thicket	*Melaleuca rhaphiophylla*	143	0.05	74
38	Thicket	*Dryandra* spp. and/or *Eucalyptus* spp.	497	0.16	18
40	Scrub	*Acacia* spp.	1109	0.36	78

41	Scrub	*Melaleuca* spp.	76	0.02	70
42	Scrub	*Agonis flexuosa, Acacia, Eucalyptus*	1636	0.53	12
44	Sparse scrub	*Acacia* spp.	121	0.04	66
45	Mallee	*Eucalyptus* spp.	46385	14.97	55
131	Mallee and woodland patches	Same	22219	7.17	40
47	Mallee-heath	*Eucalyptus tetragona*+mixed Proteaceae-Myrtaceae	17114	5.52	69
47a	Tree-heath	Heterogeneous	3626	1.17	0
48	Scrub heath	Same, Proteaceae – Myrtaceae	35070	11.32	61
49	Heath	Same as 48	2554	0.82	41
51	Sedgeland	Cyperaceae, Restionaceae	499	0.16	48
119	Succulent steppe, thickly wooded	*Atriplex, Maireana*+*Eucalyptus* spp.	4486	1.44	87
121	Same, lightly wooded	Same+*Acacia* spp.	104	0.03	68
124	Same, unwooded	Samphire: *Halosarcia*	313	0.10	64
125	Playa lakes	Unvegetated	4896	1.58	30
126	Freshwater lakes	Unvegetated	123	0.04	54
128	Rock outcrops	Unvegetated	413	0.13	54
129	Drift sand	Unvegetated	645	0.21	10
44 Units		Total:	309840	100 12.3% of State	65

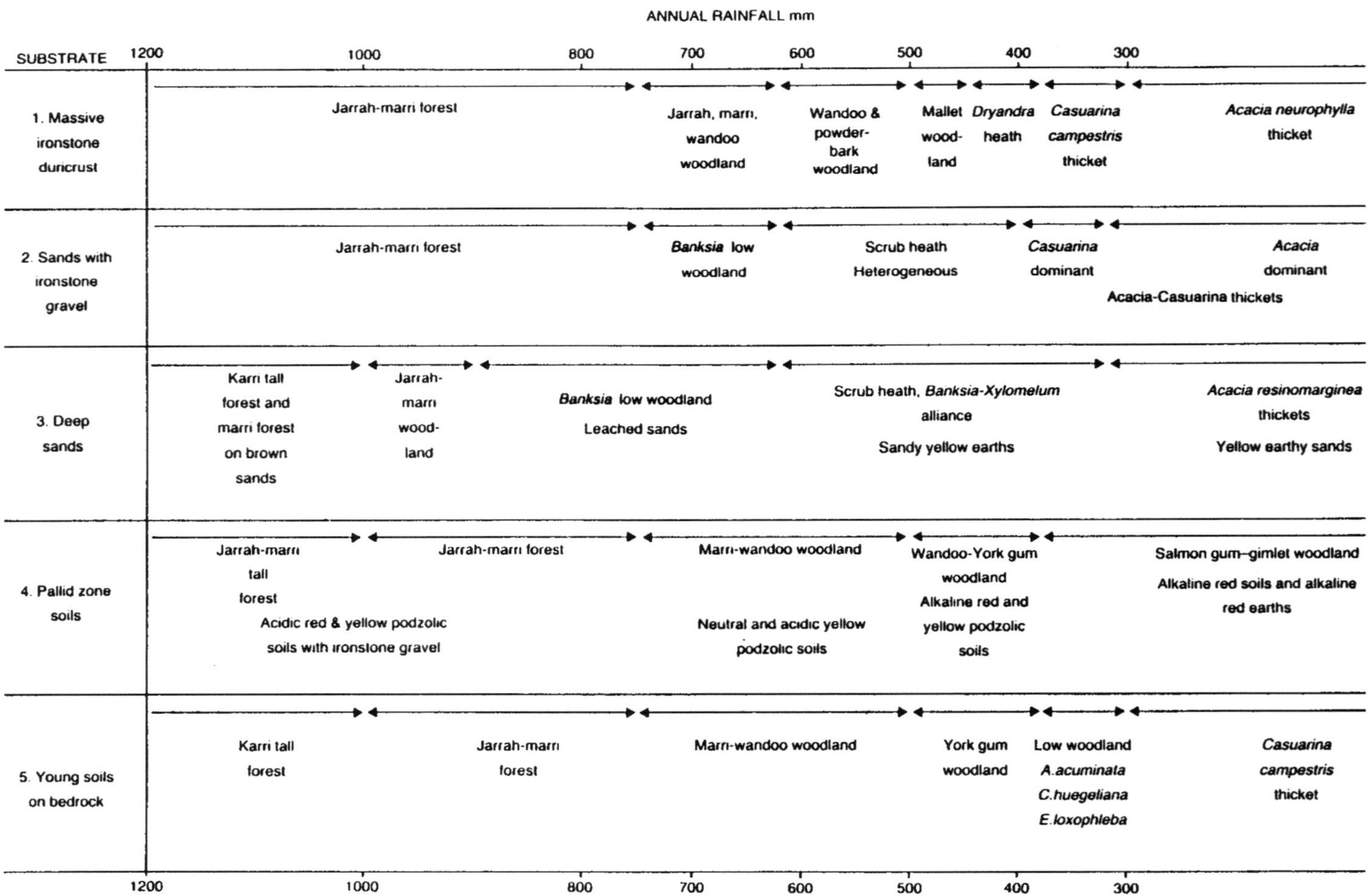

Fig. 5.3. Vegetation changes along rainfall gradients on five major substrate types in southwestern Australia. (Beard 1983). Eucalypt species indicated are: jarrah: *E. marginata*; marri: *E. calophylla*; wandoo: *E. wandoo*; powderbark: *E. accedens*; mallet: *E. astringens*; karri: *E. diversicolor*; York gum: *E. loxophleba*; salmon gum: *E. salmonophloia*; gimlet: *E. salubris*.

occupy a relatively small area, but are important ecologically. In particular, permanent and ephemeral lake systems (both salt and freshwater) are an important part of the coastal and inland landscape, and granite outcrops occur throughout the region and provide a range of local environments quite different from the surrounding areas.

5.2.3 Floristic Patterns

Southwestern Australia has a high level of plant species diversity, both within individual communities and in terms of species turnover. The flora of southwestern Australia is well known for its richness and high level of endemism. At the species level, an accurate census has yet to be completed. The pace of taxonomic and biogeographic discovery has been such that a little under three decades ago, Beard (1965) recorded just 3611 named taxa for the South West Botanical Province (Fig. 5.2a). Since then, the number of species in the region recognised in large genera such as *Eucalyptus* and *Acacia* has approximately doubled (though many names have yet to be published), and substantial increases are routine in recent revisionary work across many plant families. For example, George (1991) recognised 97 species of feather flowers (*Verticordia*), mostly from southwestern Australia, 41 of which were new species. An estimate of 8000 species in the South West Botanical Province now seems reasonable, with about 75% endemism. This compares favourably with the Cape Floristic Region of South Africa, which has 8550 named species, although these are packed into an area about a third the size (Bond and Goldblatt 1984). Both these southern mediterranean-climate regions confound the notion that global plant species richness decreases away from the tropics (Cowling et al. 1992).

There have been a number of authoritative reviews on regional and local patterns of species richness and diversity in southwestern Australia published over the past decade or so (e.g. George et al. 1979; Hopper 1979; Hopkins and Griffin 1984; Lamont et al. 1984; Griffin et al. 1990). High levels of diversity are thought to be a result of a number of factors (Hopper 1979, 1992), in particular:

1. Lack of glaciation, but major oscillations in moisture regimes over the past c. 2.5 million years.
2. Isolation from eastern Australia by the arid Nullarbor Plain.
3. Interaction of the paleotropical and temperate flora assemblages.
4. Development of a complex mosaic of landforms and soils during the Tertiary and Quaternary.

5. Interacting disturbance factors, in particular fire and drought, coupled with resource (i.e. nutrient and water) scarcity (see Hnatiuk and Hopkins 1980; Lamont 1982; Griffin et al. 1983)

The flora of the southwest is dominated by genera of woody perennials in families such as the Myrtaceae, Proteaceae, Fabaceae, and Epacridaceae. The flora has radiated greatly at the species level, while diversity at the family and genus levels is relatively low (Lamont et al. 1984). In the southwest, *Acacia* has at least 400+ species, *Eucalyptus* 300+, *Grevillea* 200+, *Stylidium* and *Melaleuca* 150+, *Hakea* and *Caladenia* 100+. This is in stark contrast to the rainforest flora of Queensland, where there is great diversity in families and genera but relatively few species within most genera. Explosive speciation occurred in the southwest in response to the environmental perturbations of the late Tertiary and Quaternary. Such speciation has been concentrated in the Transitional Rainfall Zone for the majority of woody perennial taxa (Hopper 1979; Lamont et al. 1984). Nodes of species richness have been recognised near the south coast (Stirling Range and Fitzgerald River National Parks) and in the northern sandplains north of Perth, particularly the Mt Lesueur region. While the diversity of these regions is less than in comparable regions in the Cape Province (Cowling et al. 1992) and some types of rainforest (Lamont et al. 1984), their levels of α- ,β- and γ-diversity are remarkable. For instance, Griffin et al. (1990) recorded more than 110 species in 10 x 10 m quadrats at Mt Lesueur and found that adjacent quadrats had as few as 60% of species in common, while quadrats 500 m apart could have as little as 40% in common. Lamont et al. (1984) indicated that species evenness was high in these heathland communities, with many sparsely distributed species. Similar patterns of species turnover have been found in heath vegetation further inland, but these studies also indicated that species diversity could be as high in woodland and other communities, i.e. high species richness is not restricted to the heath communities (R.J. Hobbs and L. Atkins, unpubl. data). Overall diversity in inland areas is only slightly lower than the coastal and northern areas, and this may in part be due to the level of transformation in inland parts (see Sect. 5.2.8).

At the regional level, there are major differences in the concentration of species in different genera, Lamont et al. (1984) indicated that, while genera such as *Banksia* and *Eucalyptus* are most speciose near the south coast, others such as *Acacia* are most diverse in the inland region of the wheatbelt. There are recurrent examples of allopatric replacement series of closely related, narrowly endemic species within genera. For example, genera such as *Calothamnus, Melaleuca, Hakea, Darwinia* and *Dryandra* show bimodal concentrations of species in the two major nodes mentioned above. Others are concentrated in the southern kwongan and mallee (e.g.

Adenanthos, Eucalyptus, Banksia, Leucopogon), or the Mt Lesueur-Kalbarri northern kwongan (*Grevillea, Conostylis, Lechenaultia*), or are concentrated inland through the wheatbelt (e.g. *Acacia, Verticordia*).

Although studies of variation within species have had a shorter history than those at the species level, a wealth of pattern and process has been documented (Hopper 1992). It seems likely that the concentration of species in the Transitional Rainfall Zone is a trend replicated infraspecifically in many cases. Hopper (1992) cites examples of potential divergence in ecosystem function within species. Much more research is needed, however, before the value of infraspecific variation to the maintenance of ecosystem function in southwestern Australian plants can be interpreted.

The remarkable species richness, rapid turnover of species across southwestern landscapes, and large number of rare locally endemic plants in southwestern Australia are the end-products of long and complex evolutionary processes, stimulated by environmental perturbations in the late Tertiary and Quaternary. These perturbations were most pronounced in the Transitional Rainfall Zone, the region where species, infraspecific variants, cryptic species complexes and natural hybrids are most concentrated. The evolution and persistence of so many rare locally endemic species in the Transitional Rainfall Zone flora has been interpreted as suggesting that they may have a significant role to play in dynamic environments subject to recurrent and unpredictable change (Main 1982; Pate and Hopper 1993). However, the levels of diversity at the species and infra-specific levels raise important questions concerning the link between diversity and function. For instance, do allopatric congeners or genetically and geographically distinct subpopulations of a particular species also differ functionally, or are they simply geographically separated ecological equivalents? Such levels of infra-specific variability also raise questions concerning strategies for conservation of biodiversity; maintaining representative populations of a given species is unlikely to capture all of the variation present.

5.2.4 Faunal Richness

While southwestern Australia retains elements of the mammal fauna which are now missing from the rest of the continent (e.g. Kennedy 1990), levels of faunal diversity do not, in general, match the high floristic diversity. The vertebrate fauna of Australian MTEs is characterised by a relatively low level of endemism. Many of the species which have been recorded as occurring in the region are representatives of either arid zone or more mesic faunas, and the mediterranean-climate region represents the limits of their distributions. Such species are commonly confined to the periphery of the region. This pattern reflects the transitional nature of this climatic zone between the dryer inland areas and the more mesic environments to the east and southeast.

The low level of endemism suggests that the distribution of many vertebrate species in the region may largely be a consequence of historical and biogeographic factors, rather than a result of adaptations to local ecological conditions. Much of the vertebrate fauna has broad environmental tolerances which enables it to persist under a wide range of conditions, including those presented by the marked seasonal climate of mediterranean regions. Consequently, the functional significance of vertebrate faunal biodiversity in the region is less likely to lie in specialised relationships between the various components of the biota. Rather, much of the fauna could be viewed as having a generalised role which would also apply in the other regions in which its components occur.

The low level of endemism is particularly conspicuous for birds. Only nine species are endemic to the mediterranean-climate zone (Schodde 1981), although over 190 species have been recorded in the Western Australian region alone (Saunders 1993). Of over 80 species of mammals recorded from the region, only 16 or approximately 20% are confined to it, while approximately 28% of the mediterranean-climate zone reptiles are endemic. Frogs are an exception to this trend, with 64% of the 36 local species being restricted to the region (data from distribution maps of Strahan 1983; Cogger 1986). Many invertebrate groups also show a higher degree of endemism which to a large extent reflects the dispersal characteristics of particular taxonomic groups. More vagile groups such as butterflies and grasshoppers have a higher proportion of species which extend beyond the region, whereas many terrestrial and epigaeic arthropods such as scorpions and mites have very restricted distributions (M. Harvey, pers. comm.).

An important factor to be noted concerning the Australian fauna is the absence of large herbivores The extinction of the megafauna in the late Quaternary left Australia (in general, not just the MTEs) without large herbivores and predators (Merrilees 1984; Murray 1984); see Section 5.3.1.

5.2.5 Landscape Patterns

Within the broad pattern determined by rainfall and soils, numerous smaller scales of pattern are present. Landscape heterogeneity results from the interaction of many influences (Kolasa and Rollo 1991; Milne 1991). We have illustrated how broad-scale vegetation patterns are related to landform and soils (Fig. 5.4). At a finer scale, vegetation patterning may be determined by smaller-scale variations in soil characteristics, as discussed

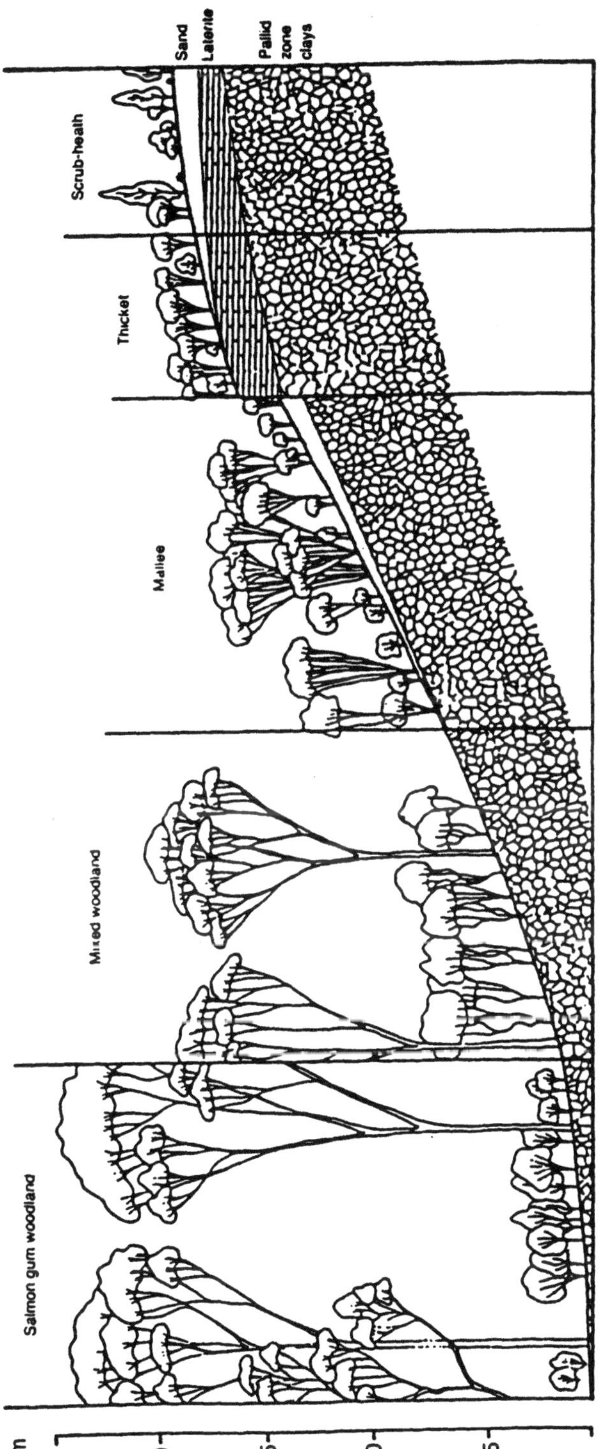

Fig. 5.4. An example of a caternary sequence from the eastern part of the South West Botanical Province, illustrating the broad determination of vegetation types by soil and landform. (Beard 1983)

for the kwongan vegetation of the northern sandplains by Froend (1987). High levels of species turnover across the landscape and patchy distribution of populations of individual species add to the complexity of vegetation patterning in this area and elsewhere (Hnatiuk and Hopkins 1980; Hopkins and Griffin 1984). This patchiness can be a result of chance dispersal events, localised disturbances or microenvironmental variation (Weiss and Murphy 1993). While some species may be restricted to particular soil types, others occur on a variety of types; for instance Hopkins and Griffin (1984) list species of *Darwinia* and *Dryandra* which are restricted to one or a few soil types, and others which are found in all soil types within its range. Similarly, while *Eucalyptus salmonophloia* is considered to occur predominantly in heavy valley soils in the Transitional Rainfall Zone and eastwards, Lantzke (1992) lists it as occurring on eight different soil types at various landscape positions. Floristic patterns within broad vegetation types can also be very complex. Thus, although the type of scheme presented in Fig. 5.3 provides a general picture of vegetation distributions, the actual landscape mosaic is generally more complex and less predictable (e.g. Hobbs et al. 1989). The effects of disturbance also complicate the landscape mosaic. The history of disturbances in a given area determines the distribution of patches of different age and successional stage across the landscape, and the disturbance regime determines the scale and pattern of variation observed; see Section 5.2.6, below.

5.2.6 Response to Disturbance

Disturbances common in Australian MTEs include fire, drought, flooding, infrequent frosts or periods of high summer temperatures, severe summer storms, localised soil disturbance by animals, treefalls and insect outbreaks (Hnatiuk and Hopkins 1980; Grubb and Hopkins 1986; Fox and Fox 1986; van Schagen et al. 1992; Main 1992). As in other mediterranean regions, fire, in particular, has received much attention as a force shaping the vegetation (Gardener 1957; Bell et al. 1984, 1989; Ford 1985). There has been considerable debate about the natural fire frequency in the ecosystems of southwestern Australia. Environmental conditions for wildfires have existed in southwestern Australia since the late Tertiary when the climate changed from humid tropical to much drier (Singh et al. 1981). The occurrence of fire has been recorded with certainty in forest vegetation over the past 5000 years, caused both by lightning strikes and Aboriginal firing (Churchill 1968; Hallam 1985). Discussion of the "natural" fire regime cannot ignore the impact of Aborigines, who were thought to set fires in most vegetation types, with the frequency of burning related to the density of the vegetation and its productivity in food terms. Hallam (1985) argues that the coastal plain and non-lateritic areas in the Avon Valley would have been burned more frequently than the lateritic uplands which support the jarrah forest. On the other hand, fuel build up is most rapid in the forest communities, and slower in the drier woodland areas (Bell et al. 1989), probably due to a combination of slower production of litter and increased utilization by termites in the woodland areas. Work by Lamont and Downes (1979) indicated that forest fire intervals ranged from 2-150 years, but that fire frequency had increased dramatically since European colonisation in 1829. Species diversity in the jarrah forest falls off dramatically after 30 years without fire (Bell and Koch 1980), but this is due to the disappearance of short-lived ephemerals, which may still be present as seed in the soil. Hobbs and Atkins (1990) have also indicated that species diversity declines after long fire-free periods in *Banksia* woodland on the coastal plain. Hopkins (1985) has discussed the relationship between fire frequency and rainfall, suggesting that frequencies are likely to be lower in areas with lower rainfall because of the longer time taken for shrub and tree species to reach maturity. For these lower rainfall areas at least, it is becoming apparent that fire is only one component of a complex disturbance regime which also includes droughts, floods and windstorms, all of which have significant impacts on the vegetation (C. Yates, unpubl. data).

Disturbances such as fire and aseasonal drought provide a layer of variation in the vegetation which can mask or override edaphic factors (Hnatiuk and Hopkins 1980; Hopkins and Griffin 1984), and Hopkins and Robinson (1981) have shown how fire can alter the vegetation from one structural type to another. In addition to the natural disturbance regime, human disturbance is now an important component of most MTEs. Human activities such as vegetation removal, timber or soil extraction and introduction of non-native grazers produce a further overlay of variation on the landscape, as well as altering the natural regime (Hobbs 1987; Scougall et al. 1993; Saunders and Hobbs 1992). Introduced and invasive organisms can significantly alter vegetation composition and structure; for instance, Scougall et al. (1993) illustrated how stock grazing significantly alters woodland composition by removing woody understorey and enhancing invasion by non-native annuals. Hobbs and Atkins (1991) further indicate that introduced annuals can prevent regeneration of woody species. The introduced pathogen *Phytophthora cinnamomi* has marked effects on the composition of many vegetation types by preferential infestation of particular groups (Dell and Malajczuk 1989; Wills 1993; Hobbs 1994).

Individual disturbance types can produce different responses depending on factors such as environmental variations within the disturbed areas, season of disturbance, weather characteristics following the disturbance, and interactions with other disturbances (Grubb and Hopkins 1986; Hobbs and Atkins 1990; Lamont et al. 1991, 1993). Majer (1992) has illustrated the effect of timing of disturbance in relation to the timing of processes such as litterfall, decomposition, growth and flowering. The seasonal demarcation of these processes has a strong influence on the resilience of natural ecosystems to disturbances at different times of year. There has been some consideration of whether different disturbances result in similar vegetation responses, and it appears that some generalised trends can be detected within particular vegetation types, but that variations occur depending on the size and intensity of the disturbance (Grubb and Hopkins 1986; Fox and Fox 1986). Hobbs (1994) illustrated how drought and invasion of heathland by *Phytophthora cinnamomi* produced similar patterns of damage since plant groups showed similar susceptibilities to both forms of disturbance. Clearly, however, a different vegetation response would be expected to fire or other disturbances which affected all species.

Fox and Fox (1986) compared the resilience of plant and animal communities in different mediterranean-type regions and also in non-mediterranean regions. They concluded that mediterranean regions are in general more resilient than non-mediterranean regions, but that there is little evidence for differences in responses between different mediterranean regions. Majer (1992), on the other hand, produced data on ant recolonisation of rehabilitated mine sites which suggested that the ant community was less resilient in southwestern Australian MTEs than in non-mediterranean areas elsewhere in Australia. Sites with lower rainfall within the mediterranean region were also less resilient than those with high rainfall. The question of the relative resilience of MTEs thus still appears to be open, at least for Australia. Grubb and Hopkins (1986) have discussed the likely effects of species richness on resilience. They suggested that increasing species richness should increase elasticity up to a point, but that very species rich communities may not have any significantly greater elasticity since many sparse species will be present. Indeed, with many species present, the conditions following a disturbance may not be suitable for them all to return. Stochastic events are likely to have a large influence on sparse species, and hence the community overall will be more vulnerable to species loss and hence less resilient.

5.2.7 Response to Environmental Constraints

The majority of plant species within heath communities are C_3 and there are few instances of C_4 or CAM plants (Winter et al. 1981). Many vegetation types have open canopies, and dense shade is rarely encountered except in some shrubland types and the tall forests. Hence in many cases, adaptations to high levels of light and heat may be more relevant than adaptation to shade (e.g. Hellmuth 1971). Shortages of water and nutrients more often limit the growth of sclerophyll vegetation in mediterranean-type regions which are, by definition, limited by regular periods of water shortage through each summer. Such vegetation generally also occurs on soils of low nutrient status (Specht and Rayson 1957a; Groves 1981). There is a diversity of adaptations shown by sclerophyllous shrublands to low water and nutrient levels (Lamont 1982, 1992b). A corollary of this is that competition will frequently be for water or nutrients rather than light. For instance, unlike dominant trees in the Mediterranean Basin, California and Chile, eucalypt canopies are generally open and hence let abundant light through to the ground layer. However, eucalypts are often very efficient competitors for water, and hence exclude other species in this way (e.g. Lamont 1985).

In areas with a mediterranean climate, rain falls mostly in winter, summers are usually hot and dry, and surface soil moisture deficits strongly limit plant growth. Because most of the dominant species in sclerophyll vegetation tend to grow mainly in summer - especially the proteaceous and eucalypt floristic groups - at a time when temperatures are high but rainfall is low, adaptations to ensure the availability of soil moisture to the shoot over the summer period are critical. Considerable variation in aboveground diversity in shoot and root form can be observed in southwestern Australian vegetation. For kwongan vegetation, Pate et al. (1984a) examined 29 growth and life form characteristics for 429 species and found a complex array of types, including some, such as the stilt plants, which are virtually restricted to this area (Pate et al 1984b; Pate 1989). A diverse array of shoot forms are found, many of which act to collect and channel water down stems, leading to stem flow being a significant pathway for water movement in some communities (Nulsen et al. 1986).

Rooting systems also reflect the need to harvest water and nutrients efficiently. This is illustrated by the root system of *Eucalyptus marginata*, which consists of a primary tap root from which an extensive horizontal system of first order laterals develop and grow up to 20 m from the lignotuber (Abbott et al. 1989). From this, riser roots ascend close to the surface and branch in the organic surface horizon (where most of the nutrients reside) to form pads of mixed ectomycorrhizal and non-mycorrhizal short roots. Also produced from the laterals are sinker roots which descend vertically and occupy semi-permanent vertical channels in the substrate. These root channels, or preferred pathways, are occupied by successive generations of trees (Abbott et al. 1989). The array of root morphologies present is illustrated by the kwongan vegetation (Specht and Rayson 1957b; Dodd et al. 1984) (Fig. 5.5). Some species have deep taproots that may penetrate as far as 2 m to the clayey B horizon with a moisture regime very different from that of the sandy top soil. There is also a group of shallow (<0.5 m) tap-rooted species. Some deep

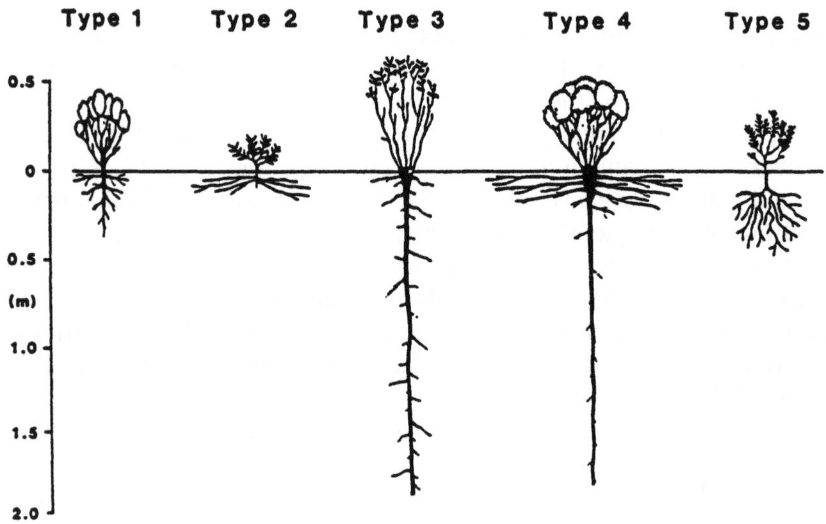

Fig. 5.5. Examples of major types of root systems in shrubs from the Swan Coastal Plain, Western Australia, illustrating taproots, surface root mats, and combinations of these. (After Dodd et al. 1984)

tap-rooted species also produce extensive lateral root systems which develop in surface soil; these roots arise from different underground organs such as caudices (in the case of *Xanthorrhoea*), stem bases, rhizomes or tubers. The great diversity of root form in sclerophyll shrublands in southern Australia results in exploitation of soil moisture down the full profile. Deep roots in summer tap the soil water stored in the heavier-textured B horizon and lateral fibrous roots tap the sandier A horizon in autumn and spring. A concentration of fine roots at the soil surface and even penetrating the litter layer may also trap scarce moisture coming from the occasional but irregular summer rainstorms that are a feature of most of southern Australia. Sometimes the concentration of surface roots may be most developed in species at the drier limits of their distribution; e.g. as in the distribution of mallee eucalypts which are confined increasingly to sand ridges in central Australia as rainfall decreases inland.

The array of root forms is even greater with respect to nutrients because of their role in fixing nitrogen, both symbiotically and non-symbiotically, and increasing surface areas for the absorption of nutrients from the less impoverished litter and upper soil layers. This enhanced absorption layer is especially important in the functioning of proteoid roots and mycorrhizae (Lamont 1982). Proteoid roots are produced in over 500 species of Proteaceae in southwestern Australia, stimulated by free-living soil bacteria.

Mycorrhizal associations probably involve hundreds of fungal species, and both vesicular-arbuscular (VA) and ectomycorrhizal associations form. Few local studies of the importance of VA mycorrhizae have been conducted, but it is known that many major tree and shrub species are ectomycorrhizal (Lamont 1992b). One functional group of species that is significantly developed in southern Australian sclerophyll communities is the symbiotic nitrogen fixers, especially the acacias and other legumes. Acacias are especially abundant following fire, where short-lived species dominate for a few years before being replaced by other slower-growing species. Acacias and other legumes possess root nodules containing nitrogen-fixing bacteria, *Bradyrhizobium* spp. Other nitrogen-fixers present include actinomycetes (*Frankia* spp) which stimulate nodule formation in Casuarinaceae, and cyanobacteria (*Nostoc* spp.) which modify coralloid root morphology in the cycad *Macrozamia riedlei*. (Lamont 1984; 1992b). Lamont (1982, 1992b) has suggested that the extent to which specialised forms of nutrient uptake operate depends on the levels of soil nutrient availability. Generally, specialised roots (nodules, mycorrhizae) will be active only at low nutrient availabilities. As nutrient levels increase, "normal" roots will take over.

5.2.8 Human Transformation

The transformation of much of the Australian MTEs from predominantly natural vegetation to predominantly agricultural has occurred in a little over a century, compared with 8000-10000 years in the Mediterranean Basin (Hobbs and Hopkins 1990). The rapid transformation has resulted in a highly fragmented landscape, with many small remnants of native vegetation embedded in a predominantly agricultural matrix (Hobbs 1992b; Saunders and Hobbs 1992; Hobbs et al. 1993) (Fig. 5.6).

This transformation, together with the introduction of domestic and feral animals and alien plant species, has had dramatic effects on the biota, both in terms of loss of habitat and extinction of species. Over 93% of the original vegetation was removed during agricultural development of the wheatbelt, with over 55% being removed after 1945 (Hobbs et al. 1993). Because vegetation types are closely allied to particular soil types, some plant communities were much more extensively modified than others. Woodlands on valley bottom soils, in particular, were cleared and utilised preferentially.

Beard and Sprenger (1984) have measured the total area of each physiognomic unit and also the percentage of that which has been cleared or alienated (Table 5.2). Some 65% of the Southwest Botanical Province has been cleared, although this proportion is much higher for certain

Fig. 5.6. Human transformation in the agricultural area of southwestern Australia has removed up to 97% of the original extent of some vegetation types, leaving native vegetation in small, often isolated remnants and as strips along road verges. (Photo R.J. Hobbs)

vegetation types. Two types appear to have virtually disappeared while others have been cleared of over 90% of their area. The main threats to the continued retention of these types, which are situated in the most densely populated part of the State, are agricultural, urban and industrial use. There are also less obvious threats to these and other types, including weed invasion and inappropriate fire regimes.

The result of clearing and fragmentation is the known extinction of 24 plant species (Leigh et al. 1984) and the possible unknown loss of many others. The agricultural area of Western Australia now has the highest numbers of rare and endangered plant species in Australia (Hartley and Leigh 1979; Leigh et al. 1984; Hopper et al. 1990). Clearing not only changed the abundance of native vegetation, but also changed the patterns and availability of the different vegetation types present. Grazing and disturbance by livestock and rabbits within remaining vegetation patches further alters the vegetation and results in changed soil composition and structure. Introduced plant species have spread into remnant areas where they can affect vegetation composition and regeneration (Hobbs and Atkins 1991; Scougall et al. 1993).

Removal of the native vegetation and the introduction of alien species (particularly rabbits and foxes) have resulted in the depletion of the native fauna. Kitchener et al. (1980) noted that only 12 of the 43 species of native mammal recorded from the wheatbelt at European settlement were still common. In fact, 9 of the original 43 species are now extinct on the Australian mainland, and mammals are still being lost from the wheatbelt (Hobbs et al. 1993). The avifauna is currently demonstrating the same pattern of species decline as the mammals (Saunders 1989; Saunders and Curry 1990; Saunders 1993). For example, of the 192 species of birds recorded from the agricultural areas since the turn of the century, nearly 50% have declined in distribution and/or abundance and at least 3 species have disappeared from the area altogether (Saunders 1993). If current land use continues, this number will certainly rise.

There is little information on the changes in status of other vertebrates as a result of development for agriculture. There is even less information on the effects on invertebrates. The few studies available indicate that development for agriculture and the subsequent agricultural practices have reduced the diversity of the soil macrofauna and significantly altered its effects on soil composition, nutrient cycling and water movement (Abbott et al. 1979; Abbott and Parker 1980; Lobry de Bruyn 1990; Hobbs et al. 1993). There are also few data available on changes in soil microfauna and flora and the potential impacts of these changes on pedological and other processes.

The jarrah forest has not been subjected to the same rapid fragmentation, but has, nevertheless, undergone dramatic modification since European settlement due to unsustainable logging of old growth, modification of fire regimes, mining activities, and the introduction of the pathogen *Phytophthora cinnamomi* (Bartle and Slessar 1989; Bell et al. 1989; Dell and Malajczuk 1989; Havel 1989; Havel et al. 1989). Mining activities also occur in some of the remaining extensive areas of non-forest vegetation, and *Phytophthora cinnamomi* is now an important problem affecting many other plant communities and threatening to become a major agent of extinction (Wills 1993).

5.3 Ecosystem Function of Biodiversity

Because of the inherent variability of mediterranean environments, the components performing each function might be expected to have characteristics which permit the performance of those functions under all conditions likely to be experienced. These requirements might be satisfied either by widely tolerant species functioning under an array of conditions, by species functioning only under a narrow environmental range, or by those with long life cycles which function only under short periods of favourable conditions. When history and environment have selected for widely tolerant species, then such an ecosystem will have few species capable of filling each functional role, e.g. under conditions of annual or regular burning when only organisms capable of withstanding extremes of light, heat and water availability will persist. Systems with the second or third set of characteristics should have species-rich assemblages and many distinct growth and activity periods, thus providing ecosystem functions under a wide array of environmental conditions. A system rich in species will have a potential pool from which substitutes able to fill essential ecosystem functions under a range of environmental conditions can be found. When substitution occurs after loss of species, the system may maintain appearances and functions for a considerable period of time. This will depend on how successful the substitution has been and whether environmental change has affected the functional efficiency of the replacement species.

There is thus the possibility that a different species can fill the same functional role as another when environmental conditions change. That this is likely is evident from the known capacity of dominant tree species to shift in distributions in response to climate changes (Churchill 1968), presumably irrespective of the influence of soil types in determining distributions under particular climatic conditions (see Sect. 5.2.2). In addition, Grubb and Hopkins (1986) point out that the species complement in a given area need not be the same after a disturbance as it was before the disturbance. For instance, intense fires can induce a marked shift in species composition, which persists presumably at least until the next fire (Grubb and Hopkins 1986). What are the implications of this for the role of diversity in ecosystem function? If the species complement changes through time, does that imply that different species are fulfilling the same role at different times? If so, are some species then functionally equivalent and hence substitutable? This can be interpreted in two ways - either all species are not needed in the system since several species can do the same thing, or the observed diversity has an important role because different species can respond to different sets of events. When one species is disadvantaged, another can take over. The open question is how reversible the changes are - do species which miss out one time round have another chance next time, or are they lost to the system? To some extent this brings in the question of whether ecosystems can have multiple alternative states, with transitions between states determined by particular combinations of climatic and disturbance events (Hobbs 1994; Hobbs et al., Chap.1, this Vol.). This question is in many ways central to the debate on the role of diversity in ecosystem function. If multiple states are possible, then not only the diversity of species in particular states needs to be considered, but also the total pool of species involved in all possible states. The maintenance of ecosystem function may depend not so much on the retention of all species per se, but on the role of these species in the retention of the capability for particular states to develop in response to particular environmental and/or disturbance cues. The question of resilience is not particularly well covered in recent discussions on the role of diversity in ecosystem function, which deals mainly with ecosystem fluxes. For instance, Vitousek and Hooper (1993) propose a set of alternative ecosystem response functions to increasing diversity, which may be linear, asymptotic or null; however, the shape of a curve based on an individual ecosystem flux may be significantly altered if ecosystem resilience requirements are factored in.

Species have been lost following environmental changes in the recent geological past, and more recently following land settlement and the introduction of non-native species. Consideration of the outcome of these events suggests that persistence of ecosystems may be affected by the causes of the extinction, the size of the pool from which substitutes may be found, and the space available for range changes. Here we examine various aspects of southwestern Australian ecosystems which may have relevance in the discussion of the role of diversity in ecosystem functioning. Firstly we discuss the consequences of the loss of the diverse megafauna present in the Pleistocene, and then we examine the consequences of extensive replacement of complex natural ecosystems with simplified agricultural systems. We then present a number of case studies which illustrate the complexity and problems involved in tackling the question of the ecosystem function of diversity, and provide examples concerning the potential use of the concepts of functional groups and keystone species raised by Hobbs et al. (Chap. 1, this Vol.).

5.3.1 Environmental Change and the Loss of Megafauna

Main (1992) has discussed the impacts of past extinctions on ecosystem structure and function in southwestern Australia. The fossil record suggests that from the late Pleistocene through the whole of the Holocene environmental change (Hope 1984) and extinctions have been occurring, so that plant (Churchill 1968; Martin 1973; Kershaw 1981), and animal (Lundelius 1983; Merrilees 1984; Murray 1984), distributions have expanded and contracted as the climate has oscillated from wet to dry periods. The present fragmented distribution of some moist habitat plants and invertebrates suggests that drying and fragmentation of distributions was continuing at the time of settlement. Nevertheless, the continued presence of species-rich assemblages indicates that under the prevailing conditions ecosystems continued to function, presumably following a series of substitutions as species became extinct. In reality the situation is more complicated. The present composition of the biota gives few clues as to how ecosystems were reorganised as a consequence of the extinctions, especially following the loss of the megafauna, which must have led to significant changes in both the structure and function of vegetation.

The megafauna consisted of large to very large animals ranging up to 1500 kg in weight and a height of 2-3 m (Murray 1984), and included large macropods in the genera *Macropus, Protemnodon, Sthenurus* and *Procoptodon*; large quadrupeds such as the monotreme *Zaglossus* (30 kg); the diprotodontids *Palorchestes, Zygomaturus, Diprotodon*; koalas, *Phascolarctos*; and wombats, *Phascolonus*. Emus were more diverse and the size range greater than in present-day members of the species (Rich and Van Tets 1984). Reconstructions and maps of fossil distribution are given in Murray (1984, 1991), Horton (1984) and Rich and Van Tets (1984). Fossil deposits in caves in southwestern Australia indicate that a diverse megafauna was present in areas now experiencing a mediterranean-type climate (Balme et al. 1978; Merrilees 1979; Porter 1979; Murray 1991). Anatomical reconstructions suggest that the feeding habits of the megafauna ranged from tall browsers to cursorial grazers, and predators such as *Thylacinus* and *Thylacoleo* (75-100 kg) appear to have been widespread.

It is clear that in southwestern Australia there was a fauna capable of browsing and grazing above- and belowground parts of plants of all ecosystems. Such a comprehensive pattern of disturbance ranging from canopy destruction to soil disturbance must have afforded many opportunities for maintaining a diverse flora adapted to regenerate in disturbed patches of various sizes. Even more significant was the potential for browsers of all sizes to affect the structural composition of any post-disturbance (including post-fire) regeneration that occurred. Grubb (1992) further argues that the extinct megaherbivores are likely to have had an ability to survive on material of very low nutritive value similar to that seen in present-day marsupials (Brown and Main 1967; Kinnear and Main 1975, 1979). The megaherbivores thus could have been a significant influence on the sclerophyllous vegetation in southwestern Australia in a way in which they were unlikely to have been in other mediterranean-type regions where the megafauna lacked such an array of adaptations. Grubb (1992) goes on to postulate that the high incidence of spininess in southwest Australian heath species (27%; Pate et al. 1984a), compared with South African fynbos species (1-2%) is related to this ability of the extinct browsers to survive on plant material of very low nutritive value. Spininess could then be seen as a defence mechanism which was necessary in southwestern Australia, but not in South Africa.

We do not know how loss of megafauna influenced ecosystem structure and function, and the effects of the loss ate liable to be confounded by the effects of changes in climate, and the influence of human activities. The decline of the megafauna has been attributed to the activities of Aboriginal populations who colonized the area (Merrilees 1968), and we have also noted that Aboriginees may have significantly altered fire regimes (Hallam 1985). In terms of ecosystem functions, therefore, it is possible that fire became the major agent of vegetation change and nutrient recycling. It could be argued that smaller vertebrates may also have taken over the major herbivorous roles, and the combination of this effect and increased fire occurrence may have resulted in little net change in ecosystem function. Certainly, we can postulate that loss of the megafauna was likely to change the competitive relationships between plant and animal species. It is possible that substitution of species took place to yield ecosystems with a different composition and hence identity. Few sites record Late Pleistocene plant communities that could have given rise to the present vegetation patterns (Hope 1984). As the megafauna became extinct, and climate and fire regimes changed, the surviving species apparently reorganised into new assemblages and ecosystems within which new interactions and feedback loops developed. This suggestion has important implications for the discussion of the importance of biodiversity for ecosystem functioning. Clearly, significant ecosystem components (in the form of the megafauna) were lost, and plant communities disassembled and reassembled in different configurations. This could have resulted in significant changes in ecosystem function, in that ecosystems developed with different structures and components. On the other hand, functioning ecosystems persisted, albeit in a different form from those present before. The coincidence of changing faunal communities with changed fire regimes makes it difficult to interpret ecosystem responses to biodiversity directly; changing disturbance regimes may have been just as, if not more, important.

The same may be true today, and any attempt to attribute system changes to diversity changes must take other, possibly confounding, factors into account. Undoubtedly further community changes are to be expected in the future, but now fragmentation of the natural environment has restricted the ability of both plants and animals to be shuffled over the landscape as in the Late Pleistocene and Holocene. Substitution may still be possible but its scope in any locality is very limited and, as potential substitutes are locally lost, radical change in function seems more likely. Consequently, attempts at preservation or retention of assemblages as they are at present is likely to be unachievable without substantial manipulative management.

5.3.2 A Comparison of Natural and Agricultural Systems

Since much of mediterranean-climate Australia has been extensively modified by human activity, it is useful to compare natural with human-modified systems. A consideration of the function of diversity might also benefit from a comparison of adjacent very diverse and very simple systems. Here we contrast a diverse natural ecosystem with an extremely simplified agricultural system. The major difference between the two is a dramatic simplification and reduction in the number of functional groups present in the agricultural system (see also Swift and Anderson 1993). In the natural system all the required functions (and associated secondary functions, see Hobbs et al., Chap. 1, this Vol.) are carried out by the functional groups present, and energy, water and nutrient transfers are thus closely interlinked (Hobbs 1992c). Natural ecosystems also exhibit considerable resilience to disturbance (See Sect. 5.2.6).

In contrast, in the agricultural system prevalent in the mediterranean regions of Australia, the functional groups present carry out only a small proportion of the sub-functions involved in water, energy and nutrient transfers which were present in the natural system (Hobbs 1992c). Hence the three major functions are poorly performed and the reduced diversity of functional groups present means that there are fewer options or pathways for each function. Thus for instance, the predominance of annual crops and pastures means that energy capture is very inefficient, since there is no plant cover for half the year. Water uptake and evapotranspiration are reduced compared to perennial-dominated communities, since rooting patterns and growth periodicities are altered. There is no longer a diversity of rooting depths and modes which take up water from different soil layers and at different times of year. This leads to rising water tables and the transport of soil-stored salt to the surface (McFarlane et al. 1993). Nutrient transfers are also significantly different since the plants mostly lack specialised roots or symbionts, decomposer communities are greatly simplified and increased leaching and soil erosion lead to greater exports from the system (Hobbs et al. 1993; Lefroy et al. 1993a). The agricultural system is thus very 'leaky' compared with the natural system, and net flows of energy, water and nutrients in and out of the system are considerably greater (Hobbs 1993a; Swift and Anderson 1993). The agricultural system also lacks resilience and is vulnerable to disturbances such as drought, flooding or insect attack.

The same considerations are valid for other human-modified systems. For instance, it is well established that timber extraction, especially in clear-felling, leads to greater nutrient and water export from forest systems (Bormann and Likens 1981; Schofield et al. 1989). Mining activities such as bauxite removal in the jarrah forest and mineral sands extraction from the northern sandplains of southwestern Australia also completely disrupt ecosystem functions by removing all biota from the extraction site. However, in all these cases, these effects may result from the simple removal of structure and/or biomass rather than the reduction in diversity per se. Certainly, agriculture, logging and mining all result in a simplification of structure and composition, which results in reduced biodiversity if we include structural aspects in the definition. The question remains whether the removal of individual species is important. A further complication is that simplification is often (or usually, in the case of agriculture) accompanied by replacement by aliens. This does not necessarily matter at the level of functional groups, but should be considered in discussions of species richness per se. There is thus a case for parallel consideration of natural communities with differing levels of species richness (see Sect. 5.4.5).

Because of the degree of fragmentation and modification of natural ecosystems experienced in mediterranean Australia, there are now few, if any, areas which are capable of maintaining functional 'integrity'. Individual reserves or remnants will not, in themselves, maintain ecosystem functions at 'normal' rates because of the increasing influence of external factors arising in the surrounding heavily modified landscape (Hobbs 1993a). The surrounding production-oriented system will continue to exert a large influence on the natural system. This is due to the alteration of the hydrologic regime over the whole landscape which resulted from the replacement of perennial vegetation with annual crops and pastures (Groves and Hobbs 1992; McFarlane et al. 1993), increased nutrient transfers in water and wind erosion, and continual influx of non-native biota.

In the agricultural system, the major management problems arise because of poor performance of ecosystem functions, due to the absence of functional groups and hence important sub-ecosystem functions, as discussed above. Land degradation problems of salinisation and erosion are directly related to the poor ability of the agricultural system to capture energy, use water and retain nutrients. Management of the agricultural system thus requires an increase in the numbers of functional groups which carry out the major functions. An approach to this has been developed by Lefroy et al. (1993a,b) which is based on increasing the amount of perennial vegetation in the agricultural landscape. The effect of this is to push the agricultural system back in the direction of the natural system and to tackle imbalances in energy, nutrient and water transfers simultaneously. The approach is thus one of increasing the complexity of the landscape (which can be viewed as increasing biodiversity at this scale) by reintroducing functional groups which had been removed during agricultural development. It remains to be seen whether the reintroduction of one set of functional groups (i.e. deep rooted perennials of varying types) will allow the establishment of other groups (e.g. decomposers) and the overall restoration or enhancement of ecosystem function (Saunders et al. 1993). Options for restoring ecosystem function range from the introduction of one perennial species, which could be native or non-native to the area, to an attempt to reconstruct a multi-species natural community with all structural components and functional groups present. The relative merits of each of these for the agricultural system will depend on how they affect the major ecosystem processes. However, if we are interested in enhancing the conservation value of the landscape too (i.e. through increasing habitat availability or providing corridors for movement), the choice of methods may have a significant impact (Lefroy et al. 1993b).

The change from indigenous to introduced vegetation has had, and continues to have, profound effects on ecosystem function in southern Australia. This change has generally been from perennial to annual, from deep-rooted to shallow-rooted and from species-rich to species-poor (i.e. crop and pasture) plant communities. The options available for water and nutrient capture in a heterogeneous and uncertain environment have thus been reduced. Reduction in landscape complexity also has implications for the conservation of species diversity. It is likely that some species require certain combinations of disturbance and environmental conditions for establishment and/or survival. Animal species may also be dependent on vegetation heterogeneity (e.g. Saunders and Ingram 1987; Lambeck and Saunders 1993). Clearly, therefore, these aspects of diversity are important over and above considerations of species diversity per se. Conservation in the mediterranean regions of Australia thus needs to consider functional and landscape diversity, as well as species or genetic diversity. A species-oriented approach is no longer enough, and overall ecosystem and landscape management will be required for the retention of biodiversity.

5.3.3 Loss of Multiple Ecosystem Components: Mistletoes in Eucalypt Woodlands

Clearing for agriculture has reduced the area of woodlands dominated by *Eucalyptus wandoo* and *E. salmonophloia*. Both species regenerate after fire or other disturbance (Burrows et al. 1990; Norton et al. 1994; C. Yates, unpubl. data) and stands of all ages can be found in even small remnants. The architectural structure of these woodlands provides the habitat for mistletoes (*Amyema* spp.) which can be common and abundant on both species (Kenneally 1973; Norton et al. 1994). As with all parasites, these probably reduce the vigour and life expectancy of their host by killing boughs and so altering canopy architecture. Local anecdotal evidence suggests that, in the past, brush-tailed possums (*Trichosurus vulpecula*) were possibly a major herbivore of mistletoes. This is supported by the observed effects of herbivory on mistletoes by possums in eastern Australia and introduced possums in New Zealand (Norton 1991; Reid and Yan 1993). Thus there was probably a strong interaction between the mistletoe and the folivores which ate them along with eucalypt leaves and other foliage. This interaction continued long after settlement and even in the presence of foxes while the poison plants (*Gastrolobium* spp.) containing monosodium fluoroacetate were abundant. Native fauna possess varying degrees of tolerance to monosodium fluoroacetate, while non-native species are highly susceptible to it (King et al. 1978; Mead et al. 1985). Following the removal of poison plants to protect livestock, foxes soon eliminated possums. It might be expected that, in the absence of the herbivore, mistletoe would become abundant and kill host trees, as in eastern Australia (Lewis 1991; Reid and Yan 1994). This has not happened.

The reason for this seems to be that other factors have conspired to limit mistletoe abundances (Norton et al. 1994). Mistletoes are widespread, but in the agricultural areas their occurrence is localised, contagious and obviously related to a source of infection (Lamont 1985). Infection is commonly mediated by the mistletoe bird (*Dicaeum hirundinaceum*) which eats the fruits and disperses the plant by defecating the sticky seeds on to boughs of trees and shrubs where they germinate and establish. The rate of passage is short, 20-40 min or slightly longer (Keast 1958; Barlow 1981). Barlow (1981) believes dispersal is most likely over short distances with long-range dispersal only if seeds stick to feathers. The mistletoe bird is nomadic but tends to breed where there is a source of fruit. This period of residence might be expected to enhance re-infection of old hosts and infection of new ones. Re-infection is a prerequisite for the persistence of mistletoe because individual plants have a life expectancy of less than 15 years (Barlow 1981).

Reid (1987) argues that mistletoe in Australia evolved in relation to autochthonous birds such as honeyeaters (Meliphagidae) and not the mistletoe bird which arrived in the Pleistocene. Honeyeaters, which might be considered as substitutes for the mistletoe bird, utilise the shrubby understorey and have different patterns of movement to mistletoe birds. Lamont and Southall (1982) showed a difference in density of mistletoes on *Acacia acuminata* along a road verge relative to the adjacent nature reserve and interpreted this as visitation pressure by bird dispersers using the road verge as a corridor. On the other hand, Norton et al. (1994) found no mistletoes in corridors of *E. salmonophloia*, and interpreted this as a lack of birds using corridors for long enough for seeds to be deposited. Saunders (1993) has shown that both mistletoe birds and many honeyeater species have declined greatly in abundance following clearing for agriculture.

Prior to settlement woodlands were extensive. At present they are small and by chance may lack sufficient mistletoe to attract seed vectors regularly either in a nomadic or breeding phase. In addition, stock grazing removes understorey shrubs, compacts the soil and alters water relations, resulting in reduced mistletoe abundance (Norton et al. 1994). One might predict a high rate of extinction of mistletoe in remnants and a low chance of re-infection should an adjacent infected remnant be distant from that suffering local extinction. The observed contagious distribution of mistletoe and a low rate of infection of regenerating trees suggests that this may be so.

The larvae of two butterflies, *Ogyris* sp. (Lycaenida) and *Delias aganippe* (Pieridae), feed on mistletoe. Formerly widespread and sometimes common, their numbers have declined in tandem with the decline in abundance of mistletoe. Pollen transfer and gene flow between flowering shrubs and herbs formerly visited by these butterflies presumably no longer occurs or occurs much less frequently.

This case study illustrates the loss of various components of the woodland system - arboreal herbivores, mistletoes, mistletoe-dispersing birds, understorey shrubs and invertebrate herbivores. There has been a complex set of changes to the woodland system as a result of fragmentation and its effects may not be readily apparent. If we were to examine the woodland ecosystem as it exists today, we might have little idea that these various components and interactions had been lost. A woodland ecosystem still exists, but without the secondary functions carried out by the various components that have been lost. The loss of these functions may not have any obvious effect on the woodland ecosystem for quite some time, since the trees are long-lived. If we wish to investigate the importance of the loss of any particular component, we must also be aware of the importance of other components. We also need some historical background which allows us to determine how the ecosystem reached its current state. Determining which components are the important links or the "keystones" in the system is liable to be very difficult.

5.3.4 Complex Interactions: Cockatoos, Banksias and Grubs, and Marsupials and Mycorrhizae

Cockatoos play an important role in enhancing seed set in *Banksia tricuspis*, a species of the northern sandplain. Most (65%) flower heads are destroyed by wood-boring grubs of the moth *Arthrophora* sp., Tortricidae (S. van Leeuwen, unpubl. data; Lamont 1992b). Carnaby's cockatoo (*Calyptorhynchus funereus latirostris*), and to a much lesser extent long-billed corellas (*Cacatua sanguinea*) feed on the grubs. In consuming

84% of these insects they make a 'mistake' only 6% of the time, i.e. breaking open extra heads that do not contain the grubs, while missing only 16% of those that contain them. The cockatoo thus acts as an agent of biological control, presumably increasing the fitness of this banksia by enabling it to set extra seeds in subsequent crops through reduced abundance of the adult insects. The cockatoo's effect is largely indirect. The moth has a direct negative effect on reproduction of *B. tricuspis*. The effects of removal of omnivorous birds on adult *B. tricuspis* populations would not be evident for hundreds of years, although the stands would function differently immediately due to a population explosion of the moths, reduced food for pollinators, negligible seed set or release and hence lack of post-fire seedlings. Carnaby's cockatoo is declining in many parts of its former range (Saunders 1990), and its loss may have important flow-on effects to the plant community. Other faunal species may have equally important, and as yet unknown, influences.

This is illustrated by a further example involving a marsupial, a poison plant and its mycorrhizal symbiont. *Gastrolobium bilobum* forms dense thickets in the southern jarrah forest where it is a major N_2-fixer and soil stabilizer and provides cover for vertebrates and food for fluoroacetate-resistant animals (Lamont 1992a). When grown in its normal soil, autoclaved seedlings were over nine times larger in the presence of ectomycorrhizas than in their absence (Lamont et al. 1985). In the absence of the symbiotic fungi, the *G. bilobum* seedlings were far too weak to have survived in the field. The underground sporocarps produced by most of these fungi are the favoured food of a small marsupial, *Bettongia penicillata* (Christensen 1980). Most spores in the faecal pellets belong to the major mycorrhizal genus *Mesophellia*. The spores are much more likely to germinate after passing through the marsupial's body than fresh spores (Lamont et al. 1985). Not only does consumption of the sporocarps increase after fire, but *Bettongia* may also travel up to 3 km overnight as it moves from one patch to another. Pellets will increase the numbers of mycorrhizal types on a given root system from an average of three to five, in a range of soils and fire histories (Lamont 1992b).

Gastrolobium bilobum takes part in a complex set of reciprocal relationships, from indirect to obligate. The species is dependent on an ectomycorrhizal guild of which some (*Mesophellia*) species are more abundant or have greater affinity for its root systems than others (C. Ralph, unpubl. data). Loss of fungal species would therefore have negligible to major impact on the presence of the plant species in the long term (the rate depending on incidence of fire and fungal identity). The loss of the marsupial may not affect population viability for hundreds of years because fires may only partially sterilise the surface soil (Malajczuk and Hingston 1981). Since this once abundant marsupial is now absent from much of the jarrah forest, this hypothesis could be tested by returning it to selected areas and monitoring any vegetation changes.

These examples illustrate the complex, and sometimes unexpected biotic interactions within natural communities. While the direct effects of these interactions on ecosystem function may be hard to define, the fact that such interactions between diverse and apparently otherwise unconnected components exist illustrates the need to consider not only diversity or components thereof, but also the biotic interactions involved. Frequently, we may be unaware of these until they have broken down due to the removal of particular ecosystem components.

5.3.5 Ecosystem Processes

While reviewing the literature on MTEs in Western Australia, we encountered only one study which provided data with which the relationship between diversity and ecosystem function could be directly examined. In this study (Springett 1976), soil microarthropod diversity and abundance and litter decomposition were compared in natural woodland and plantations of *Pinus pinaster* of three ages. Springett (1976) found no clear relationship between arthropod abundance and decomposition rates, but a significant correlation between species diversity and decomposition (Fig. 5.7). This figure indicates a large effect of increasing diversity at low levels of diversity, but this tails off at higher levels. This isolated example suggests an asymptotic relationship between diversity and function, corresponding to one of the models suggested by Vitousek and Hooper (1993). Thus a certain minimum number of species (or, more importantly, types of species) may be required for full ecosystem function, but the addition of more species may have little further effect on function (or rates thereof), except as insurance against disturbance and change, as discussed previously.

Springett (1976) found that a reduction in the number of species of litter and soil invertebrates in pine plantations, relative to adjoining native vegetation, did not result in a decline in the total number of individuals. This apparent density compensation did not, however, equate with functional compensation. Although the number of individuals remained constant, the overall rate of decomposition was lower where there were fewer species. This may simply have been due to the absence of appropriate types of species rather than because of species number per se. Clearly, care must be taken in interpreting these results since two different systems are being compared – alien pine plantations of various ages on one hand, and a native community on the other. However, the results relate to the decomposer community, not the plant component, and thus have some value to the present discussion.

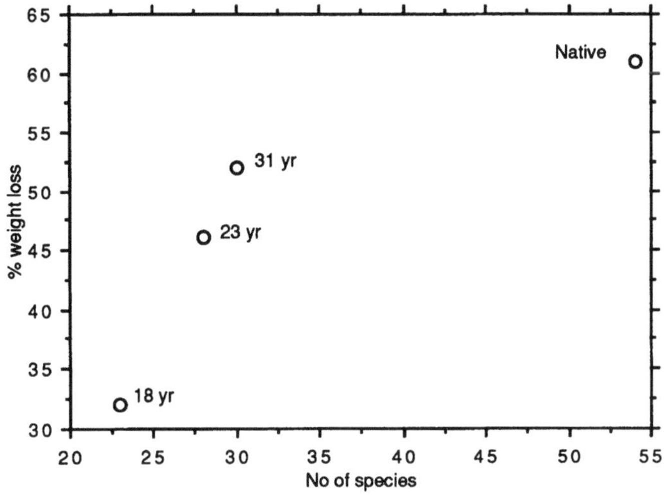

Fig. 5.7. Percentage weight loss (i.e. decomposition rate) from litter bags placed in plantations of *Pinus pinaster* of three ages (18, 23 and 31 years) and native woodland vegetation (predominantly *Banksia menziesii, B. attenuata* and *Eucalyptus todtiana*) at Gnangara, Western Australia, versus number of species of microarthropods recovered from soil cores. (Data from Springett 1976, after Hobbs 1992c)

On the other hand, species diversity is also confounded with site age, and there is the possibility that the differences in decomposition found may be related to some form of chemical leaching or accumulation totally unrelated to diversity of decomposers. Indeed, counter examples can be given which indicate that the maintenance of a given function following a decline in the number of species is not necessarily indicative of functional compensation. In contrast to Springett's (1976) results, a comparative study of two sites in jarrah woodland in Western Australia, one of which had been affected by *Phytophthora cinnamomi* and one unaffected, Postle et al. (1986) found similar rates of litter decomposition in both sites in spite of significantly fewer taxa of decomposers, and fewer individuals, in the dieback affected area. The apparent functional compensation in the area affected by *Phytophthora* was attributed to increased physical breakdown of litter in the modified site due to thinning of the canopy which left the litter more exposed to abiotic weathering processes.

Clearly, therefore, although ecosystem process rates may be linked to biotic diversity in some cases, in others this relationship may be altered or obscured by physical and/or chemical processes. Separating the two effects may be difficult, if not impossible. This needs to be taken into consideration when a link is sought between ecosystem processes and purely biotic factors.

5.3.6 Functional Groups: Useful or Not?

Since many plant species normally occupy the N_2-fixing group within a community, it is instructive to consider their ability to replace each other functionally should, for some reason, one or more of the species be eliminated. The first point to be made is that the legume-rhizobium combination is at least one to two orders of magnitude more efficient than the non-legume combinations on the basis of the C_2H_2-reduction technique (Lamont 1982, 1984). Secondly, rates of fixation may vary considerably between congenerics at the same site - for example Shea and Kitt (1976) recorded 1.7 µmol C_2H_2 reduced/g fresh nodule/h for *Acacia extensa*, 2.0 units for *A. myrtifolia* and 3.8 units for *A. pulchella* on laterite. When differences in nodule mass per plant and (actual or potential) density of plants are taken into account, this could lead to such major differences in N contribution to the community that the long-term outcomes for ecosystem maintenance could be quite different. In seeking functional equivalents, it is not just a question of whether they participate in critical pathways but the extent of *quantitative participation* in these pathways (Lamont 1992b). This is important when species redundancy or substitutability is discussed. Should we then abandon the quest for functional groupings of plants? Certainly, some care needs to be taken in developing appropriate groups, and complications such as that discussed here need to be considered. However, there is still a pressing need to develop some method of categorising the vast diversity of plants in the species rich assemblages such as those found in southwestern Australia. Hobbs (1994) has indicated that the use of easily identifiable groupings can be of some value in describing and predicting ecosystem response to environmental changes. If we want to look at the role of diversity in ecosystem function, we need to base our categories on those functions. No classification will be perfect, but hopefully, successive approaches will provide us with greater insights into the problem (see also Hobbs et al., Chap. 1, this Vol.).

5.3.7 Keystone Species: *Banksia prionotes* as a Keystone Plant

The potential importance of recognising keystone species and their roles was discussed by Hobbs et al. (Chap. 1, this Vol.), and Lamont (1992b) discussed the idea in detail for southwestern Australian examples. Plant species can play keystone roles in southwestern Australia. For example, *Banksia prionotes* provides an important nectar source for honeyeaters for a limited period of the year when there are virtually no other nectar-producing species in flower (Lambeck and Saunders 1993). Although this species is widespread and relatively abundant over much of its range, it is poorly represented in the central wheatbelt and extremely localised. Because of its asynchronous flowering relative to other nectar-producing species, it provides a critical resource for honeyeaters which, as pollinators, play an important role in the maintenance of floral biodiversity in the region. Many of the region's plants use honeyeaters as vectors for pollen transfer (Keighery 1982) and a reduction in bird numbers could have a significant impact on plant reproductive success. For example, *B. prionotes* will not set seed unless pollinated by birds (Collins and Spice 1986), and although experimental exclusion of birds from *Hakea scoparia* did not prevent seed set, it resulted in a reduction in the number of follicles formed (R.J. Lambeck, unpubl. data). The loss of such nectar-producing plants could have a significant impact on the capacity of honeyeaters to persist in the region. *B. prionotes* has become a keystone mutualist (Gilbert 1986), not solely because of its particular attributes, but because of its relationship to other species in the region. The absence of functional analogues has elevated its importance. This species does not play the same critical role in coastal parts of its range, where it occurs in sympatry with the synchronously flowering *Banksia menziesii*. Hence, when considering the status of particular species, it is important to recognise that it might act as a keystone in some parts of its range and not others. Similarly, there may be times when a species assumes a keystone role because other species from the same functional group are temporarily absent from an area (e.g. following disturbance). The classification of a species as a keystone must therefore take into account its temporal and spatial relationship with other species, and not simply its direct impact on ecosystem functions considered in isolation.

5.4 Research Opportunities

The above set of examples has illustrated both the complexity of biotic interactions to be considered in any treatment of the ecosystem role of biodiversity, and some of the difficulties involved in studying the problem. Solbrig (1991) suggested that the key questions to be addressed in the study of the ecosystem function of biodiversity centred around our ability to predict what would happen following the removal or addition of individual species, or entire functional groups, structural groups or other ecosystem components. These questions could be addressed using three approaches. Firstly, *controlled* experiments involving the deliberate removal or addition of species or groups could be undertaken. Secondly, use could be made of *inadvertent* experiments involving loss or addition of species through human activities (such as fragmentation, exploitation). Thirdly, opportunistic use could be made of *natural* experiments where episodic events such as storms or droughts have altered species or functional group compositions.

It appears likely that the first option will be available only at very small scales, although these are still very important since adequate controls can be included. While the use of natural experiments can prove fruitful (e.g. Hobbs and Mooney 1994), the main source of information would seem to be from inadvertent experiments. In addition, the rapidity and extent of change which has occurred and is still occurring in Australian MTEs means that research cannot be restricted to the few remaining "pristine" areas (if, indeed, any still exist), but must focus on the altered systems prevalent today. To be useful, research must examine questions relevant to the management and restoration of these altered systems in a dynamic and uncertain world.

Lavorel and Noble (1992) illustrated how an integrated research approach is required which couples experiments, long-term observations and modelling. The examples given above point out some areas where further work is needed. In terms of studying the role of biodiversity in ecosystem function, Main (1992) suggested that the interpretation of loss of species from southwest Australian MTEs is liable to be confounded by numerous factors, including:

1. The presence of non-natives such as foxes, rabbits and the fungal pathogen *Phytophthora* which have no strongly interacting species regulating their numbers.
2. The continued presence of a varied set of species-rich ecosystems despite the loss during the recent geological past of many species (Balme et al. 1978; Lundelius 1983; Murray 1984; Merrilees 1984).
3. Further losses of biota since European settlement.
4. Fragmentation of the natural environment following agricultural settlement
5. Lack of knowledge of the roles of species in ecosystems, their biology, life histories and life expectancy.
6. The lags within the system, related to longevity of individuals (Frank 1968).

Some of these factors are inescapable, and similar problems will be encountered in other mediterranean-climate areas. However, some of the factors listed can also be used to advantage in an adaptive approach to research on the topic. We suggest examples of potential studies which could yield useful information relevant to the question of the role of biodiversity in ecosystem function (modified from Hobbs 1992c). The topics listed address the hypotheses that removal or addition of species or functional groups will have measurable effects on ecosystem functions and related secondary functions, as outlined by Hobbs et al. (Chap. 1, this Vol.), and as treated in greater detail by Solbrig (1991).

5.4.1 Removal of a Functional Group: Jarrah Forest Fire Regimes

Understorey leguminous species such as *Acacia pulchella* contribute significant amounts of nitrogen to the jarrah forest system post-fire through symbiotic fixation (Hansen et al. 1987, 1988; Hingston et al. 1989). It is also known that these species establish from seed following fire due to heat-stimulation of germination, and that post-fire establishment could be manipulated by altering fire intensities, with low intensity fires and moist soil conditions reducing germination and establishment (Shea et al. 1979; Bell and Koch 1980; Bell et al. 1989; Portlock et al. 1990, but see also McCaw 1989). In addition, it appears that production of the zoosporangial stage of the pathogen *Phytophthora cinnamomi* is significantly lower beneath leguminous stands than under other forest species (Shea et al. 1978; Murray et al. 1985). Leguminous species may either inhibit the pathogen directly, or via associated rhizosphere fungi (Whitfield et al. 1981; Malajczuk et al. 1984; Dell and Malajczuk 1989).

Examination of nutrient cycling processes in plots where leguminous species' abundance was modified experimentally could therefore yield useful information on the importance of this group to the functioning of the jarrah forest ecosystem. This could be supplemented with similar studies on mine site rehabilitation areas, where the role of understorey legumes has already been investigated in some detail (Ward et al. 1990, and see Sect. 5.4.3 below). In addition, identification of the causal mechanisms behind the reduction in pathogen abundance could identify whether the leguminous functional group also has a further role in pathogen suppression, either directly or through its fungal associates.

5.4.2 Removal of a Functional Group: Effects of *Phytophthora cinnamomi*

The invasive fungal pathogen is currently producing considerable floristic changes in some of the most important biological reserves in southwestern Australia (Wills 1993). It does not affect all elements of the flora to a similar extent, but has a greater impact on woody species, particularly members of the Proteaceae (Wills 1993). This results in a reduction in proteaceous cover and a concomitant increase in cover of herbaceous species, particularly Cyperaceae. Similar patterns of change have been found elsewhere (Weste 1986). There are two important aspects of this change which may affect ecosystem functioning. One is the effect of removal of Proteaceae on nutrient uptake, given that this group has specialised root structures to enhance uptake (Lamont 1982, 1992b). The second is the removal of structural components of the ecosystem and of the nectar resources and seeds present in proteaceous flowers and fruits, which are important food supplies for a variety of bird, mammal and insect species (Lambeck 1992; Lamont 1992b; Newell and Wilson 1993; Wills 1993). Some of these species, in turn, are likely to be important pollinators. There is therefore the potential for studies which examine dieback-affected areas in terms of nutrient regimes and utilization by fauna.

5.4.3 Additions of Species / Functional Groups in Mine Site Rehabilitation

Rehabilitation of mine sites following extraction of bauxite in the jarrah forest and mineral sands in the northern sandplain offers great potential for the study of biodiversity and ecosystem function. Ward et al. (1990) have illustrated the effects of the presence or absence of a leguminous understorey on the nitrogen economy of rehabilitation sites in the jarrah forest. Majer (1992) has presented data sets on decomposition and on recolonisation by Collembolan species in rehabilitation sites. While these data sets are not readily amenable to cross-comparison, it should be possible with further analysis to assess the effects of diversity of decomposer organisms on decomposition rates.

Bell et al. (1990) indicate the importance of ensuring that seeds of species with different seed storage strategies are returned to rehabilitation sites in the northern sandplain heaths. Reliance on seed stored in the soil ignores the importance of a large component of the flora which relies mostly on canopy storage. Even when these species are reintroduced by way of a mulching treatment, regeneration consists mainly of non-resprouting species. Few resprouters re-establish (S. Bellairs, pers. comm.). Thus there is a situation where a highly diverse community is re-established, but a major functional group (i.e. resprouters) is absent. This has important implications for the way the restored ecosystem responds to future disturbances. Experimental manipulation of rehabilitated mine sites with different species/ functional group complements could yield significant insights into the functional importance of various aspects of biodiversity.

5.4.4 Addition and Removal of Species / Functional Groups: Fragmentation and Revegetation in Agricultural Areas

There is considerable scope for investigating the functional impact of the changes caused by fragmentation of the Western Australian wheatbelt. Grazing, selective clearing, short-term cropping and other disturbances have left areas in which species or functional groups (of plants and/or animals) have been removed. On the other hand, invasions by non-native plants and animals have added species. In addition, as indicated above, problems of land degradation and loss of conservation value in agricultural areas need to be tackled in tandem. The solution currently being pursued is one of re-establishing perennial vegetation strategically across the landscape. It is relatively well established that doing this will alleviate many of the problems associated with rising water tables (i.e. salinization, erosion, waterlogging; McFarlane et al. 1993), but there is little work on whether revegetation allows the redevelopment of functioning ecosystems (Hobbs 1993b). Much of the revegetation work has been carried out using species not native to the region (for instance, eucalypts from eastern Australia) but more recently there has been increasing impetus to use local species (e.g. Lefroy et al. 1991). While it seems likely that the species used will have little effect on the hydrological processes (except for relatively minor differences in water uptake rates), other processes and nature conservation value may be more closely linked to species composition. Studies such as those carried out by Springett (1976), comparing species diversities and ecosystem processes in areas dominated by native and non-native species in the forest region, need to be conducted in revegetated areas of the wheatbelt. If the revegetation is designed to act as habitat, food supply or movement corridors for fauna, then the types of species and functional groups used could determine the extent to which it is successful. To date, no studies have been carried out on the faunal use of revegetated areas, but this is an area which could yield useful information on the importance of diversity at a number of levels.

5.4.5 Comparisons Between Species-Rich and Species-Poor Native Plant Communities

While much of the emphasis in discussion of the vegetation of south-western Australia centres on the richness in species of the heath or kwongan vegetation, it should be noted that other vegetation types are considerably less species-rich. The relationship between diversity and ecosystem function can be explored in a non-experimental way using this natural variability in species richness. Within particular structural types, such as shrubland, species-rich and species-poor types can occur adjacent to each other (R.J. Hobbs and L. Atkins, unpubl. data). Communities dominated by single species, such as *Allocasuarina* spp. or *Melaleuca* spp., tend to have relatively few associated species. Comparison of ecosystem function in adjacent relatively rich and relatively poor communities may yield useful information to supplement the more experimental studies discussed above. For instance, Hobbs and Atkins (1991) found that shrubland dominated by *Allocasuarina campestris* responded quite differently to canopy removal than an adjacent sandplain heath, with biomass of annuals increasing significantly in the shrubland but not in the heathland. The importance of dominance by particular species in controlling ecosystem function is an important corollary to the question of the role of diversity.

Other situations such as single-species stands of *Spinifex* grass on coastal sand dunes could also be investigated in comparison with more mature multi-species dune assemblages. A further example would be a comparison of species-rich understoreys of South Australian woodlands with those simplified by frequent burning (e.g. *Epacris impressa* under *Eucalyptus obliqua*, and with naturally species-poor understoreys dominated by the grass *Themeda australis*; P.J. Grubb, pers. comm.). Such comparative studies are open to a number of problems, notably the potential importance of confounding factors including soil differences, but may nevertheless yield information which might be hard to obtain in other ways.

5.5 Conclusions

It can be argued that a major role of diversity is to provide a pool of functional substitutes following the loss of organisms which formerly filled as given role (Main 1992). The large number of extinctions and range contractions already experienced in the fragmented natural environments of the mediterranean-climate zone of Australia has massively reduced the capacity of ecosystems to engage in species substitution to fill essential roles, as probably happened following the extinction of the megafauna. The role of species diversity in ecosystem resilience, rather than simply in the maintenance of ecosystem flows at given rates, should be emphasised. Should substitution be no longer possible, further species losses may lead to the loss of any resemblance to the ecosystems now recognised. Species diversity per se may have little overall direct effect on ecosystem fluxes, but may be critical in ensuring the persistence of the ecosystem (or alternative states of the same ecosystem) in the face of varied and changing environmental and disturbance factors.

In the highly fragmented landscapes which characterize much of the Australian mediterranean regions, options for functional compensation and recolonisation, and hence much of the functional resilience, have been lost. The patchy distribution of the original vegetation (Specht 1972; Beard and Sprenger 1984) is associated with a relatively high beta diversity. Animal species tend not to be uniformly distributed throughout these spatially variable environments. Fragmentation of such a landscape results in unequal representation of the various components of the biota in the fragments which remain. The probability of extinction is increased, simply due to the reduction in the number of species which can be supported by the reduced area (Connor and McCoy 1979), and because of pressures imposed by adjacent land use practices. In addition, the stochastic risks associated with natural disturbance remain, increasing the probability of populations being reduced to levels where demographic fluctuations can jeopardise genetic integrity (Gilpin and Soulé 1986).

In these fragmented landscapes the combination of natural disturbances and human impacts has a much greater capacity to affect all of the members of a functional group than is the case for continuous, unmodified environments. Any changes in function resulting from a change in species diversity will be less likely to be compensated for in the short term, as there will be fewer analogous species, and the probability of long-term restoration of function through recolonisation will be reduced as a result of the increased distance to source populations, and the hostility of the intervening habitat.

The presence of a diverse flora and fauna in unfragmented landscapes contributes to continuity of function following local extinction, thereby maintaining conditions suitable for recolonisation. The change in function associated with fragmentation may eliminate the possibility of re-establishment of a population, even if dispersal back to these disturbed remnants is possible. Such remnants may become "sinks" (Howe et al. 1991) which dispersing organisms can colonise but in which they are either unable to survive or unable to successfully produce or raise young. Extinction will then become the dominant population process, resulting in further changes to the manner in which the ecosystem functions. In fragmented environments the probability of local extinction is higher, and such extinctions are more likely to result in permanent local changes to functioning than is the case for undisturbed habitat.

If ecosystem functioning is to be construed as the persistence of those processes necessary for the cycling of nutrients and energy, then changes in the level of biodiversity will not result in the cessation of function. By altering the number of alternative pathways via which nutrients and energy can move through a given system, changes in biodiversity will result in altered rates of function. Complete loss of function will occur only where catastrophic events or human land-use practices destroy the primary production and decomposition trophic levels (Ehrlich and Mooney 1983).

For example, the primary processes of nutrient, energy and water flux could theoretically persist in the absence of fauna. Elements extracted from the soil and air by plants could return as a result of senescence, rather than consumption by herbivores, and decomposition could result from the actions of microflora or physical weathering. However, suggesting that faunal biodiversity is not essential for ecosystem functioning is not equivalent to saying that animals are not important in the Australian MTEs. As mobilisers of nutrients and energy, and as vectors for pollen and seed dispersal, they play an important part in maintaining the "characteristic" patterns and rates of processes, and therefore influence the assemblages of species in the region. However, it is the particular attributes of the animals present, their relative abundances, and the nature of their interactions, rather than the number of species, which is most important in maintaining these rates of function. Maintenance of a given level of function is dependent upon the presence of appropriate species. The maintenance of biodiversity simply increases the probability of critical components being represented. High levels of biodiversity will not sustain a given level of function if critical components of that diversity are missing. The important factor is thus not simply species numbers, but also types of species - or, to put it in another way, species in the required functional groups.

While we have been able to produce little direct evidence on the question of the ecosystem function of biodiversity, we have indicated that the ecosystems of southwestern Australia provide many opportunities for the study of this question. Many ecosystem types are present, providing a range of variation in structure, diversity and degree of modification. Detailed studies within and between these differing ecosystems, together with comparisons of other regions with mediterranean-type climates, should yield valuable information on this difficult topic. It should be noted, however, that the examples given highlight some of the problems involved in interpreting causality between changes in diversity and altered function. The additional studies suggested are open to the same sorts of problems, and it will be a major task to take alternative explanations into account and to factor them out.

While it is still unclear whether the maintenance of biodiversity is essential for ecosystem functioning, there is no doubt that the maintenance of functioning ecosystems is essential if we want to retain biodiversity. Threats to biodiversity arise mostly from fragmentation and exploitation of natural areas, both of which disrupt or alter ecosystem functioning. The challenge for conservation is thus to find ways to maintain ecosystem processes in fragmented and extensively modified systems. The link between ecosystem function and biodiversity is thus two-way, and both directions of causality need to be investigated.

Acknowledgements. We thank participants of the Perth workshop for their input of ideas, Brian Walker for comments on sections of the chapter and for providing funds to support the Perth workshop, Peter Grubb and Ian Noble for critical and insightful reviews of the draft manuscript, and Lyn Atkins for figure preparation. We also thank Surrey Beatty and Sons for permission to reproduce material from the proceedings of the Australian workshop. The senior author also wishes to thank the South African FRD for support to attend the Cape Town meeting, and George Davis and many others in Cape Town for their assistance.

References

Abbott I, Parker CA (1980) Agriculture and the abundance of soil animals in the Western Australian wheatbelt. Soil Biol Biochem 12: 455-459

Abbott I, Parker CA, Sills ID (1979) Changes in abundance of large soil animals and physical properties of soils following cultivation. Aust J Soil Res 17: 343-353

Abbott I, Dell B, Loneragan O (1989) The jarrah plant. In: Dell B, Havel JJ, Malajczuk N (eds) The jarrah forest. A complex mediterranean ecosystem. Kluwer, Dordrecht, pp 41-51

Balme J, Merilees D, Porter J K (1978) Late Quaternary mammal remains, spanning about 30000 years, from excavations in Devil's Lair, Western Australia. J R Soc West Aust 61: 33-65

Bartle J, Slessar GC (1989) Mining and rehabilitation. In: Dell B, Havel JJ, Malajczuk N (eds) The jarrah forest. A complex mediterranean ecosystem. Kluwer, Dordrecht, pp 357-377

Barlow BA (1981) The loranthaceous mistletoes in Australia. In: Keast A (ed) Ecological biogeography of Australia. Junk, The Hague, pp 555-574

Beard JS (1965) A descriptive catalogue of West Australian plants. Society for Growing Australian Plants, Sydney

Beard JS (1983) Ecological control of the vegetation of southwestern Australia: moisture versus nutrients. In: Kruger FJ, Mitchell DT, Jarvis JUM (eds) Mediterranean-type ecosystems: the role of nutrients. Springer, Berlin Heidelberg New York, pp 66-73

Beard JS (1990) Plant life of Western Australia. Kangaroo Press, Kenthurst, NSW

Beard J S, Sprenger BS (1984) Geographical data from the vegetation survey of Western Australia. Vegetation survey of Western Australia, Occasional Paper 2. Vegmap Publications, Perth

Bell DT, Koch JM (1980) Post-fire succession in the northern jarrah forest of Western Australia. Aust J Ecol 5: 9-14

Bell DT, Hopkins AJM, Pate JS (1984) Fire in the kwongan. In: Pate JS, Beard JS (eds) Kwongan. Plant life of the sandplain. University of Western Australia Press, Nedlands, pp 178-204

Bell DT, McCaw WL, Burrows ND (1989) Influence of fire on jarrah forest vegetation. In: Dell B, Havel JJ, Malajczuk N (eds) The jarrah forest. A complex mediterranean ecosystem. Kluwer, Dordrecht, pp 203-215

Bell DT, Vlahos S, Bellairs SM (1990) Seed ecology in relation to reclamation: lessons from mined lands in Western Australia. Proc Ecol Soc Aust 16: 531-535

Bond P, Goldblatt P (1984) Plants of the Cape flora. A descriptive catalogue. J S Afr Bot Suppl Vol 13:1-455
Bormann FH, Likens GE (1981) Pattern and process in a forested ecosystem. Springer, Berlin Heidelberg New York
Bowler JM (1982) Aridity in the late Tertiary and Quaternary of Australia. In: Barker WR, Greenslade PJM (eds) Evolution of the flora and fauna of arid Australia. Peacock Publ, Adelaide, pp 35-45
Brown GD, Main AR (1967) Studies on marsupial nutrition V. The nitrogen requirement of the euro, *Macropus robustus*. Aust J Zool 15: 7-27
Burrows N, Gardner C, Ward B, Robinson A (1990) Regeneration of *Eucalyptus wandoo* following fire. Aust For 53: 248-58
Christensen PES (1980) The biology of *Bettongia penicillata* (Gray 1937) and *Macropus eugenii* (Desmarest 1817) in relation to fire. For Dep West Aust Bull 91
Churchill D (1968) The distribution and prehistory of *Eucalyptus diversicolor* F. Muell., *E. marginata* Donn. ex Sm., and *E. callophylla* R. Br. in relation to rainfall. Aust J Bot 16: 125-151
Cogger HG (1986) Reptiles and amphibians of Australia. Reed, Sydney
Collins BG, Spice J (1986) Honeyeaters and the pollination biology of *Banksia prionotes* (Proteaceae). Aust J Bot 34: 175-185
Connor EF, McCoy ED (1979) The statistics and biology of the species-area relationship Am Nat 113: 791-833
Cowling RM, Holmes PM, Rebelo AG (1992) Plant diversity and endemism. In: Cowling RM (ed) The ecology of fynbos: nutrients, fire and diversity. Oxford University Press, Cape Town, pp 62-112
Dell B, Havel JJ (1989) The jarrah forest, an introduction In: Dell B, Havel JJ , Malajczuk N (eds) The jarrah forest. A complex mediterranean ecosystem. Kluwer, Dordrecht, pp 1-10
Dell B, Malajczuk N (1989) Jarrah dieback – a disease caused by *Phytophthora cinnamomi*. In: Dell B, Havel JJ , Malajczuk N (eds) The jarrah forest. A complex mediterranean ecosystem. Kluwer, Dordrecht, pp 67-87
di Castri F (1981) Mediterranean-type shrublands of the world. In: di Casri F, Goodall DW, Specht RL (eds) Ecosystems of the world 11. Mediterranean-type shrublands. Elsevier, Amsterdam, pp 1-52
Dick RS (1975) A map of the climates of Australia according to Köppen's principles of definition. Qld Geogr J 3: 33-69
Dodd J, Heddle EM, Pate JS, Dixon KW (1984) Rooting patterns of sandplain plants and their functional significance. In: Pate JS, Beard JS (eds) Kwongan. Plant life of the sandplain. University of Western Australia Press, Nedlands, pp 146-177
Ehrlich PR, Mooney HA (1983) Extinction, substitution and ecosystem services. BioScience 33: 248-254
Ford J (1985) Fire ecology and management in Western Australian ecosystems. Western Australian Institute of Technology, Bentley
Fox BJ, Fox MD (1986) Resilience of animal and plant communities to human disturbance In: Dell B, Hopkins AJM, Lamont BB (eds) Resilience in mediterranean-type ecosystems. Junk, Dordrecht, pp 39-64
Frank PW (1968) Life histories and community stability. Ecology 49: 355-356
Froend RH (1987) Investigations into species richness patterns in the northern sandplain region of Western Australia. PhD Thesis, University of Western Australia, Nedlands
Gardner CA (1957) The fire factor in relationship to the vegetation of Western Australia. West Aust Nat 5: 166-173

Gentilli J (1989) Climate of the jarrah forest. In: Dell B, Havel JJ, Malajczuk N (eds) The jarrah forest. A complex mediterranean ecosystem. Kluwer, Dordrecht, pp 23-40

George AS (1991) New taxa, combinations and typifications in *Verticordia* (Myrtaceae: Chamelaucieae). Nuytsia 7: 231-394

George AS, Hopkins AJM, Marchant NG (1979) The heathlands of Western Australia. In: Specht RL (ed) Ecosystems of the world 9A. Heathlands and related shrublands. Elsevier, Amsterdam, pp 211-230

Gilbert LE (1986) Food web organization and the conservation of neotropical diversity. In: Soulé ME, Wilcox BA (eds) Conservation biology: an evolutionary-ecological perspective. Sinauer, Sunderland, MA, pp 11-33

Gilpin ME, Soulé ME (1986) Minimum viable populations: processes of species extinction In: Soulé ME (ed) Conservation biology: the science of scarcity and diversity. Sinauer, Sunderland, MA, pp 19-34

Griffin EA, Hopkins, AJM, Hnatiuk RJ (1983) Regional variation in mediterranean-type shrublands near Eneabba, south-western Australia. Vegetatio 52: 103-127

Griffin EA, Hopper SD, Hopkins AJM (1990) Flora. In: Burbidge AA, Hopper SD, van Leeuwen S (eds) Nature conservation, landscape and recreation values of the Lesueur area. Environmental Protection Authority, Perth, Pp 39-69

Groves RH (1981) Heathland soils and their fertility status. In: Specht RL (ed) Ecosystems of the World 9B. Heathlands and Related Shrublands, Analytical Studies. Elsevier, Amsterdam, pp 143-150

Groves RH, Hobbs RJ (1992) Patterns of plant functional responses and landscape heterogeneity. In: Hobbs RJ (ed) Biodiversity of mediterranean ecosystems of Australia. Surrey Beatty, Chipping Norton, NSW, pp 47-60

Grubb PG (1992) A positive distrust in simplicity – lessons from plant defences and from competition among plants and among animals. J Ecol 80: 586-610

Grubb PJ , Hopkins AJM (1986) Resilience at the level of the plant community. In: Dell B, Hopkins AJM, Lamont BB (eds) Resilience in mediterranean-type ecosystems. Junk, Dordrecht, pp 21-38

Hallam SJ (1985) The history of aboriginal firing. In: Ford J (ed) Fire ecology and management of Western Australian ecosystems. Western Australian Institute of Technology, Bentley, pp 7-20

Hansen AP, Pate JS, Hansen A, Bell DT (1987) Nitrogen economy in jarrah (*Eucalyptus marginata* Donn ex Sm) forest of Western Australia. J Exp Bot 38:26-41

Hansen AP, Stoneman GL, Bell DT (1988) The potential of seeder legumes as a means to increase nitrogen availability in the jarrah forest. Aust For 51:226-231

Hartley W, Leigh J (1979) Plants at risk in Australia. Australian National Parks and Wildlife Service Occasional Paper 3, Australian National Parks and Wildlife Service, Canberra

Havel JJ (1989) Land use conflicts and the emergence of multiple use. In: Dell B, Havel JJ, Malajczuk N (eds) The jarrah forest. A complex mediterranean ecosystem. Kluwer, Dordrecht, pp 281-314

Havel JJ, Dell B, Malajczuk N (1989) Concluding remarks. In: Dell B, Havel JJ, Malajczuk N (eds) The jarrah forest. A complex mediterranean ecosystem. Kluwer, Dordrecht, pp 401-403

Hellmuth EO (1971) Ecophysiological studies on plants in arid and semi-arid regions of Western Australia. J Ecol 59: 365-374

Hingston FJ, O'Connell AM, Grove TS (1989) Nutrient cycling in jarrah forest. In: Dell B, Havel JJ, Malajczuk N (eds) The jarrah forest. A complex mediterranean ecosystem. Kluwer, Dordrecht, pp 155-177

Hnatiuk RJ, Hopkins AJM (1980) Western Australian species-rich kwongan (sclerophyllous shrublands) affected by drought. Aust J Bot 28: 573-85

Hobbs RJ (1987) Disturbance regimes in remnants of natural vegetation. In: Saunders DA, Arnold GW, Burbidge AA, Hopkins AJM (eds) Nature conservation: the role of remnants of native vegetation. Surrey Beatty, Chipping Norton, NSW, pp 233-240

Hobbs RJ (ed) (1992a) Biodiversity of mediterranean ecosystems in Australia. Surrey Beatty, Chipping Norton, NSW

Hobbs RJ (1992b) Function of biodiversity in mediterranean ecosystems in Australia: definitions and background. In: Hobbs RJ (ed) Biodiversity of mediterranean ecosystems of Australia. Surrey Beatty, Chipping Norton, NSW, pp 1-25

Hobbs RJ (1992c) Is biodiversity important for ecosystem function? Implications for research and management. In: Hobbs RJ (ed) Biodiversity of mediterranean ecosystems of Australia. Surrey Beatty, Chipping Norton, NSW, pp 211-229

Hobbs RJ (1993a) Effects of landscape fragmentation on ecosystem processes in the Western Australian wheatbelt. Biol Conserv 64: 193-201

Hobbs RJ (1993b) Can revegetation assist in the conservation of biodiversity in agricultural areas? Pac Conserv Biol 1: 29-38

Hobbs RJ (1994) Can we use plant functional types to describe and predict repsonses to environmental change? In: Smith TM, Shugart HH, Woodward FI (eds) Plant functional types. Cambridge University Press, Cambridge (in press)

Hobbs RJ, Atkins L (1990) Fire-related dynamics of a *Banksia* woodland in south-western Western Australia Aust J Bot 38: 97-110

Hobbs RJ, Atkins L (1991) Interactions between annuals and woody perennials in a Western Australian nature reserve J Veg Sci 2: 643-654

Hobbs RJ, Hopkins AJM (1990) From frontier to fragments: European impact on Australia's vegetation. Proc Ecol Soc Aust 16: 93-114

Hobbs RJ, Mooney HA (1994) Effects of episodic rainfall events on mediterranean-climate ecosystems. In: Roy J, Aronoson J, di Castri F (eds) Time scales of biological responses to water constraints Springer, Berlin Heidelberg New York (in press)

Hobbs RJ, Wallace JF, Campbell NA (1989) Classification of vegetation in the Western Australian wheatbelt using Landsat MSS data. Vegetatio 80: 91-105

Hobbs RJ, Saunders DA, Lobry de Bruyn LA, Main AR (1993) Changes in biota In: Hobbs RJ, Saunders DA (eds) Reintegrating fragmented landscapes: towards sustainable production and nature conservation. Springer, Berlin Heidelberg New York, pp 65-106

Hope G (1984) Australian environmental change: timing, directions, magnitudes, and rates. In: Martin PS, Klein RG (eds) Quaternary extinctions: a prehistoric revolution. University of Arizona Press, Tuscon, pp 681-690

Hopkins AJM (1985) Fire in the woodlands and associated formations of the semi-arid region of south-western Australia. In: Ford J (ed) Fire ecology and management of Western Australian ecosystems. Western Australian Institute of Technology, Bentley, pp 83-90

Hopkins AJM, Griffin EA (1984) Floristic patterns. In: Pate JS, Beard JS (eds) Kwongan. Plant life of the sandplain. University of Western Australia Press, Nedlands, pp 69-83

Hopkins AJM, Robinson CJ (1981) Fire-induced structural change in a Western Australian woodland. Aust J Ecol 6: 177-188

Hopper SD (1979) Biogeographical aspects of speciation in the southwest Australian flora. Annu Rev Ecol Syst 10: 399-420

Hopper SD (1992) Patterns of plant diveristy at the population and species level in south-west Australian mediterranean ecosystems. In: Hobbs RJ (ed) Biodiversity of mediterranean ecosystems of Australia. Surrey Beatty, Chipping Norton, NSW, pp 27-46

Hopper SD, van Leeuwen S, Brown AP, Patrick SJ (1990) Western Australia's endangered flora and other plants under consideration for declaration. Department of Conservation and Land Management, Perth
Horton DR (1984) Red kangaroos: last of the Australian megafauna. In: Martin PS, Klein RG (eds) Quarternary extinctions: a prehistoric revolution. University of Arizona Press, Tuscon, pp 639-668
Howe RW, Davis GJ, Mosca V (1991) The demographic significance of "sink" populations Biol Conserv 57: 239-256
Keast A (1958) The influence of ecology on variation in the mistletoe-bird (*Dicaeum hirundinaceum*). Emu 58: 195-206
Keighery GJ (1982) Bird-pollinated plants in Western Australia. In; Armstrong JA, Powell JM, Richards AJ (eds) Pollination and evolution. Royal Botanic Gardens, Sydney
Kenneally KF (1973) Some observations of the stem hemiparasite, or mistletoe *Anymema miquelii* (Loranthaceae) in south-western Australia. West Aust Nat 12: 156-161
Kennedy M (1990) Australia's endangered species. The extinction dilemma. Simon and Shuster, Brookvale, NSW
Kershaw AP (1981) Quaternary vegetation and environments. In: Keast A (ed) Ecological biogeography of Australia. Junk, The Hague, pp 83-101
King DR, Oliver AJ, Mead RJ (1978) The adaptation of some Western Australian mammals to food plants containing fluoroacetate. Aust J Zool 26: 699-712
Kinnear JE, Main AR (1975) The recycling of urea nitrogen by the wild tammar wallaby (*Macropus eugenii*) – a "ruminant-like" marsupial. Comp Biochem Physiol 51A: 793-810
Kinnear JE, Main AR (1979) Niche theory and macropodid nutrition. J R Soc West Aust 62: 65-74
Kitchener DJ, Chapman A, Muir BG, Palmer M (1980) The conservation value for mammals of reserves in the Western Australian wheatbelt. Biol Conserv 18: 179-207
Kolasa J, Rollo CD (1991) Introduction: the heterogeneity of heterogeneity: a glossary. In: Kolasa J, Pickett STA (eds) Ecological heterogeneity. Springer, Berlin Heidelberg New York, pp 1-23
Kruger FJ, Mitchell DT, Jarvis JUM (eds) (1983) Mediterranean-type ecosystems. The role of nutrients. Springer, Berlin Heidelberg New York
Lambeck RJ (1992) The role of faunal diversity in ecosystem function. In: Hobbs RJ (ed) Biodiversity of mediterranean ecosystems in Australia. Surrey Beatty, Chipping Norton, NSW, pp 129-148
Lambeck RJ, Saunders DA (1993) The role of patchiness in reconstructed wheatbelt landscapes. In: Saunders DA, Hobbs RJ, Ehrlich PR (eds) Nature conservation 3. reconstruction of fragmented ecosystems, global and regional perspectives. Surrey Beatty, Chipping Norton, NSW, pp 153-161
Lamont BB (1982) Mechanisms for enhancing nutrient uptake in plants, with particular reference to mediterranean South Africa and Western Australia. Bot Rev 48: 597-689
Lamont BB (1984) Specialized modes of nutrition. In: Pate JS, Beard JS (eds) Kwongan. Plant life of the sandplain. University of Western Australia Press, Nedlands, pp 126-145
Lamont BB (1985) Gradient and zonal analysis of understorey suppression by *Eucalyptus wandoo*. Vegetatio 63: 49-66
Lamont BB (1992a) Triangular trophic relationships in mediterranean Western Australia. In: Thanos CA (ed) Proc 6th Int Conf Mediterranean Ecosystems. University of Athens, Athens, pp 171-176

Lamont BB (1992b) Functional interactions within plants – the contribution of keystone and other species to biological diversity. In: Hobbs RJ (ed) Biodiversity of mediterranean ecosystems of Australia. Surrey Beatty, Chipping Norton, NSW, pp 95-127

Lamont BB, Connell SW, Bergl SM (1991) Seed bank and population dynamics of *Banksia cuneata*: the role of time, fire and moisture. Bot Gaz 152: 114-122

Lamont BB, Downes S (1979) The longevity, flowering and fire history of the grasstrees *Xanthorrhoea preissii* and *Kingia australis*. J Appl Ecol 16: 893-899

Lamont BB, Hopkins AJM, Hnatiuk RJ (1984) The flora – composition, diversity and origins. In: Pate JS, Beard JS (eds) Kwongan. Plant life of the sandplain. University of Western Australia Press, Nedlands, pp 27-50

Lamont BB, Ralph CS, Christensen PES (1985) Mycophagous marsupials as dispersal agents for ectomycorrhizal fungi on *Eucalyptus calophylla* and *Gastrolobium bilobum*. New Phytol 101: 651-656

Lamont BB, Southall K (1982) Biology of the mistletoe *Amyema preissii* on road verges and undisturbed vegetation. Search 13:87-88

Lamont BB, Witkowski ETF, Enright NJ (1993) Post-fire litter microsites: safe for seeds, unsafe for seedlings. Ecology 74: 501-512

Lantzke N (1992) Soils of the Northam advisory district. The zone of ancient drainage. Department of Agriculture, Perth

Lavorel S, Noble IR (1992) Ecosystem function of biodiversity: experimentation, long-term observations and modelling. In: Hobbs RJ (ed) Biodiversity of mediterranean ecosystems of Australia. Surrey Beatty, Chipping Norton, NSW, pp 149-167

Lefroy EC, Hobbs RJ, Atkins LA (1991) A revegetation guide to the central wheatbelt. Western Australian Department of Agriculture, CSIRO and Greening Australia, Perth

Lefroy EC, Salerian J, Hobbs RJ (1993a) Integrating ecological and economic considerations: a theoretical framework. In: Hobbs RJ, Saunders DA (eds) Reintegrating fragmented landscapes: towards sustainable production and nature conservation. Springer, Berlin Heidelberg New York, pp 209-244

Lefroy EC, Hobbs RJ, Scheltema M (1993b) Reconciling agriculture and nature conservation: towards a restoration strategy for the Western Australian wheatbelt. In: Saunders DA, Hobbs, RJ, Ehrlich PR (eds) Nature conservation 3: reconstruction of fragmented ecosystems, global and regional perspectives. Surrey Beatty, Chipping Norton, NSW, pp 243-257

Leigh JH, Boden R, Briggs J (1984) Extinct and endangered plants of Australia. Macmillan, Melbourne

Lewis SA (1991) The conservation and management of roadside vegetation in South Australia. In: Saunders DA, Hobbs RJ (eds) Nature conservation 2: The role of corridors. Surrey Beatty, Chipping Norton, NSW, pp 313-318

Lobry de Bruyn LA (1990) The role of ants and termites in modifying soil properties in naturally vegetated and agricultural environments. PhD Thesis, University of Western Australia, Nedlands

Lundelius EL (1983) Climatic implications of Late Pleistocene and Holocene faunal associations in Australia. Alcheringa 7: 125-149

Main AR (1982) Rare species: precious or dross? In: Groves RH, Ride WDL (eds) Species at risk: research in Australia. Australian Academy of Science, Canberra, pp 163-174

Main AR (1992) The role of diversity in ecosystem function: an overview. In: Hobbs RJ (ed) Biodviersity of mediterranean ecosystems in Australia. Surrey Beatty, Chipping Norton, NSW, pp 77-93

Majer JD (1992) Maintaining and restoring diversity. In: Hobbs RJ (ed) Biodiversity of mediterranean ecosystems in Australia. Surrey Beatty, Chipping Norton, NSW, pp 169-191

Malajcuk N, Hingston FJ (1981) Ectomycorrhizae associated with jarrah. Aust J Bot 29: 453-462

Malajczuk N, Pearce M, Litchfield RT (1984) Interactions between *Phytophthora cinnamomi* and *Rhizobium* isolates. Trans Br Mycol Soc 80: 305-312

Martin H (1973) Palynology and historical ecology of some cave excavations in the Australian Nullarbor. Aust J Bot 21: 283-316

McArthur WM (1993) History of landscape development. In: Hobbs RJ, Saunders DA (eds) Reintegrating fragmented landscapes: towards sustainable production and nature conservation. Springer, Berlin Heidelberg New York, pp 10-22

McArthur WM, Bettenay E (1979) The land. In: O'Brien BJ (ed) Environment and science. University of Western Australia Press, Nedlands

McCaw WL (1989) Regeneration of *Acacia* and *Kennedia* from soil-stored seed following an autumn fire in jarrah (*Eucalyptus marginata*) forest. J R Soc West Aust 71: 1-6

McFarlane DJ, George RJ, Farrington P (1993) Changes in the hydrologic cycle. In: Hobbs RJ, Saunders DA (eds) Reintegrating fragmented landscapes: towards sustainable production and nature conservation. Springer, Berlin Heidelberg New York, pp 146-86

Mead RJ, Oliver AJ, King DR, Hubach PH (1985) The coevolutionary role of fluoroacetate in plant-animal interactions in Australia. Oikos 44: 63-76

Merrilees D (1968) Man the destroyer: Later Quaternary changes in the Australian marsupial fauna. J R Soc West Aust 51: 1-24

Merrilees D (1979) The prehistoric environment in Western Australia. J R Soc West Aust 62: 109-128

Merrilees D (1984) Comings and goings of Late Quarternary mammals in extreme Southwestern Australia. In: Martin PS, Klein RG (eds) Quaternary extinctions: a prehistoric revolution. University of Arizona Press, Tuscon, pp 629-638

Milne BT (1991) Heterogeneity as a multiscale characteristic of landscapes. In: Kolasa J, Pickett STA (eds) Ecological heterogeneity. Springer, Berlin Heidelberg New York, pp 69-84

Murray DIL, Darling DD, McGann LR (1985) Indirect effect of florisitic composition on production of sporangia by *Phytophthora cinnamomi* in jarrah forest soils. Aust J Bot 33: 109-113

Murray P (1984) Extinctions downunder: a bestiary of extinct late pliestocene monotremes and marsupials. In: Martin PS, Klein RG (eds) Quaternary extinctions. University of Arizona Press, Tuscon, pp 600-627

Murray P (1991) The pleistocene megafauna of Australia. In: Vickers-Rich P, Monaghan JM, Baird RF, Rich TH (eds) Vertebrate paleontology of Australasia. Pioneer Design Studio, Melbourne, pp 1071-1164

Newell GR, Wilson BA (1993) The relationship between cinnamon fungus (*Phytopthora cinnamomi*) and the abundance of *Antechinus stuartii* (Dasyuridae: Marsupialia) in the Brisbane Ranges, Victoria. Aust Wild Res 20: 251-259

Noble JC, Bradstock RA (eds) (1989) Mediterranean landscapes in Australia. Mallee ecosystems and their management. CSIRO, Melbourne

Norton DA (1991) *Trilepidea adamsii*; an obituary for a species. Conserv Biol 5: 52-57

Norton DA, Hobbs RJ, Atkins L (1994) Fragmentation, disturbance and plant distribution: mistletoes in woodland remnants in the Western Australian wheatbelt. Conserv Biol in press

Nulsen RA, Bligh KJ, Baxter IN, Solin EJ, Imrie DH (1986) The fate of rainfall in a mallee- and heath vegetated catchment in southern Western Australia. Aust J Ecol 11: 361-371

Pate JS (1989) Australian micro stilt plants. Trends Ecol Evol 4: 45-49

Pate JS, Hopper SD (1993) Rare and uncommon plants in ecosystems, with special reference to the South West Australian flora. In: Schulze E-D, Mooney HA (eds) Ecosystem function of biodiversity. Ecological studies 99. Springer, Berlin Heidelberg New York, pp 293-325

Pate JS, Dixon K, Orshan G (1984a) Growth and life form characteristics of kwongan species. In: Pate JS, Beard JS (eds) Kwongan. Plant life of the sandplain. University of Western Australia Press, Nedlands, pp 84-100

Pate JS, Weber G, Dixon KW (1984b) Stilt plants – extraordinary growth form of the kwongan. in: Pate JS, Beard JS (eds) Kwongan. Plant life of the sandplain. University of Western Australia Press, Nedlands, pp 101-125

Porter J (1979) Vertebrate remains from a stratified Holocene deposit in Skull Cave, Western Australia, and a review of their significance. J R Soc West Aust 61: 109-117

Portlock CC, Shea SR, Majer JD, Bell DT (1990) Stimulation of germination of *Acacia pulchella*: laboratory basis for forest management options. J Appl Ecol 27: 319-324

Postle AC, Majer JD, Bell DT (1986) Soil and litter invertebrates and litter decomposition in Jarrah (*Eucalyptus marginata*) forest affected by Jarrah dieback fungus (*Phytophthora cinnamomi*). Pedobiologia 29: 47-69

Reid N (1987) The mistletoe bird: coincidence or coevolution. Emu 87: 130-131

Reid N, Yan Z (1994) Mistletoe biology, pathology and management. In: Kile GA, Old K, Keane PJ (eds) Eucalypt diseases. CSIRO, Melbourne (in press)

Rich PV, Van Tets G (1984) What fossil birds contribute towards an understanding of origin and development of the Australian avifauna. In: Archer M, Clayton G (eds) Vertebrate zoogeography and evolution in Australia. Hesperian Press, Carlisle, Western Australia, pp 421-446

Saunders DA (1989) Changes in the avifauna of a region, district and remnant as a result of fragmentation of native vegetation: the wheatbelt of Western Australia. A case study. Biol Conserv 50: 99-135

Saunders DA (1990) Problems of survival in an intensely cultivated landscape: the case of Carnaby's cockatoo *Calyptorhynchus funereus latirostris*. Biol Conserv 54: 277-290

Saunders DA (1993) A community-based observer scheme to assess avian responses to habitat reduction and fragmentation in south-western Australia. Biol Conserv 64: 203-218

Saunders DA, Curry PJ (1990) The impact of agricultural and pastoral industries on birds in the southern half of Western Australia: past and future. Proc Ecol Soc Aust 16: 303-321

Saunders DA, Hobbs RJ (1992) Impact on biodiversity of changes in land-use and climate. In: Hobbs RJ (ed) Biodiversity of mediterranean ecosystems in Australia. Surrey Beatty, Chipping Norton, NSW, pp 61-75

Saunders DA, Ingram JA (1987) Factors affecting survival of breeding populations of Carnaby's cockatoo *Calyptorhynchus funereus latirostris* in remnants of native vegetation. In: Saunders DA, Arnold GW, Burbidge AA, Hopkins AJM (eds) Nature conservation: the role of remnants of native vegetation. Surrey Beatty, Chipping Norton, NSW, pp 249-258

Saunders DA, Hobbs RJ, Ehrlich PR (eds) (1993) Nature conservation 3: reconstruction of fragmented ecosystems, global and regional perspectives. Surrey Beatty, Chipping Norton, NSW

Schodde R (1981) Bird communities of the Australian mallee: composition, derivation, distribution, structure and seasonal cycles. In: di Castri F, Goodall DW, Specht RL (eds) Ecosystems of the world 11. Mediterranean-type shrublands. Elsevier, Amsterdam, pp 387-416

Schofield NJ, Stoneman GL, Loh IC (1989) Hydrology of the jarrah forest. In: Dell B, Havel JJ, Malajczuk N (eds) The jarrah forest. A complex mediterranean ecosystem. Kluwer, Dordrecht, pp 179-201

Scougall A, Majer JD, Hobbs RJ (1993) Edge effects in grazed and ungrazed Western Australian wheatbelt remnants in relation to ecosystem reconstruction. In: Saunders DA, Hobbs RJ, Ehrlich PR (eds) Nature conservation 3: reconstructing fragmented ecosystems, global and regional perspectives. Surrey Beatty, Chipping Norton, NSW, pp 163-178

Shea SR, Kitt RJ (1976) The capacity of jarrah forest native legumes to fix nitrogen. For Dept West Aust Res Pap 21: 1-34

Shea SR, Gillen KJ, Kitt RJ (1978) Variation in sporangial production of *Phytophthora cinnamomi* Rands on jarrah (*E. marginata* Sm.) forest sites with different understorey compositions. Aust For Res 8: 219-226

Shea SR, McCormick J, Portlock CC (1979) The effects of fires in regeneration of leguminous species in the northern jarrah (*Eucalyptus marginata* Sm.) forest of Western Australia. Aust J Ecol 4: 195-205

Singh G, Kershaw AP, Clark R (1981) Quaternary vegetation and fire history in Australia. In: Gill AM, Groves RH, Noble IR (eds) Fire and the Australian biota. Australian Academy of Science, Canberra, pp 23-54

Solbrig OT (1991) From genes to ecosystems: a research agenda for biodiversity. IUBS, Paris

Specht RL (1972) The vegetation of South Australia. Government Printer, Adelaide

Specht RL (1981) Mallee ecosystems in southern Australia. In: di Castri F, Goodall DW, Specht RL (eds) Ecosystems of the world 11. Mediterranean-type shrublands. Elsevier, Amsterdam, pp 203-231

Specht RL, Rayson P (1957a) Dark Island heath (Ninety Mile Plain, South Australia) I. Definition of the ecosystem. Aust J Bot 5: 52-85

Specht RL, Rayson P (1957b) Dark Island heath (Ninety Mile Plain, South Australia) III. The root systems. Aust J Bot 5: 103-114

Springett BP (1976) The effect of planting *Pinus pinaster* Ait. on populations of soil microarthropods and on litter decomposition at Gnangara, Western Australia. Aust J Ecol 1: 83-87

Strahan RE (1983) The Australian Museum complete book of Australian mammals. Angus and Robertson, London

Swift MJ, Anderson JM (1993) Biodiversity and ecosystem function in agricultural systems. In: Schulze E-D, Mooney HA (eds) Ecosystem function of biodiversity. Ecological studies 99. Springer, Berlin Heidelberg New York, pp 15-41

van Schagen JJ, Hobbs RJ, Majer JD (1992) Defoliation of trees in roadside corridors and remnant vegetation in the Western Australian wheatbelt. J R Soc West Aust 75: 75-81

Vitousek PM, Hooper DU (1993) Biological diversity and terrestrial ecosystem biogeochemistry. In: Schulze E-D, Mooney HA (eds) Ecosystem function of biodiversity. Ecological studies 99. Springer, Berlin Heidelberg New York, pp 3-14

Walker BA, Pate JS (1986) Morphological variation between seedling progenies of *Viminaria juncea* (Schrad. & Wendl.) Hoffmans (Fabaceae) and its physiological significance. Aust J Plant Physiol 13: 305-319

Ward SC, Koch JM, Nichols OG (1990) Bauxite mine rehabilitation in the Darling Range, Western Australia. Proc Ecol Soc Aust 16: 557-565

Weiss SB, Murphy DD (1993) Climatic considerations in reserve design and ecological restoration. In: Saunders DA, Hobbs RJ, Ehrlich PR (eds) Nature conservation 3: reconstruction of fragmented ecosystems, global and regional perspectives Surrey Beatty, Chipping Norton, NSW, pp 89-107

Weste G (1986) Vegetation changes associated with the invasion by *Phytophthora cinnamomi* of defined plots in the Brisbane Ranges, Victoria, 1975-1985. Aust J Bot 34: 633-648

Whitfield FB, Shea SR, Gillen KJ, Shaw KJ (1981) Volatile compounds from the roots of *Acacia pulchella* RBr and their effect on *Phytophtora cinnamomi* Rands. Aust J Bot 29: 195-208

Wills RT (1993) The ecological impact of *Phytopthora cinnamomi* in the Stirling Range National Park, Western Australia. Aust J Ecol 18: 145-159

Winter K, Osmond B, Pate JS (1981) Coping with salinity. In: Pate JS, McComb AJ (eds) The biology of Australian plants. University of Western Australia Press, Nedlands, pp 88-114

6 Links Between Biodiversity and Ecosystem Function in the Cape Floristic Region

D. M. Richardson, R. M. Cowling, W. J. Bond, W. D. Stock and G. W. Davis

6.1 Introduction

6.1.1 The Link Between Biodiversity and Ecosystem Function

Natural landscapes do not satisfy human requirements for food and fibre, and are being transformed at an increasing rate to make way for intensive agriculture, forestry and other activities. This wave of transformation has increased the rate of reduction in biological diversity at all scales, from genes to landscapes. Other agents of change, including global warming, threaten to accelerate this impoverishment. It has been suggested that this magnitude of loss of biodiversity could threaten the continued functioning of certain vital processes that provide the biophysical foundation for human life. The reasoning behind this prediction is that "biotic" effects could influence the functioning of ecosystems via "ecological" effects. Biotic effects include reductions in the numbers of phenotypes and genotypes per species, species per genus, individuals per population, populations per community and communities per landscape. Ecological effects include reduced variation in age structure and levels of sharing of resources between taxa, fewer taxa in pathways, and fewer pathways for resource transfer (Lamont 1992).

Ecologists have only recently started to explore the ways in which taxa and groups of taxa (guilds or functional groups) affect the functioning of ecosystems. Theoretical paradigms concerning the role of biodiversity in controlling ecosystem processes are lacking. There are, thus, no clear rules for pronouncing one taxon or one group of taxa more important than others – we do not know whether some species are ecologically equivalent to others and therefore, if need be, expendable (Walker 1992).

If the support base for human life is under imminent threat because of the loss of biodiversity, then ecology has several important questions to answer. For example, how would the Earth change with a different living cover (with fewer, more, or a different combination of species)? Biologists, interested in detail and diversity, might argue that large changes in

G.W. Davis and D.M. Richardson (Eds.)
Ecological Studies, Vol.109

ecosystem function would result from changes in diversity and composition of the biota. Atmospheric modellers, struggling to simplify complex processes into predictive models, prefer the "green slime approach". They view the living cover of the planet as essentially homogeneous and simply parameterised – see Smith et al. (1992) for discussion on this topic. The fundamental question is whether system processes are particularly sensitive to variations in the composition of the plant cover.

6.1.2 Defining Biodiversity, Ecosystem Function and Scales of Measurement

What is biodiversity, and at what temporal and spatial scale should it be measured for linking it to ecosystem function? The easiest and most common measure of biodiversity is a count of the number of taxa in a given area. Biogeographers usually distinguish between the number of biological species within a habitat (alpha diversity), the number of habitats and turnover of species between them (beta diversity), and the turnover within a habitat from one area to the next (gamma or delta diversity) (Whittaker 1972; Cody 1986b; Bond 1989). Even if it was possible to document all the species in all taxonomic groups in an ecosystem, it is not clear why biodiversity thus defined should relate directly to the functioning of that ecosystem (see Hobbs et al., Chap. 1. this Vol. for a definition of ecosystem functioning). Different functions of ecosystems are driven by different mechanisms and processes, and by different components of biodiversity (see Fig. 1.3 in Hobbs et al., Chap. 1, this Vol.). For example, primary productivity depends largely on functional attributes such as the growth form and rooting depth of constituent plants, their fecundity and sociability characteristics, growth rates, phenology and nitrogen-fixing behaviour. Water yield is influenced mainly by interception qualities and transpiration rates. Arranging taxa into functional groups is intuitively more appropriate than using species as a measure of biodiversity when addressing such ecosystem processes.

For some groups, components of biodiversity fluctuate markedly over time. Since functionally-important taxa may be present for short periods, the time span of assessment of diversity must be sufficiently long to capture such transient components. For systems prone to stand-replacing disturbances such as fire, the shortest practical time scale for addressing the role of biodiversity in ecosystem functioning is the interval between disturbance events. Because systems respond differently to each event, it is also important to consider a time scale of several cycles (Richardson and van Wilgen 1992).

The process or function under consideration dictates the appropriate spatial scale. It is at the scale of the ecological community that one can most meaningfully test for specific ecological equivalence within functional groups. Community-level studies are also useful for identifying keystone taxa whose presence or absence controls the occurrence of other taxa and the operation of important processes (Lamont 1992; Bond 1993). The role of biodiversity in functions involving the transfer of matter and nutrients is best examined at the scale of ecological communities or landscapes.

Our aim was to seek meaningful links between variously defined measures of biodiversity and function at appropriate scales in mediterranean-climate South Africa.

6.1.3 Opportunities at the Cape

The part of South Africa with mediterranean-type climate is a good area for exploring the link between biodiversity and ecosystem function. The region has a rich biota that has been relatively well studied at various scales (see reviews in Deacon et al. 1983; Cowling 1992; van Wilgen et al. 1992b). There is a rich flora with large genera (10 with > 100 species, including *Erica* with 526 and *Aspalathus* with 245) and large groups of species that *appear* to be functionally similar with respect to several important processes. This is ideal for exploring notions of functional equivalence (Richardson and Cowling 1993). Several very different vegetation formations (e.g. species-rich fynbos, species-poor forest patches, stands of alien trees and shrubs) occur in areas with roughly similar environmental conditions. These provide ideal situations for exploring the determinants of ecosystem boundaries and the consequences of different functional groups and species assemblages on ecosystem processes.

The region has been home to three major ethnic groups of humans, with different levels of dependency on biodiversity (Donaldson and Scott 1994). The Khoi-San, present since about 21 000 years B.P., were hunter-gatherers or pastoralists and exemplified the "subsistence" extreme of total dependence on local biodiversity for survival. The Southern Nguni (Xhosa), present on the eastern fringe of the region since the 16th century, had a developed agriculture based on a small number of species. European colonists brought with them, in 1652, a complex substitution culture that relied largely on introduced plants and animals. It is important to realize that the scarcity of resources such as nutrients, water, fuel and fibre precluded the establishment of sedentary human settlement without the "substitution technology" introduced by the European colonists. Three centuries of intensive substitution culture have created a diversity of landscapes, ranging from totally transformed to almost pristine (the latter almost exclusively where nutrients or topography have precluded substitution), with a wide array of natural fragmentation "experiments". In the lowlands of the western Cape, less than 6% of renosterveld and 14% of fynbos are untransformed (Boucher 1981). The region shares many features with other mediterranean-type ecosystems (MTEs), most notably those in Australia (Kruger et al. 1989).

This chapter provides a brief review of what is known about the biodiversity of South African MTEs. We discuss the patterns and determinants of diversity for plants, birds and mammals, and summarize what little is known about arthropods. We then explore the links between biodiversity (the independent variable, defined in various ways) and ecosystem functions (dependent variables) which we define as the processes that maintain natural systems through the transfer of matter, nutrients and energy (see Hobbs et al., Chap. 1 this Vol.). Our choice of ecosystem functions to study was governed largely by the availability of data. Most ecological studies in the region have addressed biodiversity at spatial and temporal scales too fine for linking with ecosystem function. Also, almost all studies have focussed on either the population/community level *or* ecosystem process – few have attempted to link these levels (Richardson and van Wilgen 1992). Nevertheless, we synthesise available information and attempt to determine whether biodiversity affects ecosystem functions in a predictable way, and whether there is evidence that biodiversity buffers ecosystem function. We address the question of whether plant cover can be viewed as essentially homogeneous for system processes (the "green slime approach" outlined above) by asking whether changes in biodiversity influence salient ecosystem functions.

6.2 Geography of the Region

For convenience we have chosen the Cape Floristic Region (CFR) (Fig. 6.1) as the domain for discussion. Although not all of the CFR experiences a true mediterranean-type climate as defined by Aschmann (1973), we consider the entire region since it is at this scale that diversity patterns are most informative. Also, no vegetation formations are confined to the true mediterranean-climate zone.

The CFR landscape is dominated by rugged and steep fold mountains of sandstone or quartzite that reach moderate elevations (ca. 2000 m) (Campbell 1983). The mountain soils are extremely nutrient-poor, whereas soils in the lowlands, mostly derived from shales, are moderately fertile. The coastal margin has a complex sequence of acidic and alkaline sands of marine and aeolian origin (Deacon et al. 1992). The western part of the

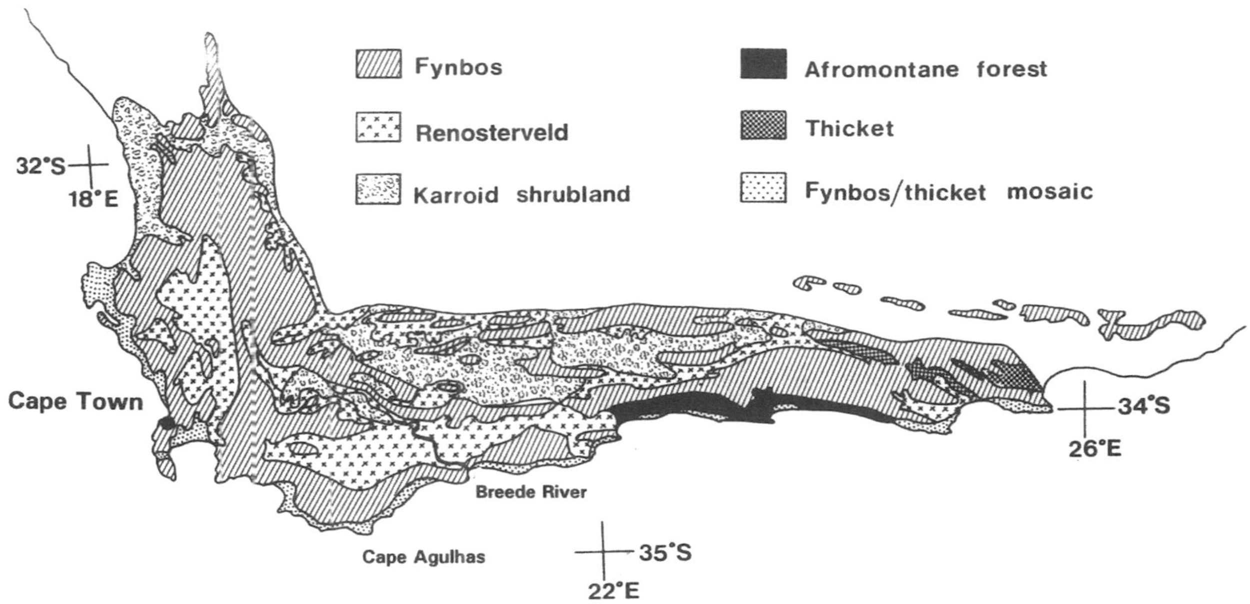

Fig. 6.1. The Cape Floristic Region (CFR) showing the major vegetation formations. The Breede River defines the boundary between the southwestern and southeastern CFR as discussed in the text

region has a mediterranean-type climate with cool, wet winters with low solar irradiance, and hot dry summers with high solar irradiance (Deacon et al. 1992; but see Versfeld et al. 1992, Hobbs et al., Chap. 1. 1.5.1, this Vol. for comparisons with other MTEs). The change from predominantly winter rainfall in the west to non-seasonal rainfall in the east is relatively abrupt at about the Breede River which defines the boundary between the south-western and south-eastern CFR (Fig. 6.1). Annual rainfall varies from <200 mm in the inland valleys, to about 400 mm on the broad coastal forelands, and from about 800 to over 3000 mm in the mountains, where rainfall gradients are very steep (Campbell 1983; Deacon et al. 1992; Versfeld et al. 1992).

6.3 Patterns of Biodiversity in the Cape Floristic Region

6.3.1 Broad-Scale Patterns of Biodiversity

The CFR, which is recognised as a distinct plant kingdom (Goldblatt 1978; Bond and Goldblatt 1984), covers about 90 000 km^2 at the south-western tip of Africa (Fig. 6.1). Regional richness of vascular plant species in Cape shrublands of the CFR is especially high (e.g. 2256 species in 470 km^2 of the Cape Peninsula; 1383 in 240 km^2 at Cape Hangklip) (Cowling et al. 1992). Interestingly, this diversity resides within relatively few genera, giving species-to-genus ratios (e.g. 4.2 for the Cape Peninsula, 8.6 for the entire CFR; Cowling and Holmes 1992) more typical of islands than of a continental flora (Goldblatt 1978). About 70% of the 8504 plant species in the CFR are endemic (Bond and Goldblatt 1984); many highly localised taxa occur in small populations (Cowling et al. 1992). There is a high degree of endemism in several other taxonomic groups, including archaic invertebrates, terrestrial molluscs, Coleoptera, Lepidoptera, amphibia and fishes (references in Macdonald and Richardson 1986). Although a distinct biotic zone is recognised for mammals (Rautenbach 1978), the species richness and endemism in this and other higher vertebrate classes does not parallel that of the flora (Bigalke 1979 for mammals; Siegfried and Crowe 1983 for birds). Large mammalian predators (Stuart et al. 1985) and a few large predatory birds (Boshoff et al. 1983) have been eliminated or greatly reduced in numbers in the region. On the other hand, many birds have invaded the region from adjacent biomes in response to habitat modification (Macdonald 1991), thus reducing the effects of the region's insularisation on bird distribution patterns. This phenomenon is also evident for dung beetles (Coleoptera: Scarabaeidae, Aphodiidae, Staphylinidae,

Histeridae, Hydrophilidae; Davis 1993), and probably for many other animal groups.

The characteristic vegetation of the CFR is fynbos, a sclerophyllous shrubland which covers about 47% of the area, mainly on sandy nutrient-poor soils in the mountains and lowlands (Figs 6.1 and 6.2). Fynbos is prone to fire at intervals of between 4 and 40 years (Kruger 1983; van Wilgen et al. 1992b). Seedling recruitment is virtually confined to the immediate post-fire period (Kruger 1983; but see Le Maitre 1992). Other major vegetation formations are karroid shrublands and renosterveld, Afromontane forest and various thicket types (Table 6.1). Several species of alien plants, notably species of *Acacia*, *Hakea* and *Pinus*, have invaded natural vegetation over large parts of the region (Richardson et al. 1992). These species now form dense self-sown stands that have changed vegetation structure significantly (van Wilgen and Richardson 1985).

6.3.2 Patterns and Determinants of Diversity: Plant Species and Functional Groups

The CFR is unique among the world's centres of plant diversity in that patterns of alpha diversity and species turnover have been relatively well documented (Table 6.2). Most research has been done in fynbos which harbours about 80% of the CFR species (Goldblatt 1978). Fynbos communities are moderately rich in plant species by global standards (Cowling et al. 1992). Mean species richness was 16 (range: 7-27) in 1 m^2 plots and 66 (31-126) in 1000 m^2 plots at 40 sites throughout the region (Cowling et al. 1992). These values are about twice the number recorded in Californian chaparral (Naveh and Whittaker 1979; Bond 1983; Cowling 1983), and of the same order as southwestern Australian kwongan or heathland (Naveh and Whittaker 1979; Rice and Westoby 1983). Species richness in tropical rainforest communities is about double that of fynbos (Gentry 1988). Other temperate communities such as certain European grasslands and grazed woodlands in Israel are also richer in species than fynbos at this scale (Bond 1983). Forest patches in the CFR, usually separated from fynbos by a distinct boundary, are relatively species-poor. The number of major woody plant species in forests decreases from 50 to 31 between Knysna in the east to Cape Town in the west. This is mainly the result of increasing isolation and decreasing stand size (McKenzie 1978). See Table 6.2 for a summary of the components of diversity for the other major vegetation types.

The high regional richness of plant species in the CFR is due to the very high turnover of moderately rich communities over long ecological gradients (beta diversity) and the very high turnover within communities

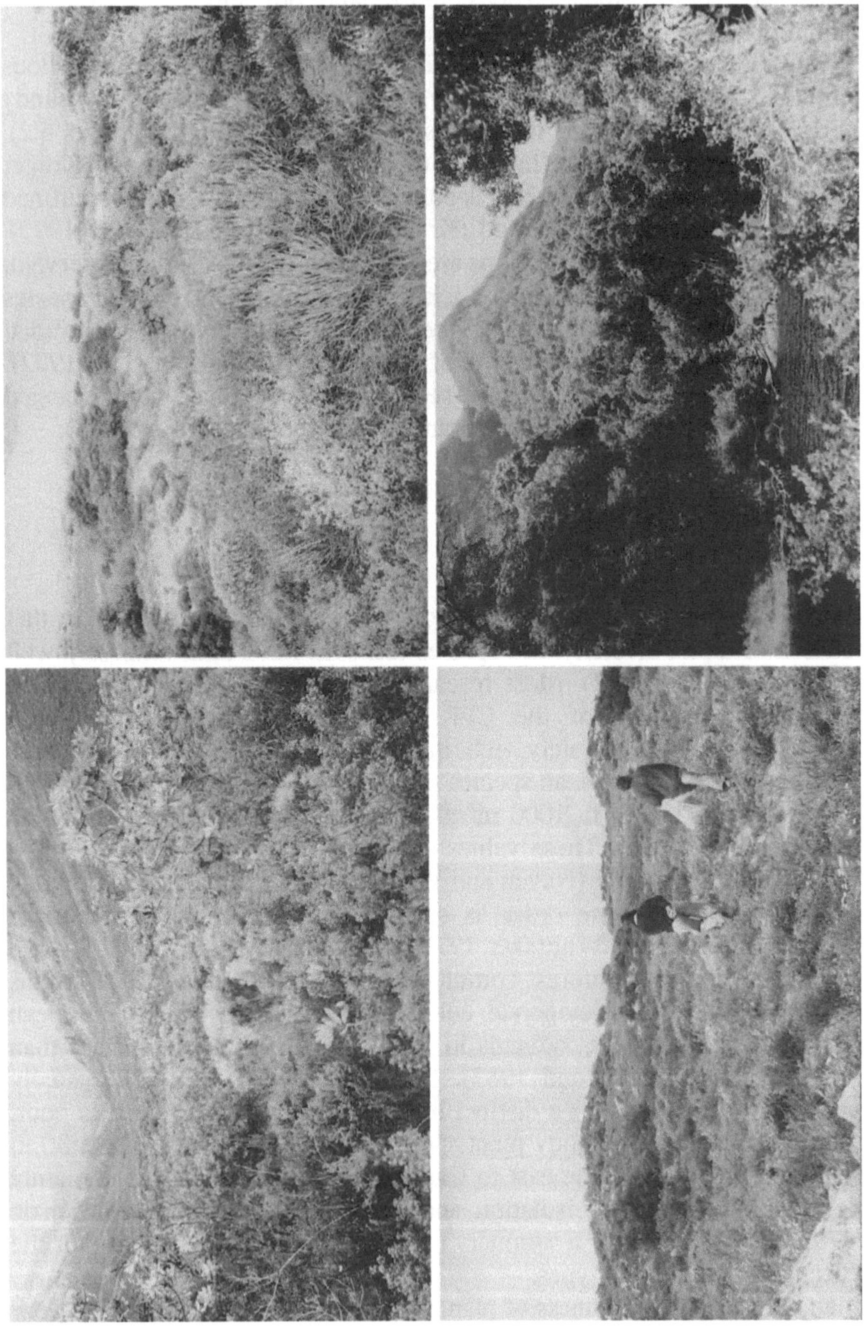

Fig. 6.2a-f. Major vegetation formations in the Cape Floristic Region (clockwise, starting top left): **a** Fynbos (mountain fynbos at Swartboskloof). **b** Karroid shrubland (karroid broken veld near Robertson). **c** Afromontane forest (Natures Valley). **d** Dense stand of alien trees (*Pinus halepensis* at Miller's Point, Cape Peninsula). **e** Sub-tropical thicket (Valley Bushveld, Bontebok National Park). **f** Renosterveld (near Malmesbury). (Photos c,d,e R.M. Cowling; a,b,f D.M. Richardson)

Table 6.1. Features of major vegetation formations in the Cape Floristic Region

Formation	Area (% of CFR)	Altitude (m)	Rainfall (mm a^{-1})	Soils	Dominant growth forms (growth-form diversity)	Flam-mability
Fynbos	47	0 - 2325	210 - 3000	Sandy, acid, infertile	Sclerophyll shrubs, evergreen graminoids, geophytes (low)	High
Renoster-veld	20	10 - 1500	250 - 550	Sandy loams, weakly acid, moderately fertile	Sclerophyll shrubs, deciduous graminoids, geophytes (moderate)	Moderate
Karroid shrubland	26	0 - 1500	50 - 250	Sand-clay, alkaline, fertile	Succulent and deciduous shrubs, geophytes, annuals (moderate)	Low
Sub-tropical thicket	5	0 - 1000	400 - 650	Sand-clay, alkaline-weakly acid, fertile	Sclerophyll, deciduous and succulent shrubs and low trees, lianas, geophytes (high)	Low
Afromon-tane forest	2	0 - 1000	650 - 1200	Sand-clay, acid, moderately fertile – infertile	Sclerophyll trees and shrubs, lianas, forbs (low to moderate)	Low

Table 6.2. Diversity of species and growth forms in major vegetation formations in the Cape Floristic Region

Formation	Alpha diversity (species)	Beta diversity (species)	Beta diversity (functional groups)	References
Fynbos	Moderate	Very high	Low	Cowling and Campbell (1983); Cowling (1990)
Karroid shrublands and renosterveld	Moderate to high	Moderate to high	High	Jurgens (1986); Cowling et al. (1989)
Thicket and forest	Low to high	Moderate	Moderate (forest) to high (thicket)	Cowling and Campbell (1983)

and between floras along geographical gradients (gamma diversity) (Cowling et al. 1992). Alpha diversity is remarkably constant across the CFR, but both beta and gamma diversity are lower in the southeast; regions in the southwest support more than double the number of species than equivalent regions in the southeast (Fig. 6.3). Alpha diversity in fynbos is unrelated to regional richness at this scale, and is best explained by non-equilibrium models that invoke low nutrient soils and recurrent fire to maintain coexistence (Bond 1983; Cowling 1987; Kruger 1987).

Plant diversity in fynbos communities is divided among four major growth forms (Cowling 1983; Kruger 1983; Cowling 1990; Le Maitre and Midgley 1992). The relatively open overstorey (1-3 m) usually comprises between one and four proteoid (Proteaceae) shrub species. More than half the species in fynbos are morphologically similar, small-leaved, understorey shrubs. Evergreen graminoids and geophytes are common, the latter being most prominent after fire (Kruger 1983; Le Maitre and Brown 1992). Most fynbos species (e.g. about 86% of a sample of 210 species at Swartboskloof; van Wilgen and Forsyth 1992) resprout after fire, and appear to be resilient under a range of fire regimes. However, in most communities the dominant species (those making up most biomass) are killed by fire. Although there may be very large changes in species composition

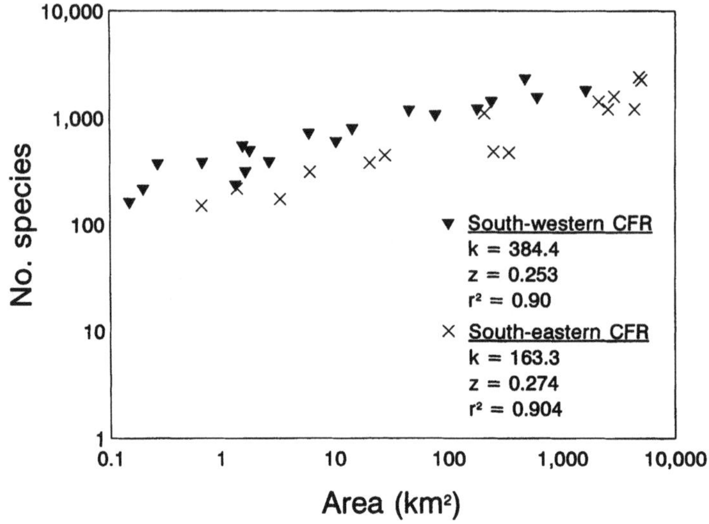

Fig. 6.3. Species-area relationships for regional fynbos floras from the south-western and the southeastern Cape Floristic Region (west and east of the Breede River respectively - see Fig. 6.1)

of fynbos communities along environmental gradients (Cowling and Campbell 1984; Cowling 1990), the mix of growth forms remains relatively uniform (Cowling and Campbell 1983; Cowling 1990). For example, on the Agulhas Plain on the southwestern Cape lowlands, adjoining fynbos communities on different substrata may share no species but have an almost identical mix of growth forms (Cowling 1990). A similar pattern was evident along a rainfall gradient (450-1000 mm a^{-1}) in a southeastern Cape fynbos landscape (Cowling and Campbell 1983). Shrubs and graminoids predominate throughout the fynbos - in bioclimates which support tall woodland or forest in other mediterranean-climate regions (Richardson and Cowling 1992). The mix of growth forms in fynbos is similar to that of Australian kwongan (Milewski and Cowling 1985) but differs from other mediterranean-climate shrublands: mature chaparral comprises a single closed stratum of shrubs (Keeley 1992), whereas annuals contribute most species to the diverse, heavily grazed and open shrublands and woodlands of Israel (Naveh and Whittaker 1979). Fynbos vegetation lacks the structural complexity (or growth-form diversity) required to contribute plant species richness such as that of some tropical rainforest communities.

Non-fynbos vegetation on more fertile soils ranges from low succulent thicket to tall evergreen forest along rainfall gradients (e.g. 450-1000 mm a^{-1} in the southeastern Cape; Cowling and Campbell 1983). The dominant growth forms in karroid shrublands are evergreen shrubs and leaf succulent shrubs, both < 1 m tall. Stem succulent shrubs, deciduous shrubs (both < 1 m) and grasses are also important at some sites (Cowling et al. 1994).

6.3.3 Patterns and Determinants of Diversity: Animal Species and Functional Groups

The study of patterns and determinants of diversity for animal groups in the CFR has received relatively little attention. Bird distribution has been well documented at the scale of quarter-degree squares, at least for the western part of the region (Hockey et al. 1989). Alpha, beta and gamma diversity for birds is low in the region, as are typical population sizes. The small population sizes (Siegfried 1983; Fraser 1989) are probably due to the scarcity of resources, especially insects (Siegfried 1979; Cody 1983b). The low species richness is probably due to the relatively low structural diversity of the vegetation within and between CFR landscapes. Recent changes in distribution patterns and abundance of many taxa and the composition of bird assemblages provide useful clues concerning the determinants of community assembly.

Fynbos shrublands throughout the region have a small distinctive resident avifauna, the specific composition of which depends on vegetation structure and resource availability. It typically comprises at least several nectarivores, one or more granivore, one or more "grass warbler" (sensu Garcia 1985) and one or more "bush warbler" (Richardson and Fraser 1994). As is true for forest tree species, there is a marked pattern of species impoverishment for birds in forest patches as one moves west and away from large forests (Cody 1983a,b). The avifauna of small forest patches in the southwestern part of the CFR typically comprises a mix of true forest birds and several other species with wide ecological niches that track pulses of resource availability; Cody (1983a) defined guilds of foliage insectivores, sallying flycatchers and slow-searching omnivores. Plantations and self-sown stands of alien trees in fynbos support sparse species-poor avifaunas (Armstrong 1993), but may increase bird diversity in treeless areas (Macdonald 1989).

A feature of avifaunas of representative patches of natural vegetation, at least in the western part of the region, is the small contribution of sedentary (resident) species. Many common and widespread species show marked intra-region movements linked to resource requirements (Cody 1983a; Siegfried 1983; Rebelo et al. 1984; Rebelo 1987; Fraser et al. 1989, 1990; Hockey et al. 1989). For example, at Swartboskloof (29-year-old mountain fynbos with small forest patches; 375 ha; 34°00'S; 18°57'E), only 28% of species recorded over 12 months were resident (Richardson and Fraser 1994; see also Fraser 1989); this reflects the transient nature of resources such as insects, fruit and especially nectar associated with the strongly seasonal rainfall (see Cody 1983a p 213). These factors are important determinants of bird diversity patterns, but influence different guilds in different ways. Cody (1983a) has invoked interspecific competition and density compensation explanations to explain bird diversity patterns in forests in the CFR.

Food for mammals is scarce in fynbos shrublands. Palatable grasses are rare and productivity is comparatively low apart from the first few years after fire, so that large mammal densities are low (Bigalke 1979). However, seeds of some fynbos species are quite large and protein-rich; these support several small mammal species. Small mammals are thought to be important seed predators potentially controlling plant populations (Bond 1984; Bond and Breytenbach 1985; Botha 1988; van Hensbergen et al. 1992), and possibly acting as a major selective force in the evolution of serotiny and myrmecochory. Alpha diversity of small mammals in fynbos is comparable to that of chaparral, and higher than that for Australian heath (Fox et al. 1985), but lower than African grasslands and savanna. There is noticeable species turnover along altitudinal gradients in mountain fynbos (Bond et al. 1980; Breytenbach 1982), but little variation in species within the CFR (Botha 1988; van Hensbergen et al. 1992). Small mammal densities vary with habitat structure and season, averaging about 20 animals ha^{-1} (Breytenbach 1982; Botha 1988).

There have been few studies of small mammals in the sandy soils of the fynbos lowlands where both species composition and system effects may be very different from the mountains. Mole-rats (family Bathyergidae) are common in lowland fynbos and can create impressive cratered landscapes with (presumably) large effects on soil-forming processes, plant productivity and species composition. Several species are endemic to the region but their interrelationships, functional equivalence and system-wide effects have yet to be studied.

Small mammal distribution is remarkably insensitive to plant diversity or floristic composition. Small mammals respond more to physical structure of the vegetation than its floristic composition or diversity (Bond et al. 1980; Breytenbach 1982; Botha 1988; van Hensbergen et al. 1992). The striped field mouse (*Rhabdomys pumilio*), for example, prefers "grassy" habitats but densities vary little between "grass" made up of fynbos Restionaceae or the true grasses of grassland or bushveld (Bond et al. 1980). Major changes in floristic composition or diversity may therefore not be matched by comparable changes in small mammals.

To summarise, vegetation structure is far more important than plant species richness in determining the composition of bird and small mammal communities. The lack of structural complexity in the fynbos vegetation of the CFR and the scarcity of food resources accounts for the low levels of species richness in these groups. No bird or small mammal species are dependent on individual plant species (or vice versa). Loss of plant species or a change in composition with no concomitant change in vegetation structure is therefore likely to have little effect on bird and small mammal diversity. Loss of plant guilds, such as tall non-sprouting shrubs, would, however, result in large changes in animal communities because of the importance of the former in structuring animal habitats and/or providing specific resources (e.g. for nectar-feeding birds; Richardson and Fraser 1994). Increased structural diversity through habitat modification has increased bird and mammal diversity at local and regional scales. All animal groups that have been studied in fynbos (i.e. birds, small mammals and ants) show marked changes in community composition that are clearly correlated with vegetation recovery after fire, as shown for example by Donnelly and Giliomee (1985a), van Hensbergen et al. (1992) and Wright and Donaldson (1994).

6.4 Biodiversity and Ecosystem Function: Community-Level Perspectives

6.4.1 Plant Growth Forms as Functional Groups

A community is a group of organisms, many of which interact within a framework of horizontal and vertical linkages (Giller and Gee 1987). Classical competition theory predicts that every species in a community occupies a distinct niche which is defined relative to a set of limiting resources, to which species respond differentially (MacArthur and Levins 1967). This theory has its roots in animal ecology and was only recently extended explicitly to plant communities (Cody 1986a,b,1989,1991). Cody argues that each plant growth form represents a particular strategy for resource utilisation which can be envisaged as a structural niche. The quantity and quality of resources, and species interactions, determine the length of the structural niche axis and, hence, the number of growth forms (or functional types) that may coexist in a community. Under conditions of a strong limiting resource (e.g. summer soil moisture in MTEs), each growth form represents a unique solution (as shown for example by Davis and Mooney 1986), and the diversity of growth forms parallels that of species: species belonging to similar growth forms (functional analogues) are segregated by habitat (Cody 1989, 1991).

Classical competition theory predicts little or no functional equivalence in plant communities and suggests that each species plays a fundamental role in ecosystem processes. However, many plant ecologists have found the classical competition theory of limited value in predicting plant community structure (Table 6.3). Other models of community structure allow for many functional analogues, suggesting that species can be lost with little effect on ecosystem processes (Solbrig 1991; Chapin et al. 1992; Walker 1992).

Cowling et al. (1994) examined the relationship between species richness and growth form richness (functional diversity) in phytosociological plots from the southwestern Cape (winter rainfall) and the southeastern Cape (non-seasonal rainfall). For fynbos and non-fynbos in both regions, the slopes of the regressions between growth form richness and species-richness exceeded unity (Fig. 6.4), suggesting the existence of many functionally analogous taxa (at least in terms of growth forms as defined). The slopes were considerably steeper for fynbos than non-fynbos communities, but significantly different ($P<0.05$) only between fynbos and non-fynbos in the southwest. Contrary to prediction, there were no significant differences in both species richness and growth forms in plots from fynbos in the southwest and southeast. The southeastern non-fynbos plots had significantly more growth forms ($P<0.005$) and species ($P<0.001$) than those from the southwest. As indicated by the steeper (but not significantly so) slope, there are more functional analogues in the southwest.

6.4.2 Regeneration Traits of Plants

The large number of taxa within certain growth forms suggests that the removal of some analogues should have little impact on ecosystem-level processes. Much of the diversity in the fynbos flora is associated with post-fire regeneration traits that facilitate competition for space (Le Maitre and Midgley 1992). Lottery and regeneration niche models can explain the coexistence of functionally analogous proteoid and ericoid shrubs (Bond et al. 1992). The diversity of regeneration traits (notably age to maturity, seed production or sprouting ability, seed size, seed storage and dispersal syndromes, germination cues, pollination and breeding systems) within growth forms seems to provide a "backup" in the event of stochastic population fluctuations induced by fire (Yeaton and Bond 1991; Kilian and Cowling 1992). This ensures long-term stability of community-level functional diversity and low variance in ecosystem processes (review in Richardson and van Wilgen 1992).

Table 6.3. A summary of reasons for the limited value of classical competition theory for predicting plant community structure

Reasons	Key references
Frequent disturbance limits the role of competition	Huston (1979); Denslow (1985)
Plant communities are rarely, if ever, saturated with species, thus rendering niche limitation meaningless	Whittaker (1977); Wilson et al. (1987)
Resources are too few or too similar for niche differentiation to explain diversity	Silvertown and Law (1987)
Coexistence of functional analogues can be explained by regeneration niche differences	Grubb (1977)
For sessile organisms such as plants, similarity in mean lifetime propagule production and, covariantly, in growth form, will promote long-term coexistence (the so-called lottery model)	Fagerstrom (1988)

6.4.3 An Ecophysiological Approach to Functional Grouping in Plants

The approaches to understanding the functional relationships between plants and their environments described in sections 6.4.1 and 6.4.2 are probably extremes which bracket reality. On the one hand it is unlikely that the high taxonomic diversity of the characteristic fynbos families (Ericaceae, Proteaceae and Restionaceae) represent a highly complex, exhaustive, and non-overlapping set of unique niches, in spite of Cody's (1986a,b) demonstration of distinct leaf morphologies in co-existing Proteaceae. The work of Cowling et al. (1994) referred to in section 6.4.1 probably represents a more realistic alternative in suggesting the existence of broad functional plant groups which reflect the limited growth form diversity of the main fynbos vegetation types. In many respects this inferred equivalence of growth-form and functional character may be reasonably accurate. It should, at least, provide a reasonable framework for understanding many broad-scale processes in South African MTEs.

Intensive management of land-use is likely to be necessary in the future. This will demand ecological insights based on some mechanistic understanding of system processes if stability and sustainability of imposed system shifts are to be reasonably predicted. The basis of that understanding resides in the ecophysiological information gained over the past few

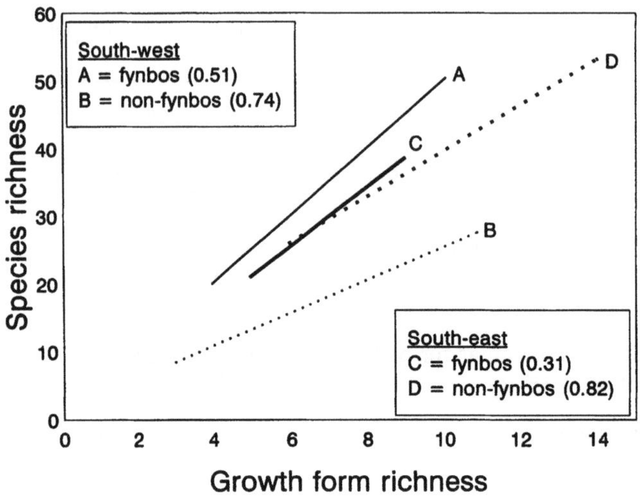

Fig. 6.4. The relationship between growth form richness and species richness in fynbos and non-fynbos phytosociological plots from the southwestern and southeastern Cape (west and east of the Breede River respectively – see Fig. 6.1). The growth forms used in the analysis are: geophytes, non-geophytic forbs, evergreen graminoids, deciduous graminoids, small-leaved evergreen shrubs <1 m and >1 m tall, large-leaved (nanophyllous and larger) evergreen shrubs <1 m and >1 m, deciduous shrubs <1 m and >1 m, succulent shrubs <1 m and <1 m, evergreen trees, and vines. (Data from Cowling et al. 1992)

decades. This must be incorporated into the systematic search for the links between diversity and system function.

Adopting a more detailed perspective of functional diversity in fynbos plants reveals that the criteria for the grouping of functional attributes cannot be identical for all processes. Plants that utilise deeper soil strata for water (broadly epitomised by the proteoid growth form) will not necessarily all acquire nutrients in the same manner, nor will they necessarily provide the same production basis for higher trophic levels. Similarly, the contribution to community gene flow by different species of the same growth form may also be very different. Within the proteoid class, some members will only occur on calcareous soils, whereas structurally similar congenerics are restricted to quartzitic substrata (Mustart and Cowling 1993). Some members of this group also produce fruits or nectar which provide a food resource for birds. Others host arthropods which contribute to the diet of animals. The proteoid growth form also embraces plants with a range of reproductive strategies, for example monoecy versus dioecy, wind pollination versus pollination by insects and birds, and seed dispersal by vectors ranging from wind to ants. In general, it is the set of selective pressures, acting under the constraints of phylogeny, that shape the functional relationships between taxa and their environment. These various adaptations to environmental pressures in turn aggregate to contribute to overall system function. As has been shown, these contributions are not uniform. As a further example, species with medium rooting depth, such as those in the genera *Erica* and *Aspalathus*, may play similar roles in the hydrological cycle, whereas the ericoid mycorrhizae of the former and the nitrogen-fixing symbionts of the latter are very different in their contribution to the capture and recycling of nutrients.

A conceptual model for extending these organism-level functions to an understanding of system-level processes in fynbos may be derived from the notion of a flowpath, which is the linked set of patches that comprise a system (Woodmansee 1988). Using the hydrological process within a plant community as an illustration of a fynbos flowpath, the movement of water from precipitation is influenced by the canopy architecture, the litter layer and other soil strata, subterranean faunal disturbances, a complex set of plant rooting systems, and the transpirational interface between the vegetation and the atmosphere. This flowpath is clearly influenced by its structural heterogeneity, or its diversity. More importantly, the configuration, and hence the characterisation of the flowpath changes depending on what resource flow we are considering. The concept of functional grouping is therefore a relative one; its utility is dependent upon a quantitative knowledge of the system components. The contribution of the ecophysiological perspective of biodiversity to conservation and management in fynbos systems is therefore important, and needs to be a part of the overall strategy for research into ecological complexity in MTEs.

Ecophysiological research into fynbos systems has mainly addressed the factors that limit primary productivity. Environmental stresses in the region, as in other MTEs, are epitomised by nutrient limitations, especially N and P, and by summer drought. As a result, much of the work that could contribute to the understanding of functional diversity within systems relates to the acquisition and cycling of N and P, and to the patterns of water use by typical components of the vegetation. The reproductive biology of fynbos plants has also received some attention. Pollination, dispersal and germination have been reasonably well studied in the Proteaceae, but little or nothing is known about these processes for the remaining taxa. In Table 6.4 we summarise current knowledge of functional phenomena at the plant/organism level, and attempt to relate these to selective environmental pressures that gave rise to them, and to the key system processes, or flowpaths, of which they form part. Clearly, the notion of functional grouping needs considerable sophistication to accommodate

Table 6.4. Fynbos plants are adapted to environmental stresses typical of a mediterranean-type region, with functional consequences that contribute to processes at the ecosystem level. The table provides a synopsis of adaptations to environmental stresses, their functional stresses, and the processes that are affected

Environmental stresses	Adaptation	Functional consequences	Process[a]	Prominent fynbos taxa[b]
Summer drought	Epidermal trichomes	Increased boundary layer resistance	Hy	Pr; Rh; Er
	Sunken stomata	Increased boundary layer resistance	Hy	Er; Re
	Reduced leaf area	Reduced heat loading and water loss	R	mft
	Deep rooting	Access to perennial water	Hy	Pr
	Poikilohydry	Tolerance of extreme stress	Hy	Re; Er
Nutrient limitation	Symbioses	Bacterial facilitation of nutrient uptake	N	Fa
		Fungal facilitation of nutrient uptake	N	Er
	Cluster roots	Increased exchange surface	N	mft
	Slow growth at maturity	Reduced nutrient requirements	N	mft
				Er; Re
	Vigorous post-fire growth	Nutrient sequestration	N	Pr; As

Table 6.4 *continued*

Environmental stresses	Adaptation	Functional consequences	Process[a]	Prominent fynbos taxa[b]
	Cumulative set of persistent seed	Slow build-up of seed store	G	-
Fire	Serotiny	Fire avoidance	G	Pr
	Myrmecochory	Fire avoidance	G	Pr
	Resprouting	Rapid re-establishment	P	Pr; Re; Br
	Geophytic habit	Rapid re-establishment	P	Ir
	Soil-stored seed	Maintenance of viable populations	G	As
Summer heat stress				
1. Seedling	Rapid root growth	Adequate water supply	Hy	Pr
2. Mature plant	Reduced or dissected leaves	Sensible heat removal	R	mft
	Open canopy architecture	Sensible heat removal	R	mft
	Epidermal trichomes	Increased reflectivity	R	Pr; Rh; Er
Predation				
1. Seeds	Serotiny	Avoidance of inter-fire predation	T	Pr

Table 6.4 *continued*

Environmental stresses	Adaptation	Functional consequences	Process[a]	Prominent fynbos taxa[b]
	Myrmecochory	Avoidance of post-fire predation	T	Pr
	Low seed set	Minimization of predator resource base	T	Pr
	Prolific seed set	Predator satiation	T	Er; Re; As
2. Shoots	Sclerophylly	Low nutrient content for herbivores	T; F	mft
	Secondary compounds	Toxicity and unpalatability	T	Ru
Wind	Reduced leaves	Reduced physical resistance and damage	Ha	mft
	Low stature of plants	Protection from landscape boundary layer	Hy	mft

[a] Ecosystem processes (not necessarily mutually exclusive): F - fire; G - gene flow; Ha - habitat structure; Hy - hydrology; N - nutrient cycling; P - pioneer establishment; R - transfer of radiant energy; T - trophic linking
[b] Taxa: As - Asteraceae; Br - Bruniaceae; Er - Ericaceae; Fa - Fabaceae; Ir - Iridaceae; mft - most fynbos tax; Pr - Proteaceae; Re - Restionaceae; Rh - Rhamnaceae; Ru - Rutaceae.

the functional diversity that exists in the mediterranean-type ecosystems of South Africa.

With regard to hydrological processes in fynbos systems, the assumed equivalence of growth form and functional character (Cowling et al. 1994) may be reasonably accurate. The broad classes of rooting depth represented by the different growth forms – typically depicted as restioid (shallow rooting), ericoid (medium depth rooting), and proteoid (deeper rooting) (Higgins et al. 1987) – have been shown to behave differently with respect to their ability to utilise soil water (see review in Stock et al. 1992). If a system is subject to perturbation in which the distribution pattern of these groups is altered by fire, disease or physical disturbance, it follows that the hydrological properties of the system must change. This is illustrated by a study of changes induced in a mountain fynbos system where the substratum was disturbed by tilling (Davis 1992). The system shifts included a reduction in species richness and Shannon-Wiener diversity (Davis 1990), and increases in surface albedo and summer soil water content, and a reduction in soil temperature (Davis 1992). Nutrient pools were unaltered. Simplification (i.e. a reduction in biodiversity) caused by physical disturbance therefore had a significant effect on energy flow and hydrological processes in this system at the patch level. This may or may not have repercussions at the landscape level.

6.4.4 Empty Functional Groups

The causes and consequences of the marked under-representation of trees in fynbos, relative to environmentally similar regions in other MTEs, have been debated by several authors (Campbell et al. 1979; Moll et al. 1980; Lamont 1982; Richardson and Cowling 1992). Several invasive alien plants have filled this "vacant niche", thereby *increasing* the species diversity and functional diversity of many fynbos communities in the *short term*, but decreasing these parameters in the long term (i.e. over several fire cycles) as native elements are eliminated. These invaders have had marked effects on ecosystem function (Macdonald and Richardson 1986; Richardson and van Wilgen 1986a,b). Species of the genera *Acacia, Hakea* and *Pinus* have been particularly successful, and dense self-sown stands of these taxa cover large parts of the region (Richardson et al. 1992) (see Sect. 6.4.6).

6.4.5 Animal Communities

Very little is known about the structure of animal communities in the CFR. It is thus difficult to address issues such as functional equivalence at the same level of detail as for plants. In this section we attempt to synthesise salient information for birds, small mammals and arthropods. Fynbos birds could feasibly influence ecosystem function through their roles as consumers (especially of seeds), pollinators and seed dispersers. The trophic structure of fynbos birds has been poorly studied but there are few specialists and many generalists. Even in a relatively well-studied group like the canaries (*Serinus* spp.), we cannot begin to guess how adding or removing species would affect ecosystem functioning. There is considerable overlap in the diets of seven *Serinus* spp. that occur in the CFR. Closer examination, however, reveals quantitative differences in the frequency of feeding on food types (Milewski 1978). These differences are not a good foundation for dividing canaries into functional groups since taxa show different patterns of local and regional movement and abilities to track resources. Some granivorous birds, including canaries, consume substantial amounts of seed between or soon after fire (D.M. Richardson, unpubl. data). The effect of this seed loss on plant community dynamics has not been assessed. Nectar-feeding specialists are important (but apparently not essential) pollinators of many fynbos plant species (Coetzee and Giliomee 1985; Wright et al. 1991; Visser 1992). Very few fynbos plants have fleshy fruits (Le Maitre and Midgley 1992), but seed dispersal by birds is common in forest and thicket formations (Knight 1988). Several frugivorous birds move regularly across the forest/fynbos ecotone and play an important role in mediating the dynamics of this ecotone in the absence of fire (Manders and Richardson 1992); this feeds forward to precipitate ecosystem-level effects (see Sect. 6.5.2). The role of birds in regulating fynbos insect populations has not been studied.

About 120 bird species have invaded the region from adjacent biomes since 1850, increasing the number of species in the region from around 220 to 340. This increase in species richness is probably the result of the increased abundance and diversity of insects in the region as a result of the increasing extent of irrigated pastures and various crops and weeds, and the increase in the occurrence of trees (Macdonald 1991). None of these "new" species can survive in natural vegetation formations, although some do utilise resources in natural formations opportunistically (Richardson et al. 1992). We do not know whether the addition of these species (mainly insectivores) has altered the functioning of these ecosystems. At least 36 of the 224 native birds recorded from terrestrial ecosystems of the fynbos biome have invaded in response to the new habitats created by woody alien plants (Macdonald and Richardson 1986).

The question of functional equivalence in small mammal communities is also not simply answered. From a dietary point of view, small mammals show considerable functional overlap in fynbos; most species, including shrews, eat substantial amounts of seeds (Breytenbach 1982). The exceptions are *Otomys* spp. which consume green matter, and the elephant shrews (Macroscelididae), which are almost exclusively insectivorous (Skinner and Smithers 1990). There is some degree of dietary specialization. For example, the fynbos endemics *Praomys verreauxi* and *Acomys subspinosus* consume mostly protea seeds and restio seeds respectively (Breytenbach 1982). Most species show distinct cover preferences; this tends to segregate species with similar food preferences in space. We do not know whether these habitat preferences are immutable. *Aethomys namaquensis*, typical of open rocky sites, differs from other species in its early colonisation of sites after fire (Breytenbach 1987; van Hensbergen et al. 1992). The common species of unburnt fynbos are presumably unable to tolerate low-cover conditions. It remains to be seen whether *Aethomys* could survive in high cover areas should the high cover "specialists" be eliminated. In a study of islands in a human-made lake in karroid shrubland, *Aethomys* was found to leave its typical rocky habitats for atypical grassy flats where the grassland species had become extinct (H. Erasmus 1988, pers. comm.). Similar insular situations have yet to be studied in the CFR.

Ants are functionally important in fynbos, primarily because of their role as seed dispersers of many plant species (Bond and Slingsby 1983). Ant species richness is comparable to that in MTE shrublands elsewhere (Koen and Breytenbach 1988), though species number declines with increasing time after fire (Donnelly and Giliomee 1985a). There seems to be limited ecological equivalence with respect to seed dispersal, since most seeds are taken by *Anoplolepis custodiens*, the very similar *A. steingroeveri*, or *Pheidole* spp. These species are widespread but are absent from very moist fynbos, such as the south slope of the Outeniqua mountains, where seeds are moved by a different suite of species (Bond and Breytenbach 1985). An interesting omission from the fynbos ant fauna is *Messor capensis*. In the karoo this common central-place forager harvests large quantities of seeds (Kerley 1991) – up to a third of the annual seed production (Dean and Yeaton 1992). This species is occasionally present in renosterveld, but its absence in fynbos may be indicative of the scarcity of a regular supply of medium-sized seeds.

Anoplolepis and *Pheidole* are eliminated by the invasive Argentine ant, *Linpithema humile* (formerly *Iridomyrmex humilis)* (Donnelly and Giliomee 1985a; de Kock et al. 1992), in this way disrupting the seed dispersal process (Bond and Slingsby 1984). One or two indigenous ant species that collect seeds do, however, tolerate the presence of *L. humile* and may play a limited role in seed dispersal (Giliomee 1992).

Ant diversity declines in fynbos invaded by hakeas or pines (Donnelly and Giliomee 1985b; Koen and Breytenbach 1988); elaiosome-gathering ants are virtually eliminated in the latter habitat (Donnelly and Giliomee

1985b). Ant species composition is also very different in forest stands, and there are no seed-gatherers among the ground-foraging species (Koen and Breytenbach 1988). Shifts in vegetation composition driven by one or a few plant species can therefore have potentially major effects on the survival of ant-dependent plant species. The results of shifts in ant community composition are completely unknown.

Though studies so far imply only limited ecological equivalence in ant species with respect to seed dispersal, we know very little about dietary specialisation of fynbos ants or the fate of smaller myrmecochorous seeds which are ignored by *Anoplolepis*. For example, *Pheidole* is supposedly a granivore but Proteaceae seedlings frequently emerge from *Pheidole* nests. Experimental studies, such as those of Hughes and Westoby (1990) on the fate of seeds carried by different ant species in Australian heath, are urgently needed.

Coetzee and Giliomee (1985) identified 91 insect taxa associated with the inflorescences of *Protea repens*, many of which are potentially important pollinators or seed predators. One hundred and eight insect taxa, representing 63 families, were collected from the foliage and infloresences of *P. nitida*. Up to 1411 insects, and as many as 17 taxa were counted in single *P. nitida* infloresences. All 75 taxa collected in the infloresences carried pollen grains, and appear to be potential pollinators. One species (*Chirodica wollastoni*, Coleoptera, Chrysomelidae) made up 60% of all insects collected and is the principal pollinator of *P. nitida*; it is also the dominant visitor to several other *Protea* species (Gess 1968; Mostert et al. 1980; Coetzee and Giliomee 1985; Visser 1992). MacGillivray (1990) identified 37 hymenopteran visitors to the Ericaceae, most of which are potential pollinators. Many natural enemies, including 33 spider taxa (18 families) were found on *P. nitida*. The spiders, when grouped into families, immatures/adults or male/female showed differences in microhabitat preference, seasonal distribution and abundance (Visser 1992). The level of ecological equivalence in these groups has not been studied.

6.4.6 Keystone Taxa

Some organisms are clearly more important than others in mediating the structure of the communities of which they form part. Those with a disproportionate effect on the persistence of a large proportion of other taxa are termed keystone species (Bond 1993). The ecological literature on fynbos reveals several taxa from a wide range of taxonomic groups that could be termed keystones (Table 6.5). Some of these are single biological species, whereas others are groups of species that perform similar functions. Two broadly defined keystone guilds are evident among the successful

Table 6.5. Examples of keystone taxa in the vegetation formations of the Cape Floristic Region

Keystone taxa	Keystone process	(Probable) impact of removal of keystone taxon	Reference
PLANTS			
Overstorey proteoid shrubs	Spatial and temporal population fluctuations resulting in dynamic resource availability, and variable mortality and recruitment of understorey species	Community-wide changes in composition, probably resulting in a loss of species	Cowling and Gxaba (1990); Bond and Stock (1989)
Mat-forming *Aspalathus* species	Several *Aspalathus* species form dense mats after intense fires in fynbos. These mats sometimes cover large areas and suppress most other plants. Because of their ability to fix nitrogen, these mono-specific thickets have a marked influence on soil nutrient status, thereby facilitating invasion by components that would otherwise be unable to colonize these sites when the *Aspalathus* spp. senesce 4-5 years post-fire	A marked change in soil nutrient status, with implications for succession and therefore ecosystem functioning	D.M. Richardson (unpubl. data)

Table 6.5. *continued*

Keystone taxa	Keystone process	(Probable) impact of removal of keystone taxon	Reference
Forest pioneer trees (e.g. *Olea europaea* ssp. *africana, Kiggelaria africana, Rapanea melanophloeos, Rhus* spp.)	Seeds of these species are dispersed across the forest/fynbos ecotone by birds. Forest foci develop under perch sites. In long fire-free intervals, these foci grow and become fire-proof, leading to forest expansion and suppression of fynbos species	Reduction in the dynamic nature of the forest/fynbos ecotone, and eventual shrinkage of forests as edges are eroded in successive fires	Manders and Richardson (1992); Manders et al. (1992)
Metalasia muricata	Facilitates establishment of thicket pioneers on coastal dunes	Establishment of dune thicket retarded	R M Cowling (unpubl. data)
Serotinous alien trees and shrubs (*Hakea* and *Pinus* spp.)	Populations of these alien species are well buffered against fire-induced population crashes (because of their short juvenile periods and large reserves of viable seeds). The prevailing non-equilibrium status of natural communities is disrupted, resulting in new depauperate steady-states and local extinction of indigenous taxa	Restoration of natural cyclical replacement processes	Richardson and Cowling (1992); Richardson et al. (1992)

Table 6.5. *continued*

Keystone taxa	Keystone process	(Probable) impact of removal of keystone taxon	Reference
ANIMALS:			
Anoplolepis custodiens and *Linpithema humile*	Seeds of over 1300 fynbos plants are dispersed by ants. One ant species, *A. custodiens*, which is common throughout the fynbos, is especially important – it buries the seed, thus protecting it from predation and facilitating germination. Other ants, including the introduced *L. humile* consume the elaiosome on the surface and leave the seed exposed to predation and the elements	*L. humile* displaces *A. custodiens*, thus disrupting the seed dispersal of many fynbos plants	Bond and Slingsby (1983, 1984); De Kock et al. (1992)
Small mammal seed predators	Cones of serotinous shrubs protect seeds from rodents before dispersal. Almost all seeds released between fires are consumed by rodents. Seeds released en masse when shrubs are killed by fire are an important resource for granivorous mammals. The degree to which seed stocks are consumed depends on the size of the seed bank and the season of seed release. In the absence of ants, rodents consume most seeds with elaiosomes	Increased seedling to parent ratios, leading to a greater abundance of over-storey shrubs. A dense cover of over-storey shrubs has a marked influence on the dynamics of under-storey species	Bond (1984); Bond and Breytenbach (1985); Cowling and Gxaba (1990)

Table 6.5. *continued*

Keystone taxa	Keystone process	(Probable) impact of removal of keystone taxon	Reference
MICROORGANISMS			
Vesicular-arbuscular (VA) mycorrhizae	These organisms, which grow and reproduce only when in symbiosis, exchange carbon fixed by green plants for enhanced uptake of phosphorus	Loss of small-seeded species of Asteraceae, Bruniaceae, Rutaceae, Rhamnaceae and Thymeliaceae	Allsopp and Stock (1992, 1993)
Ericoid mycorrhizae	As for VA mycorrhizae. There is possibly a high degree of microsymbiont-host (Ericaceae) specificity	Loss of Ericaceae – an exceptionally important group in terms of cover and species diversity (526 spp. in the CFR)	Straker et al. (1989); Cowling et al. (1990)
Rhizobia	Nitrogen fixation in legumes, including *Aspalathus* (245 spp. in the CFR). There is evidence of specificity between *Aspalathus* and Rhizobium strains	Loss of *Aspalathus* species and other legumes; these taxa play important roles in nutrient dynamics	Staphorst and Strijdom (1975); Cowling et al. (1990)

alien trees and shrubs in fynbos: the serotinous, wind-dispersed species of *Hakea* and *Pinus*, and the hard-seeded, animal-dispersed *Acacia* species (Richardson et al. 1992). The ecology of these taxa has been well studied. Efforts directed at achieving a predictive understanding of plant invasions in the region (Richardson et al. 1990; Richardson and Cowling 1992) have shed light on the characteristics of these taxa that make them keystones. The serotinous, wind-dispersed species will be used to illustrate this. Individual plants of species such as *Hakea sericea* and *Pinus pinaster* initially behave much like the indigenous serotinous proteoid shrubs. However, once established in fynbos communities, the superior fire resilience of populations (largely a result of the greater number of highly mobile, fire-protected seeds) disrupts the prevailing non-equilibrium system. As the invaders proliferate, competition with fynbos elements intensifies, eventually leading to the local extinction of many indigenous species (Richardson et al. 1989; Richardson and Cowling 1992). The new steady state is depauperate in terms of species and structure, and this alters several ecosystem processes, including hydrology and soil nutrient status.

6.5 Biodiversity and Ecosystem Function: Landscape-Level Perspectives

6.5.1 Within Formations

What are the links between biodiversity and ecosystem function at the level of a landscape dominated by a single formation such as fynbos? There is generally little change in the composition of functional types along environmental gradients in the CFR which would in other MTEs support several formations (see Sect. 6.3.2.). However, Campbell (1985,1986) defined a large number of fynbos communities in the mountains of the CFR on the basis of the frequency of a limited number of growth forms or functional types, or else the presence of structural traits of unknown functional significance. The high cover (biomass) of certain growth forms (e.g. proteoid shrubs, restioids) was clearly associated with a suite of primary resource variables (Campbell and Werger 1988). Therefore, although fynbos landscapes are not diverse in terms of functional groups (as currently defined by researchers – see Sect. 6.4.1), there is a predictable relationship between their abundance and resource factors. Very little is known of how the constituents of these communities affect ecosystem functioning. Phytosociologically defined communities within major formations such as fynbos appear to sustain similar ecosystem processes. For example, fires burn with ease across boundaries between proteoid shrublands, ericoid-restioid shrublands and hygrophilous communities in the fynbos at Swartboskloof, indicating that these communities have reasonably similar fuel properties (van Wilgen and van Hensbergen 1992).

6.5.2 Between Formations

Several vegetation formations that are distinct in terms of species composition and structure occur under roughly similar environmental conditions in the CFR. These provide ideal situations for exploring the consequences of different species assemblages on ecosystem processes such as primary productivity, water production, fire behaviour and nutrient cycling (Midgley et al. 1994).

6.5.2.1 Net Primary Productivity and Standing Phytomass

Aboveground biomass accumulates at 1-4 t ha^{-1} a^{-1} in fynbos (Rutherford et al. 1986) and at about 9 t ha^{-1} a^{-1} in indigenous forest on similar nutrient-poor soils (calculated from increment data in van Daalen 1991). Aboveground biomass in typical stands of self-sown *Pinus* spp., *Acacia saligna* and *Hakea sericea* can be up to five, three and two times that of the pre-invasion fynbos, respectively (van Wilgen and Richardson 1985; Versfeld and van Wilgen 1986).

6.5.2.2 Water production

Catchments covered in fynbos have a higher water yield over short periods than those covered with indigenous or alien trees. Streamflow from fynbos catchments also shows greater variance over longer periods (> 40 years), due to short-lived pulses of increased flow after fire (Scott and van Wyk 1992). Fynbos shrubs and indigenous tree species of riparian forests have similar transpiration rates per unit leaf area (Richardson and Kruger 1990). However, since fynbos achieves a relatively low biomass, it transpires and intercepts relatively little rainfall per unit land area (interception estimated at 5-10%; Versfeld and van Wilgen 1986). Indigenous forests intercept and use more water because of their greater biomass; interception in the Knysna forest is about 26% (Lubbe and Versfeld 1991). Differences in water production between fynbos and forest catchments are clearly attributable to the differences in structure between the two formations. Different fynbos catchments also show different hydrological responses to fire, depending on the relative representation of resprouting plants. The magnitude of response is dependent on pre-fire biomass, which determines levels of interception and transpiration; the rate of return to pre-fire levels is dependent on the rate of recovery of the vegetation, which in turn depends on the ratio of resprouting to non-sprouting plants (Bosch et al. 1986). Pine monocultures, with higher transpiration and interception rates, reduce streamflow far more than mature fynbos (van Wyk 1987).

6.5.2.3 Fire behaviour

Differences in the vertical structure of plants and dead material, rather than differences in fuel loads or fuel chemistry between species, explain differences in fire behaviour between fynbos, indigenous forests and stands of alien plants (van Wilgen and Richardson 1985; van Wilgen et al. 1990; van Wilgen and van Hensbergen 1992).

6.5.2.4 Fire-induced soil water-repellency and sediment yield

The hydrological consequences of fire in MTEs are often dramatic. These hydrological responses to fire have been linked to fire-induced water repellency in the soil. Soils are usually completely wettable, and water repellency is an abnormality resulting from the coating of soil particles with hydrophobic organic substances which reduce the attraction between soil particles and water. Plant litter is the source of hydrophobic substances which move into the soil by incorporation of organic matter, by leaching through the soil profile, or by vapour transport during burning. The degree of repellency shown by the soil is positively correlated with the period of contact between plant litter and soil. Water-repellent soils inhibit infiltration and percolation of water, resulting in overland flow. In the CFR, fire-induced water repellency in soils has been recorded in proteoid shrublands and riparian forest (Scott and van Wyk 1992) and in a catchment afforested with *Pinus radiata* (Scott and van Wyk 1990). Vegetation characteristics, soil type and fire intensity influence the level and position in the soil profile of water repellency (Scott and van Wyk 1992). Alteration of vegetation characteristics (e.g. through invasion of alien trees and shrubs) can therefore have a major effect on water repellency and sediment movement. Such an effect was observed in a fynbos catchment in the southwestern Cape where a dense stand of *Hakea sericea* had been felled and burnt under severe weather conditions. These conditions and the massive fuel load (estimated at 7180 g m^{-2}) and altered fuel properties (notably the amount of dead fuel and the reduced fuel bed depth) resulted in a much more intense fire than any that has been recorded for fynbos (estimated total heat release = 33 241 vs 13 338 kJ m^{-2}). The heat shattered large boulders and created a water-repellent layer 0.12-0.30 m below the surface, resulting in severe erosion after the first heavy rains (Richardson and van Wilgen 1986b; DM Richardson, unpubl. data). An even more intense fire and soil loss of 200 t ha^{-1} a^{-1} was documented for a site in the southern Cape where hakea was felled and burnt (Breytenbach 1989). The conversion of species-rich fynbos to dense monospecific stand (reduced species diversity and reduced structural diversity) and the influence of humans (felling the shrubs), therefore, has a marked influence on the ecosystem processes of soil formation and sediment transfer. High-intensity fires through unfelled stands of alien trees have similar but less dramatic effects. For example, wildfires in pine plantations cause much higher first-year soil losses than from adjacent fynbos (e.g. see Scott and van Wyk 1990).

6.5.2.5 Nutrient cycling

Fynbos produces very little litter (0.78 t ha^{-1} a^{-1} in 9-year-old arid fynbos; 2.17 t ha^{-1} a^{-1} in 21-year-old mesic fynbos; Versfeld and van Wilgen 1986). Litter also decomposes extremely slowly (Mitchell et al. 1986), probably because of the high degree of sclerophylly and high levels of secondary compounds. Fire is thus the main nutrient cycler in fynbos (Stock and Lewis 1986; Stock and Allsopp 1992; van Wyk et al. 1992). Coarse litter accumulates at about 1.2 t ha^{-1} a^{-1} (Milton 1991) and fine litter at up to 6 t ha^{-1} a^{-1} in the Knysna forest (Geldenhuys 1991), giving an annual litter production of about 7 t ha^{-1}. Litter decomposition in forest is extremely rapid; Clough (1991) found only between 4.1 and 11.5 t ha^{-1} of litter on the surface. The rapid nutrient cycling has implications for productivity and soil-forming processes. Clough noted that soil under pines and fynbos had similar characteristics (e.g. acidic and nutrient-poor) whereas soils in indigenous forest had higher pH and quantities of most nutrients. Van Daalen (1981) considered differences between soils under Knysna forest and adjacent fynbos to be plant-induced. Similar differences between soils under forest and fynbos have been reported from the southeastern (Cowling 1984) and southwestern Cape (Manders 1990).

The fact that soil nutrients are strongly affected by the species assemblage they support is further illustrated by comparisons between soils under fynbos and alien trees and shrubs. Since all the most important invaders have mechanisms for enhancing nutrient acquisition (including N_2-fixing symbionts, vesicular-arbuscular and sheathing mycorrhizae, and extensive proteoid root production), one would expect altered nutrient status at invaded sites. The most notable feature in stands of alien *Acacia* and

Pinus species is the altered distribution of nutrients (Musil and Midgley 1990; Witkowski 1991; Musil 1993). Enhanced productivity (see above) results in increased biomass production which influences litterfall dynamics and thus the input of organic matter and nutrients to the soil (Stock and Allsopp 1992). Impacts on soil nutrient budgets depend on soil type and the invading species involved. On clay-rich soils that support renosterveld, *A. saligna* greatly enhances organic matter and the concentration of total N, Ca, Mg and P in litter and topsoil. These effects are less evident on sandy soils that support fynbos and dune thicket (Low 1988). The enhanced levels of organic matter are due to the higher carbon-gaining capacities of the acacias, whereas increased nitrogen levels presumably result from increased fixation of N_2 by the acacia symbionts. With regular fires and the volatilisation of N and the enrichment of the soil surface with cations and P, it appears that the alien acacias create an environment which ensures the survival of early-succession N_2-fixing species such as themselves (Stock and Allsopp 1992) (Fig. 6.5). Invading species of *Hakea* and *Pinus* do not fix nitrogen, and appear to have a less marked effect on soil properties (organic matter, pH, total N, total P) than *A. cyclops* or *A. saligna* (H. Engledow, cited by Stock and Allsopp 1992).

Midgley et al. (1994) suggested that the composition of species assemblages influences the long-term productivity of a site. They argued that fynbos assemblages could drive soil fertility downwards (few additions to the soil and regular loss through volatilisation during fires), thus increasing the rate of soil impoverishment through podsolisation. This reduces the chances of other vegetation types, such as forest, becoming established at the same site (Fig. 6.5).

The only detailed study of chemical budgets and nutrient capital at the catchment level in the CFR (at Swartboskloof near Stellenbosch) revealed that fire had little overall effect on budgets or capital. This was ascribed to the intact riparian vegetation (i.e. structural diversity at the patch scale) which acted as a nutrient filter or trap (van Wyk et al. 1992).

6.6 The Ecology of Formation Shifts

Until recently, the prevailing view was that the boundaries between major vegetation formations in the CFR were controlled by edaphic factors and moisture availability (see review in Cowling and Holmes 1992). However, recent studies have revealed the dynamic nature of boundaries between forest and fynbos, fynbos and grassland, fynbos and renosterveld, and renosterveld and grassland. In all cases, changed disturbance regimes

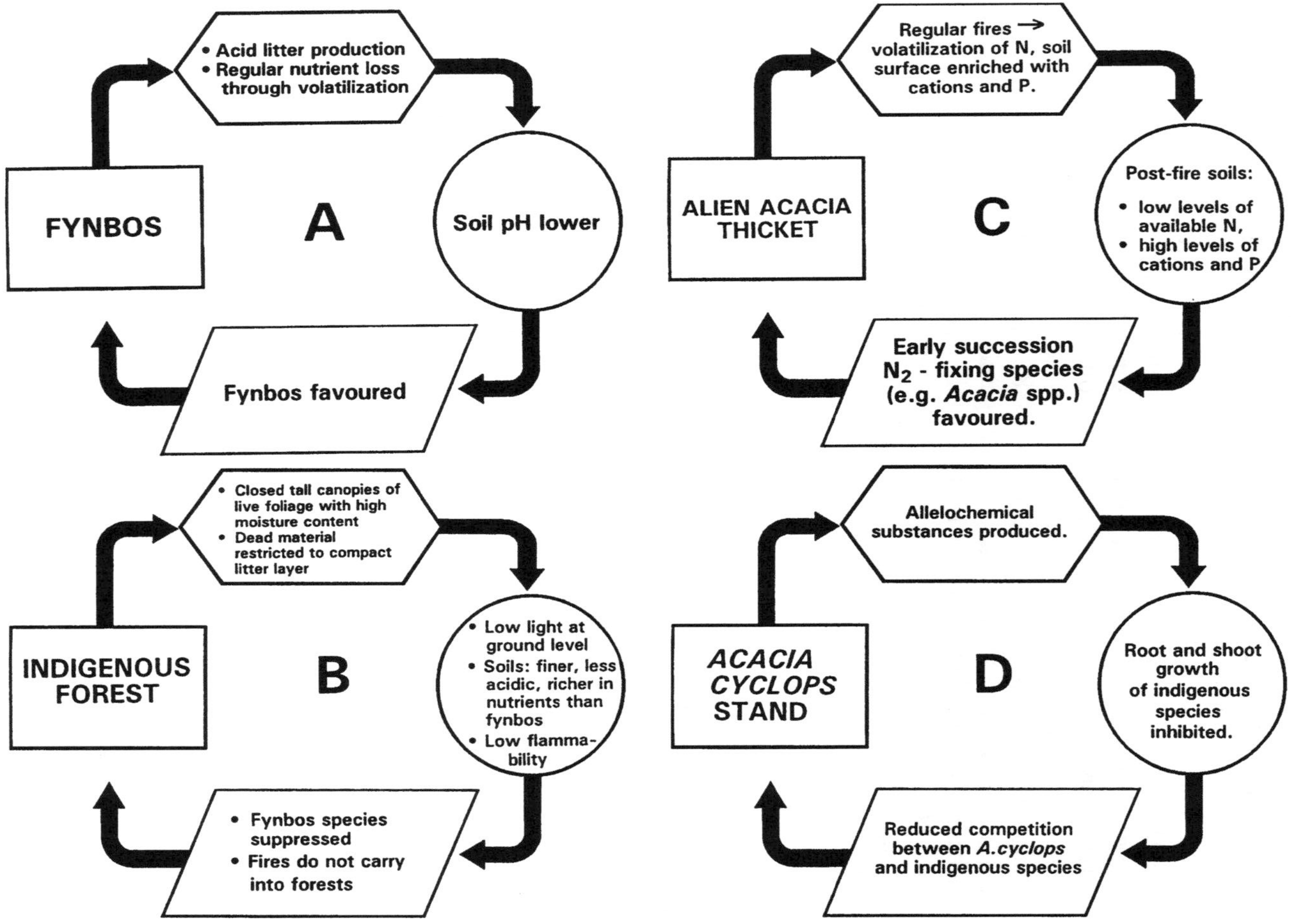

Fig. 6.5. Examples of biologically induced positive-feedback switches in fynbos plant communities. A positive-feedback switch is a process whereby a community modifies the environment, making it more suitable for that community (Wilson and Agnew 1992). Switch A is discussed by Grubb (1992, p. xi) and Midgley et al. (1994) but has yet to be verified. Primary sources of information for other switches are: B – Manders (1990); van Wilgen et al. (1990); Manders et al. (1992); C – Stock and Allsopp (1992), see

(notably fire and grazing intensity) have caused shifts in the boundaries between formations.

6.6.1 Forest-Fynbos

Forest and fynbos communities intermingle virtually throughout the CFR, but are distinct in terms of structure, the composition of species and functional groups, and community processes. Fires occur frequently in fynbos, but burn into forests only in exceptional conditions. Seeds of forest species are dispersed into the shrublands, and seedlings sometimes establish in mature fynbos communities, especially in sheltered sites. Colonising forest species are, however, usually killed by fire before forest patches have developed sufficiently to become fire-proof. Where fire-free intervals are long (> 50 years), colonising forest elements form dense thickets which eventually coalesce to form continuous forests that exclude fire, thus perpetuating the forest (Manders and Richardson 1992; Manders et al. 1992).

6.6.2 Fynbos-Renosterveld

Disturbance has been widely implicated as a determinant of the boundary of renosterveld. Boucher (1987) and Cowling et al. (1986) documented the replacement of fynbos by renosterveld as a result of post-colonial burning and grazing on the southern and southwestern coastal forelands of the CFR. Isotope evidence from soil carbon (Stock et al. 1993) does not support Levyns's (1956) notion that renosterveld on fine-textured soils in the southwestern Cape were converted from C_4-dominated grassland.

6.6.3 Fynbos-Grassland

Disturbance, especially fire and grazing, has long been regarded as the major determinant of the boundary between fynbos and grassland in the eastern part of the CFR (Acocks 1953). In the Afromontane Region of the eastern Cape, fynbos can be converted to grassland under certain fire regimes (Trollope 1973).

6.6.4 Fynbos-Alien Trees

The alien trees and shrubs that invade mountain fynbos are fire-adapted species. Whereas fires at intervals of less than about 35 years eliminate indigenous forest elements from fynbos, fires are essential for invasion and population growth of alien trees and shrubs (Richardson et al. 1992). The aliens initially behave much like the indigenous shrubs but they are well buffered against fire-induced population crashes because of their short juvenile periods and large reserves of viable seeds. Invasion thus disrupts the non-equilibrium status of natural communities and results in new depauperate steady states, which have further impacts at community and ecosystem levels (Richardson and Cowling 1992).

6.6.5 Formation Shifts and Ecosystem Function

In Section 6.5.2, we showed that many ecosystem functions are very different in the major vegetation formations of the CFR. These differences were attributed to structural and functional characteristics of the assemblages, rather than the biodiversity of these formations per se. Changing patterns of land use can and do influence the boundaries between the major formations in the CFR. The replacement of species-rich fynbos by species-poor indigenous forest or stands of a few alien trees has major effects on ecosystem function. Such vegetation changes are often rapid and irreversible (Bond and Richardson 1990), and their effects are pervasive. For example, the recent invasion of the region by many bird species from adjacent biomes is an indicator of profound alterations to many ecosystem features caused by man-induced vegetation change. Intentional or naturally occurring formation shifts offer excellent opportunities for studying the effects of changes in biodiversity on ecosystem function.

6.7 Conclusions

Two related issues were explored in an attempt to throw new light on the nature of the link between biodiversity and ecosystem function. We examined whether diversity at the level of species, functional groups, communities and landscapes affects the flux of resources through ecosystems, and whether there was any evidence that biodiversity buffers ecosystem function. Although keystone taxa have been identified in the mammal and insect faunas, very little is known of how the composition of animal assemblages affects ecosystem function. Most of our evidence deals with the effect of the composition of the plant cover on ecosystem function. Changes in other components of biodiversity could well affect ecosystem functioning indirectly – through their influence on vegetation. We note, however, that some marked changes in the composition of animal communities (e.g. the elimination of many large mammals and avian predators) have apparently had remarkably little effect on ecosystem functioning. It is likely that, *in terms of ecosystem functioning as defined in this chapter*, these components of biodiversity were redundant – their roles in important flowpaths are now performed by humans.

Diversity at the level of vegetation formations (i.e. fynbos, indigenous forest, alien forest, grassland) has a profound influence on the way that ecosystems processes energy, nutrients and matter in the CFR. This is evident for such important ecosystem processes as productivity, water production, fire behaviour and nutrient cycling (and probably many more that have not been studied). Effects on ecosystem function can be ascribed to *structural* (e.g. height, biomass, canopy cover) and *functional* (e.g. nitrogen-fixing ability) features of the vegetation, rather than to species richness. Structural (or growth form) diversity at the landscape level is important for maintaining ecosystem stability in fynbos catchments. For example, riparian forests are relatively fire-proof and form filters for nutrients and matter.

Current evidence suggests that adjoining species assemblages with structurally similar but taxonomically distinct taxa in fynbos sustain very similar ecosystem processes. This begs the question: are many plant taxa in fynbos communities "redundant links" in flowpaths affecting services such as productivity, hydrology, fire and nutrient cycling? It is very unlikely that each of the many species in the four main growth forms in fynbos could perform unique functions in these processes. We have shown that functional attributes of some taxa accord them greater importance than other structurally similar taxa with respect to certain functions. Evidence from invasive alien plants has been useful in this regard: components of diversity (suites of life history traits) that can change ecosystem functioning have been identified. There is a need to refine the categorisation of plants into functional groups by incorporating ecophysiological criteria.

The search for links between biodiversity and ecosystem function in systems prone to stand-replacing disturbances such as fire must cover a time scale of several cycles. In fynbos there are many species in each major growth form category that apparently perform similar functions. The high diversity of regeneration traits within growth forms provides a backup in the event of stochastic recruitment failures. This provides a buffer to ensure long-term stability of processes at the scale of communities and landscapes.

The boundaries between some major vegetation formations in the CFR are not static, and changed disturbance regimes have caused boundary shifts in many parts of the region. These shifts, and other changes in the species composition and structure of vegetation (e.g. through afforestation and invasion) are probably the best source of information on how changes in diversity affect ecosystem function in the CFR. Wherever the changes involve a significant alteration in the mix of growth forms in the plant cover, they have a marked effect on ecosystem functioning. Several species assemblages induce positive-feedback switches that favour the persistence of those assemblages. Studies on the mechanisms of these switches have just begun and further work would be profitable. Much more work needs to be done to determine the role of animals in ecosystem functioning.

Acknowledgements. We thank John Lawton, Mike Rutherford and Peter Ryan for useful comments on various drafts of the manuscript.

References

Acocks JPH (1953) Veld types of South Africa. Mem Bot Surv S Afr 28:1-192

Allsopp N, Stock WD (1992) Mycorrhizas, seed size and seedling establishment in a low nutrient environment. In: Read DJ, Lewis DH, Fitter AH, Alexander IJ (eds) Mycorrhizas in ecosystems. CAB International, Wallingford, pp 59-64

Allsopp N, Stock WD (1993) Mycorrhizal status of plants growing in the Cape Floristic Region, South Africa. Bothalia 23: 91-104

Armstrong AJ (1993) Is the "inhospitable sea" asumption of island biogeographic theory applicable in the context of natural habitat patches within southwestern Cape pine plantations? Environmental Forum Report, Foundation for Research Development, Pretoria

Aschmann H (1973) Distribution and peculiarity of mediterranean ecosystems. In: Di Castri F, Mooney HA (eds) Mediterranean-type ecosystems: origin and structure. Chapman & Hall, London, pp 11-19

Bigalke RC (1979) Aspects of vertebrate life in fynbos, South Africa. In: Specht RL (ed) Heathlands and related shrublands. Descriptive studies. Elsevier, Amsterdam, pp 81-95

Bond P, Goldblatt P (1984) Plants of the Cape flora – a descriptive catalogue. J S Afr Bot Suppl Vol 13:1-455

Bond WJ (1983) On alpha diversity and the richness of the Cape flora: a study in southern Cape fynbos. In: Kruger FJ, Mitchell DT, Jarvis JUM (eds) Mediterranean-type ecosystems: the role of nutrients. Ecological Studies 43. Springer, Berlin, Heidelberg, New York, pp 225-243

Bond WJ (1984) Fire survival of Cape Proteaceae – influence of fire season and seed predators. Vegetatio 56: 65-74

Bond WJ (1989) Describing and conserving biotic diversity. In: Huntley BJ (ed) Biotic diversity in southern Africa: concepts and conservation. Oxford University Press, Cape Town, pp 2-18

Bond WJ (1993) Keystone species. In: Schulze E-D, Mooney HA (eds) Biodiversity and ecosystem function. Ecological studies 99. Springer, Berlin Heidelberg New York pp 237-253

Bond WJ, Breytenbach GJ (1985) Ants, rodents and seed predation in Proteaceae. S Afr J Zool 20: 150-154

Bond WJ, Richardson DM (1990) What can we learn from extinctions and invasions about the effects of climate change? S Afr J Sci 86:429-433

Bond WJ, Slingsby P (1983) Seed dispersal by ants in shrublands of the Cape Province and its evolutionary implications. S Afr J Sci 79:231-233

Bond WJ, Slingsby P (1984) Collapse of an ant-plant mutualism: the Argentine ant (*Iridomyrmex humilis*) and myrmecochorous Proteaceae. Ecology 65:1031-1037

Bond WJ, Stock WD (1989) The costs of leaving home: ants disperse myrmecochorous seeds to low nutrient sites. Oecologia 81:412-417

Bond WJ, Ferguson M, Forsyth G (1980) Small mammals and habitat structure along altitudinal gradients in the southern Cape mountains. S Afr J Zool 15:34-43

Bond WJ, Cowling RM, Richards MB (1992) Competition and coexistence. In: Cowling RM (ed) The ecology of fynbos: Nutrients, fire and diversity, Oxford, Cape Town, pp 206-225

Bosch JM, van Wilgen BW, Bands DP (1986) A model for comparing water yields from fynbos catchments burnt at different intervals. Water SA 191-202

Boshoff AF, Vernon CJ, Brooke RK (1983) Historical atlas of the diurnal raptors of the Cape Province. Ann Cape Prov Mus 4:173-297

Botha SA (1988) Die invloed van diere op die populasiebiologie van die Clanwilliam seder (*Widdringtonia cedarbergensis* Marsh) met spesiale verwysing na die rol van klein soogdiere. MSc Thesis, Stellenbosch University, Stellenbosch

Boucher C (1981) Floristic and structural features of the coastal foreland vegetation south of the Berg river, western Cape Province, South Africa. In: Moll E (ed) Proc symp on coastal lowlands of the western Cape, University of Western Cape, Bellville, pp 21-26

Boucher C (1987) A phytosociological study of transects through the western Cape coastal foreland, South Africa. PhD Thesis, University of Stellenbosch, Stellenbosch

Breytenbach GJ (1982) Small mammal responses to environmental gradients in the Groot Swartberg of the southern Cape. MSc Thesis, University of Pretoria, Pretoria

Breytenbach GJ (1987) Small mammal dynamics in relation to fire. S Afr Nat Sci Prog Rep 135, CSIR, Pretoria pp 56-58

Breytenbach GJ (1989) Alien control: can we afford to slash and burn hakea in fynbos ecosystems? S Afr For J 151:6-16

Campbell BM (1983) Montane plant environments in the fynbos biome. Bothalia 14:283-298

Campbell BM (1985) A classification of the mountain vegetation of the fynbos biome. Mem Bot Surv S Afr 50:1-115

Campbell BM (1986) Montane plant communities of the fynbos biome. Vegetatio 66:3-16

Campbell BM, Werger MJA (1988) Plant form in the mountains of the Cape, South Africa. J Ecol 76:637-653

Campbell BM, McKenzie B, Moll EJ (1979) Should there be more tree vegetation in the mediterranean climate region of South Africa. J S Afr Bot 45:543-557

Chapin FS, Schulze E-D, Mooney HA (1992) Biodiversity and ecosystem processes. Trends Ecol Evol 7:107-108

Clough ME (1991) Comparison of aluminium, basic cations and acidity in the soil of indigenous forests and pine stands in the southern Cape. S Afr J Plant Soil 8:177-183

Cody ML (1983a) Bird diversity and density in South African forests. Oecologia 59:201-215

Cody ML (1983b) Continental diversity patterns and convergent evolution in bird communities. In: Kruger FJ, Mitchell DT, Jarvis JUM (eds) Mediterranean-type ecosystems: the role of nutrients. Ecological studies 43. Springer, Berlin Heidelberg New York, pp 357-391

Cody ML (1986a) Diversity, rarity, and conservation in mediterranean-climate regions. In: Soule ME (ed) Conservation biology: the science of scacity and diversity. Sinauer, Sunderland, MA, pp 122-152

Cody ML (1986b) Structural niches in plant communities. In Diamond J, Case TJ (eds) Community ecology, Harper and Row, New York, pp 381-405

Cody ML (1989) Growth form diversity and community structure in desert plants. J Arid Environ 17:199-209

Cody ML (1991) Niche theory and plant growth form. Vegetatio 97:39-55

Coetzee JH, Giliomee JH (1985) Insects in association with the inflorescence of *Protea repens* (L)(Proteaceae) and their role in pollination. J Entomol Soc South Afr 48:303-314

Cowling RM (1983) Diversity relations in Cape shrublands and other vegetation in the south-eastern Cape, South Africa. Vegetatio 54:103-127

Cowling RM (1984) A syntaxonomic and synecological study in the Humansdorp region of the fynbos biome. Bothalia 15:175-227

Cowling RM (1987) Fire and its role in coexistence and speciation in Gondwanan shrublands. S Afr J Sci 83:106-112

Cowling RM (1990) Diversity components in a species-rich area of the Cape Floristic Region. J Veg Sci 1:699-710

Cowling RM (ed)(1992) The ecology of fynbos: nutrients, fire and diversity. Oxford University Press, Cape Town

Cowling RM, Campbell BM (1983) A comparison of fynbos and non-fynbos coenoclines in the lower Gamtoos River Valley, south-eastern Cape, South Africa. Vegetatio 53:161-178

Cowling RM, Campbell BM (1984) Beta diversity along fynbos and non-fynbos coenoclines in the lower Gamtoos River Valley, south-eastern Cape. S Afr J Bot 50:187-189

Cowling RM, Gibbs Russel GE, Hoffman MT, Hilton-Taylor C (1989) Patterns of plant species diversity in southern Africa. In: Huntley BJ (ed) Biotic diversity in southern Africa: concepts and conservation. Oxford University Press, Cape Town, pp 19-50

Cowling RM, Gxaba T (1990) Effects of a fynbos overstorey shrub on understorey community structure: implications for the maintenance of community-wide species richness. S Afr J Ecol 1:1-7

Cowling RM, Holmes PM (1992) Flora and vegetation. In: Cowling RM (ed) The ecology of fynbos: nutrients, fire and diversity, Oxford University Press, Cape Town, pp 23-61

Cowling RM, Holmes PM, Rebelo AG (1992) Plant diversity and endemism. In: Cowling RM (ed) The ecology of fynbos: nutrients, fire and diversity, Oxford University Press, Cape Town, pp 62-112

Cowling RM, Mustart PJ, Laurie H, Richards MB (1994) Species diversity, functional diversity and functional redundancy in fynbos communities. S Afr J Sci (90:333-337)

Cowling RM, Pierce SM, Moll EJ (1986) Conservation and utilization of South Coast Renosterveld, an endangered South African Vegetation type. Biol Conserv 37:363-377

Cowling RM, Straker CJ, Deignan MT (1990) Does microsymbiont-host specificity determine plant species turnover and speciation in Gondwanan shrublands? A hypothesis. S Afr J Sci 86:118-120

Davis ALV (1993) Alpha-diversity patterns of dung beetle assemblages (Coleoptera: Scarabaeidae, Aphodiidae, Staphylinidae, Histeridae, Hydrophilidae) in the winter rainfall region of South Africa. Afr Entomol 1:67-80

Davis W (1990) Commercial wildflower production in the fynbos biome and its role in the management of land-use. PhD Thesis, University of Cape Town, Cape Town

Davis W (1992) Effects of tillage on heathland soil: disturbance of a natural mountain fynbos system in the southwestern Cape, South Africa. Soil Tillage Res 24:29-45

Davis SD, Mooney HA (1986) Water use patterns of four co-occurring chaparral shrubs. Oecologia (Berl) 70:172-177

Deacon HJ, Hendey QB, Lambrechts JJN (eds)(1983) Fynbos palaeoecology: a preliminary synthesis. S Afr Nat Sci Prog Rep 75, CSIR, Pretoria

Deacon HJ, Jury MR, Ellis F (1992) Selective regime and time. In: Cowling RM (ed) The ecology of fynbos: nutrients, fire and diversity, Oxford University Press, Cape Town, pp 6-22

Dean WRJ, Yeaton RI (1992) The importance of harvester ant (*Messor capensis*) nest mounds as germination sites in the southern Karoo, South Africa. Afr J Ecol 30:335-345

De Kock AE, Giliomee JH, Pringle KL, Majer JD (1992) The influence of fire, vegetation age and Argentine ants (*Iridomyrmex humils*) on ant communities in Swartboskloof. In: van Wilgen BW, Richardson DM, Kruger FJ, van Hensbergen HJ (eds) Fire in South African mountain fynbos. Ecological studies 93. Springer, Berlin Heidelberg New York, pp 203-215

Denslow JS (1985) Disturbance-mediated coexistence of species. In: Pickett STA, White TS (eds) The ecology of natural disturbance and patch dynamics. Academic Press, New York, pp 307-324

Donaldson JS, Scott G (1994) Aspects of human dependence on plant diversity in the Cape mediterranean-type ecosystem. S Afr J Sci 90:338-342

Donnelly D, Giliomee JH (1985a) Community structure of epigaeic ants (Hymenoptera: Formicidae) in fynbos vegetation of different ages in the Jonkershoek Valley. J Entomol Soc South Afr 48:247-257

Donnelly D, Giliomee JH (1985b) Community structure of epigaeic ants in a pine plantation and in newly burnt fynbos. J Entomol Soc South Afr 48:259-265

Fagerstrom T (1988) Lotteries in communities of sessile organisms. Trends Ecol Evol 3:303-306

Fox BJ, Quinn RD, Breytenbach GJ (1985) A comparison of small mammal succession following fire in shrublands of Australia, California, and South Africa. Proc Ecol Soc Aust 14:179-198

Fraser MW (1989) Short-term responses of birds to fire in old mountain fynbos. Ostrich 60:172-182

Fraser MW, McMahon L, Underhill LG, Underhill GD, Rebelo AG (1989) Nectarivore ringing in the southwestern Cape. Safring News 18:3-18

Fraser MW, Underhill GD, McMahon L (1990) Cape Weaver movements in the southwestern Cape. Safring News 19:17-25

Garcia EFJ (1985) Warbler. In: Campbell B, Lack E (eds) A dictionary of birds, Poyser, Calton, pp 639-641

Geldenhuys C (1991) Woud afval en voeding siklus in Groenkopwoud, Suid Kaap. Saasveld Forestry Research Centre, George, unpublished report

Gentry AH (1988) Change in plant community diversity and floristic composition on environmental and geographic gradients. Ann Mo Bot Gard 75.1-34

Gess FW (1968) Insects found on proteas. J Bot Soc S Afr 54:29-33

Giliomee JH (1992) Factors disturbing insects in a fynbos reserve. In: Thanos C (ed) MEDECOS VI: Plant animal interactions in mediterranean type ecosystems, University of Athens, Athens, pp 133-139

Giller PS, Gee JHR (1987) The analysis of community organization: The influence of equilibrium, scale and terminology. In: Gee JHR, Giller PS (eds) Organization of communities past and present. Blackwell, Oxford, pp 519-542

Goldblatt P (1978) An analysis of the flora of southern Africa: its characteristics, relationships, and origins. Ann Mo Bot Gard 65:369-436

Grubb PJ (1977) The maintenance of species-richness in plant communities: the importance of the regeneration niche. Biol Rev 52:107-145

Grubb PJ (1992) Foreword. In: Cowling RM (ed) The ecology of fynbos: nutrients, fire and diversity, Oxford University Press, Cape Town, pp x-xi

Higgins KB, Lamb AJ, van Wilgen BW (1987) Root systems of selected plant species in mesic mountain fynbos in the Jonkershoek Valley, south-western Cape Province. S Afr J Bot 53: 249-257

Hockey PAR, Underhill LG, Neatherway M, Ryan PG (1989) Atlas of the birds of the southwestern Cape. Cape Bird Club, Cape Town

Hughes L, Westoby M (1990) Removal rates of seeds adapted for dispersal by ants. Ecology 71:138-148

Huston M (1979) A general hypothesis of species diversity. Am Nat 133: 81-101

Jurgens N (1986) Untersuchungen zur Okologie sukkulenter Pflanzen des sudlichen Afrika. Mitt Inst Allg Bot Hamb 21:129-365

Keeley JE (1992) A Californian's view of fynbos. In: Cowling RM (ed) The ecology of fynbos: nutrients, fire and diversity. Oxford University Press, Cape Town, pp 372-388

Kerley GIH (1991) Seed removal by rodents, birds and ants in the semi-arid Karoo, South Africa. J Arid Environ 20:63-69

Kilian D, Cowling RM (1992) Comparative seed biology and co-existence of two fynbos shrub species. J Veg Sci 3:637-646

Knight RS (1988) Aspects of plant dispersal in the southwestern Cape with particular reference to the roles of birds as dispersal agents. PhD Thesis, University of Cape Town, Cape Town

Koen JH, Breytenbach W (1988) Ant species richness of fynbos and forest ecosystem in southern Cape. S Afr J Zoo 23:184-8

Kruger FJ (1983) Plant community diversity and dynamics in relation to fire. In: Kruger FJ, Mitchell DT, Jarvis JUM (eds) Mediterranean-type ecosystems: the role of nutrients. Ecological Studies 43. Springer, Berlin Heidelberg New York, pp 466-472

Kruger FJ (1987) South African heathlands. In: Specht RL (ed) Heathlands of the world. 9A. Heathlands and related shrublands: descriptive studies. Elsevier, Amsterdam, pp 19-80

Kruger FJ, Breytenbach GJ, Macdonald IAW, Richardson DM (1989) Characteristics of invaded mediterranean-climate regions. In: Drake JA, Mooney HA, Di Castri F, Groves RH, Kruger FJ, Rejmanek M, Williamson MH (eds) Ecology of biological invasions: a global perspective. Wiley, Chichester, pp 181-213

Lamont BB (1982) Mechanisms for enhancing nutrient uptake in plants, with particular reference to mediterranean South Africa and Western Australia. Bot Rev 48:597-689

Lamont BB (1992) Functional interactions within plants – the contribution of keystone and other species to biological diversity. In: Hobbs RJ (ed) Biodiversity of mediterranean ecosystems in Australia. Surrey Beatty, Chipping Norton, pp 95-127

Le Maitre DC (1992) The relative advantages of seeding and sprouting in fire-prone environments: a comparison of life histories of *Protea neriifolia* and *Protea nitida*. In: van Wilgen BW, Richardson DM, Kruger FJ, van Hensbergen HJ (eds) Fire in South African mountain fynbos. Ecological studies 93. Springer, Berlin Heidelberg New York, pp 123-144

Le Maitre DC, Brown PJ (1992) Life cycles and fire-stimulated flowering in geophytes. In: van Wilgen BW, Richardson DM, Kruger FJ, van Hensbergen HJ (eds) Fire in South African mountain fynbos. Ecological studies 93. Springer, Berlin Heidelberg New York, pp 145-160

Le Maitre DC, Midgley JJ (1992) Plant reproductive ecology. In: Cowling RM (ed) The ecology of fynbos: Nutrients, fire and diversity, Oxford, Cape Town, pp 135-174

Levyns MR (1956) Notes on the biology and distribution of the rhenoster bush. S Afr J Sci 52:141-143

Low AB (1988) Phytomass and major nutrient pools in an 11-year post-fire coastal fynbos community. S Afr J Bot 2:98-104
Lubbe WA, Versfeld DB (1991) Rainfall interception in the Groenkop Indigenous Forest, Southern Cape. Forestek report 91299. CSIR, Pretoria
MacArthur RH, Levins R (1967) The limiting similarity, convergence and divergence of coexisting species. Am Nat 101:377-385
Macdonald IAW (1989) Man's role in changing the face of southern Africa. In: Huntley BJ (ed) Biotic diversity in southern Africa: concepts and conservation. Oxford University Press, Cape Town, pp 51-77
Macdonald IAW (1991) Conservation implications of the invasion of southern Africa by alien organisms, PhD Thesis, University of Cape Town, Cape Town
Macdonald IAW, Richardson DM (1986) Alien species in terrestrial ecosystems of the fynbos biome. In: Macdonald IAW, Kruger FJ, Ferrar AA (eds) The ecology and management of biological invasions in southern Africa. Oxford University Press, Cape Town, pp 77-91
MacGillivray DB (1990) The insect pollination of selected *Erica* species. MSc Thesis, University of Stellenbosch, Stellenbosch
Manders PT (1990) Fire and other variables as determinants of forest/fynbos boundaries in the Cape Province. J Veg Sci 1:483-490
Manders PT, Richardson DM (1992) Colonization of Cape fynbos communities by forest species. For Ecol Manage 48:277-293
Manders PT, Richardson DM, Masson PH (1992) Is fynbos a stage in succession to forest? Analysis of the perceived ecological distinction between two communities. In: van Wilgen BW, Richardson DM, Kruger FJ, van Hensbergen HJ (eds) Fire in South African mountain fynbos. Ecological studies 93. Springer, Berlin Heidelberg New York, pp 81-107
McKenzie B (1978) A quantitative and qualitative study of the indigenous forests of the southwestern Cape. MSc Thesis, University of Cape Town, Cape Town
Midgley JJ, Bond WJ, Richardson DM (1994) Different ecosystems in the same environment; determinants, consequences and implications. S Afr J Sci (in press)
Milewski AV (1978) Diet of *Serinus* species in the southwestern Cape, with special reference to the Protea Seedeater. Ostrich 49:174-184
Milewski AV, Cowling RM (1985) Anomalies in the plant and animal communities in similar environments at the Barrens, Western Australia and the Caledon coast, South Africa. Proc Aust Ecol Soc 14:199-212
Milton SJ (1991) Estimated woody litterfall in broadleaved, evergreen forest, Southern Cape, South Africa. S Afr For J 158:29-32
Mitchell DT, Coley PGF, Webb S, Allsopp N (1986) Litterfall and decomposition processes in the coastal fynbos vegetation, south-western Cape, South Africa. J Ecol 74:977-993
Moll EJ, McKenzie B, McLachlan D (1980) A possible explanation for the lack of trees in the fynbos, Cape Province, South Africa. Biol Conserv 17:221-228
Mostert DP, Siegfried, WR Louw GN (1980) Protea nectar and satellite fauna in relation to the food requirements and pollinating role of the Cape Sugarbird. S Afr J Sci 76:409-412
Musil CF (1993) Effect of invasive Australian acacias on the regeneration, growth and nutrient chemistry of South African lowland fynbos. J Appl Ecol 30: 361-372
Musil CF, Midgley GF (1990) The relative impact of invasive Australian acacias, fire and season on the soil chemical status of a sand plain lowland fynbos community. S Afr J Bot 56:419-427

Mustart PJ, Cowling RM (1993) The role of regeneration stages in the distribution of edaphically restricted fynbos Proteaceae. Ecology 74: 1490-1499
Naveh Z, Whittaker RH (1979) Structural and floristic diversity of shrublands in northern Israel and other mediterranean areas. Vegetatio 41:171-190
Rautenbach IL (1978) A numerical re-appraisal of southern African biotic zones. Bull Carnegie Mus Nat Hist 6:175-187
Rebelo AG (1987) Bird pollination in the Cape Flora. In Rebelo AG (ed) A preliminary synthesis of pollination biology in the Cape flora. S Afr Nat Sci Prog Rep 141, pp 109-125
Rebelo AG, Siegfried WR, Crowe TM (1984) Avian pollinators and the pollinating syndromes of selected mountain fynbos plants. S Afr J Bot 3:285-296
Rice B, Westoby M (1983) Plant species richness at the 0.1 ha scale in Australian vegetation compared to other continents. Vegetatio 52:129-140
Richardson DM, Cowling RM (1992) Why is mountain fynbos invasible and which species invade? In: van Wilgen BW, Richardson DM, Kruger FJ, van Hensbergen HJ (eds) Fire in South African mountain fynbos. Ecological studies 93. Springer, Berlin Heidelberg New York, pp 161-181
Richardson DM, Cowling RM (1993) Biodiversity and ecosystem processes: opportunities in mediterranean-type ecosystems. Trends Ecol Evol 8:79-81
Richardson DM, Cowling RM, Le Maitre DC (1990) Assessing the risk of invasive success in *Pinus* and *Banksia* in South African mountain fynbos. J Veg Sci 1:629-642
Richardson DM, Fraser M (1994) The birds of Swartboskloof, Jonkershoek Valley, Stellenbosch. South Birds (in press)
Richardson DM, Kruger FJ (1990) Water relations and photosynthetic characteristics of selected trees and shrubs of riparian and hillside habitats in the south-western Cape Province, South Africa. S Afr J Bot 56:214-225
Richardson DM, van Wilgen BW (1986a) Effects of thirty-five years of afforestation with *Pinus radiata* on the composition of mesic mountain fynbos near Stellenbosch. S Afr J Bot 52:309-315
Richardson DM, van Wilgen BW (1986b) The effects of fire in felled *Hakea sericea* and natural fynbos and the implications for weed control in mountain catchments. S Afr For J 139:4-14
Richardson DM, van Wilgen BW (1992) Ecosystem, community and species response to fire in mountain fynbos: Conclusions from the Swartboskloof experiment. In: van Wilgen BW, Richardson DM, Kruger FJ, van Hensbergen HJ (eds) Fire in South African mountain fynbos. Ecological studies 93. Springer, Berlin Heidelberg New York, pp 273-284
Richardson DM, Macdonald IAW, Holmes PM, Cowling RM (1992) Plant and animal invasions. In: Cowling RM (ed) The ecology of fynbos: nutrients, fire and diversity, Oxford University Press, Cape Town, pp 271-308
Richardson DM, Macdonald IAW, Forsyth GG (1989) Reductions in plant species richness under stands of alien trees and shrubs in the fynbos biome. S Afr For J 149:1-8
Rutherford MC, Powrie LW (1993) Allelochemic control of biomass allocation in interacting shrub species. J Chem Ecol 19: 893-960
Rutherford MC, Pressinger FM, Musil CF (1986) Standing crops, growth rates and resource efficiency in alien invaded ecosystems. In: Macdonald IAW, Kruger FJ, Ferrar AA (eds) The ecology and management of biological invasions in southern Africa. Oxford University Press, Cape Town, pp 189-199
Scott DF, van Wyk DB (1990) The effects of wildfire on soil wettability and hydrological behaviour of an afforested catchment. J Hydrol 121:239-256

Scott DF, van Wyk DB (1992) The effects of fire on soil water repellency, catchment sediment yields and streamflow. In: van Wilgen BW, Richardson DM, Kruger FJ, van Hensbergen HJ (eds) Fire in South African mountain fynbos. Ecological studies 93. Springer, Berlin Heidelberg New York, pp 216-239

Siegfried WR (1979) Animal ecology. In: Day J, Siegfried WR, Louw GN, Jarman ML (eds) Fynbos ecology: A preliminary synthesis, S Afr Nat Sci Prog Rep 52, CSIR, Pretoria, pp 127-132

Siegfried WR (1983) Trophic structure of some communities of fynbos birds. J S Afr Bot 49:1-43

Siegfried WR, Crowe TM (1983) Distribution and species diversity of birds and plants in fynbos vegetation of mediterranean climate-zone South Africa. In: Kruger FJ, Mitchell DT, Jarvis JUM (eds) Mediterranean-type ecosystems: the role of nutrients. Ecological studies 43, Springer, Berlin Heidelberg New York, pp 403-416

Silvertown J, Law R (1987) Do plants need niches? Trends Ecol Evol 2:24-26

Skinner JD, Smithers RHN (1990) Mammals of the Southern African subregion. University of Pretoria, Pretoria

Smith TM, Shugart HH, Bonan GB, Smith JB (1992) Modelling the potential response of vegetation to global climate change. Adv Ecol Res 22:93-116

Solbrig OT (ed) (1991) From genes to ecosystems: a research agenda for biodiversity. IUBS, Cambridge, MA

Staphorst JL, Strijdom BW (1975) Specificity in the rhizobium symbiosis of *Aspalathus linearis* (Burm. Fil.) R. Dahlgr. ssp. *linearis*. Phytophylactica 7:95-96

Stock WD, Allsopp N (1992) Functional perspectives of ecosystems. In: Cowling RM (ed) The ecology of fynbos: nutrients, fire and diversity, Oxford University Press, Cape Town, pp 241-259

Stock WD, Bond WJ, Le Roux D (1993) Isotope evidence from soil carbon to reconstruct vegetation history in the south-western Cape Province. S Afr J Sci 89:153-154

Stock WD, Lewis OAM (1986) Soil nitrogen and the role of fire as a mineralizing agent in a South African coastal fynbos ecosystem. J Ecol 74:317-328

Stock WD, van der Heyden F, Lewis OAM (1992) Plant structure and function. In: Cowling RM (ed) The ecology of fynbos: nutrients, fire and diversity, Oxford University Press, Cape Town, pp 226-240

Straker CJ, Gianinazzi-Pearson V, Gianinazzi S, Cleyet-Marel J-C, Bosquet N (1989) Electrophoretic and immunological studies on acid phosphatase from a mycorrhizal fungus of *Ericea hispidula* L. New Phytol 111:215-221

Stuart CT, Macdonald IAW, Mills MGL (1985) History, current status and conservation of large mammalian predators in the Cape Province, Republic of South Africa. Biol Conserv 31:9-17

Trollope WSW (1973) Fire as a method of controlling macchia (fynbos) vegetation on the Amatole mountains of the eastern Cape. Proc Grassl Soc S Afr 8:35-41

van Daalen JC (1981) The dynamics of the indigenous forest-fynbos ecotone in the southern Cape. S Afr For J 119:14-23

van Daalen JC (1991) Forest growth: a 35 year Southern Cape case study. S Afr For J 159:1-10

van Hensbergen HJ, Botha SA, Forsyth GG, Le Maitre DC (1992) Do small mammals govern vegetation recovery after fire in fynbos? In: van Wilgen BW, Richardson DM, Kruger FJ, van Hensbergen HJ (eds) Fire in South African mountain fynbos. Ecological studies 93. Springer, Berlin Heidelberg New York, pp 182-202

van Wilgen BW, Forsyth GG (1992) Regeneration strategies in fynbos plants and their influence on the stability of community boundaries after fire. In: van Wilgen BW,

Richardson DM, Kruger FJ, van Hensbergen HJ (eds) Fire in South African mountain fynbos. Ecological studies 93. Springer, Berlin Heidelberg New York, pp 54-80
van Wilgen BW, Richardson DM (1985) The effects of alien shrub invasions on vegetation structure and fire behaviour in South African fynbos shrublands: a simulation study. J app Ecol 22:955-966
van Wilgen BW, van Hensbergen HJ (1992) Fuel properties of vegetation in Swartboskloof. In: van Wilgen BW, Richardson DM, Kruger FJ, van Hensbergen HJ (eds) Fire in South African mountain fynbos. Ecological studies 93. Springer, Berlin Heidelberg New York, pp 37-53
van Wilgen BW, Higgins KB, Bellstedt DU (1990) The role of vegetation structure and fuel chemistry in excluding fire from forest patches in the fire-prone fynbos shrublands of South Africa. J Ecol 78:210-222
van Wilgen BW, Bond WJ, Richardson DM (1992a) Ecosystem management. In: Cowling RM (ed) The ecology of fynbos: nutrients, fire and diversity, Oxford University Press, Cape Town, pp 344-371
van Wilgen BW, Richardson DM, Kruger FJ, van Hensbergen HJ (eds)(1992) Fire in South African mountain fynbos. Ecological studies 93, Springer, Berlin Heidelberg New York
van Wyk DB (1987) Some effects of afforestation on streamflow in the Western Cape Province, South Africa. Water SA 13:31-36
van Wyk DB, Lesch W, Stock WD (1992) Fire and catchment chemical budgets. In: van Wilgen BW, Richardson DM, Kruger FJ, van Hensbergen HJ (eds) Fire in South African mountain fynbos. Ecological studies 93. Springer, Berlin Heidelberg New York, pp 240-257
Versfeld DB, van Wilgen BW (1986) Impacts of woody aliens on ecosystem properties. In: Macdonald IAW, Kruger FJ (eds) The ecology and control of biological invasions in South Africa. Oxford University Press, Cape Town, pp 239-246
Versfeld DB, Richardson DM, van Wilgen BW, Chapman RA, Forsyth GG (1992) The climate of Swartboskloof. In: van Wilgen BW, Richardson DM, Kruger FJ, van Hensbergen HJ (eds) Fire in South African mountain fynbos. Ecological studies 93. Springer, Berlin Heidelberg New York, pp 21-36
Visser D (1992) The arthropods associated with *Protea nitida*. MSc Thesis, University of Stellenbosch, Stellenbosch
Vitousek PM (1986) Biological invasions and ecosystems properties: can species make a difference? In: Mooney HA, Drake JA (eds) Ecology of biological invasions of North America and Hawaii. Ecological studies 58. Springer, Berlin Heidelberg New York, pp 163-176
Walker BH (1992) Biodiversity and ecological redundancy. Conserv Biol 6:18-23
Whittaker RJ (1972) Evolution and measurement of species diversity. Taxon 21:213-351
Whittaker RJ (1977) Evolution of species diversity in land communities. Evol Biol 10:1-67
Wilson JB, Agnew ADQ (1992) Positive-feedback switches in plant communities. Adv Ecol Res 23:263-336
Wilson JB, Gitay H, Agnew ADQ (1987) Does niche limitation exist? Funct Ecol 1:391-397
Witkowski ETF (1991) Effects of invasive alien acacias on nutrient cycling in the coastal lowlands of the Cape fynbos. J Appl Ecol 28:1-15
Woodmansee RG (1988) Ecosystem processes and global change. In: Rosswell T, Woodmansee RG, Risser PG (eds) Scales and global change. John Wiley, Chichester, pp 11-27
Wright MG, Donaldson JS (1994) Insect diversity and ecosystem function in the Cape mediterranean-type ecosystem. S Afr J Sci (in press)

Wright MG, Visser D, Coetzee JH, Giliomee JH (1991) Insect and bird pollination of *Protea* species in the western Cape: further data. S Afr J Sci 87:214-215
Yeaton RI, Bond WJ (1991) Competition between two shrub species: dispersal differences and fire promote co-existence. Am Nat 138:328-341

7 Ecosystem Function of Biodiversity: Can We Learn From the Collective Experience of MTE Research?

G.W. Davis and M.C. Rutherford

7.1 Introduction

The central problem tackled in this Volume involves separating the consequences of biodiversity from all other system relations and attributes. It is a question which impinges on both theoretical ecology, and the application of ecological knowledge to maintenance of the human environment (Fig. 7.1). The mediterranean-type ecosystem (MTE) project is part of a far-reaching initiative formulated within the IUBS-SCOPE-UNESCO *Programme on Ecosystem Function of Biodiversity* (see Preface; Fig. 7.2), and has the broad mandate to address the following questions:

- "How is system stability and resistance affected by species diversity, and how will global change affect these relationships?
- What is the role of biodiversity in ecosystem processes, including feedbacks, over short and long term spans, and in the face of global change?" (Younés 1992).

There are some large gaps in the theoretical understanding of links between biodiversity and ecosystem function, many of which were addressed at the initial SCOPE symposium on this topic in Bayreuth, Germany, in 1991. The proceedings of that meeting (Schulze and Mooney 1993a) contains chapters by highly experienced ecologists. It concludes, however, that it is "... unclear exactly how and why a change in biodiversity might alter the functioning of ecosystems, i.e. the transfer of carbon, water and nutrients, and the maintenance of ecosystem stability" (Schulze and Mooney 1993b). These words suggest that theoretical understanding of the problems associated with the topic is not imminent. The IUBS-SCOPE-UNESCO programme does also, however, have a pragmatic basis. It recognizes that policy makers cannot wait for the resolution of theoretical issues or the evolution of new ecological paradigms, and will need to take decisions based on the best available information long before major results from formal research programmes are obtained (Younés 1992). It is therefore important that interim results be made accessible to politicians and managers, and that channels of communication be developed to allow new

G.W. Davis and D.M. Richardson (Eds.)
Ecological Studies, Vol.109

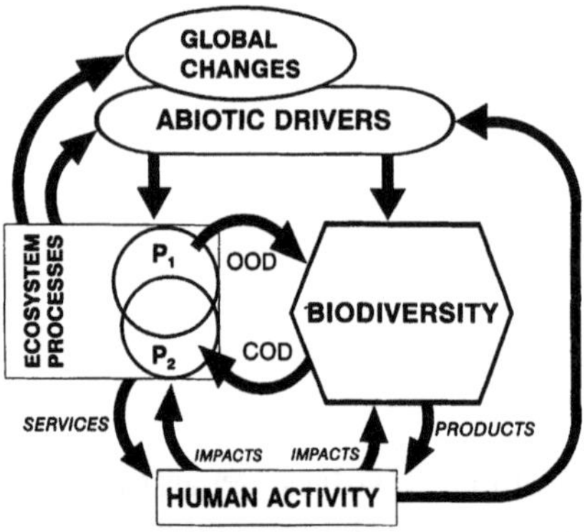

Fig. 7.1. Certain ecosystem processes (P_1) contribute to the origins of diversity (*OOD*) over evolutionary time, while others (P_2) are in part the consequences of diversity (*COD*). Both biodiversity and ecosystem processes are utilized by humanity, which in turn affects the natural world by direct elimination of biological species and transformation of process pathways. Global changes come under the influence of system processes (e.g. trace gas emission can affect climate, and altered hydrology will influence land use practice), and are also driven by human activity

ecological insights to find their way into the planning arena with as little delay as possible (Pitelka 1993).

Ecologists are perhaps more acutely aware than many planners of the biological consequences of management without sound scientific input. This concern often emerges in the pages of the ecological literature, and is possibly indicative of a drive towards integrating theoretical and management aspects of ecology. Ehrlich (1993), for instance, distinguished between the amount of scientific knowledge needed to understand linkage of biodiversity and ecosystem function, and the insight which justifies conservation of biodiversity for the probable safeguarding of ecosystem functions. This call for appropriate conservation action can be interpreted as scientific advocacy of the "precautionary principle" (IUCN 1991), and an attempt to build a bridge between basic and applied ecological research. A second challenging formulation is more explicit in its call for contributions to conservation management from the scientific side. Western (1992) has devised an axiomatic framework that outlines the relationship between biological theory and management practice for the conservation of biodiversity. This schema describes the development of theory and associated methods of application along a developmental path from *identification*, through *safeguard*, to *rescue*. An adaptation of this model to the current topic of diversity and system function is presented in Table 7.1. A third article germane to this discussion is an overview by Grime (1993),

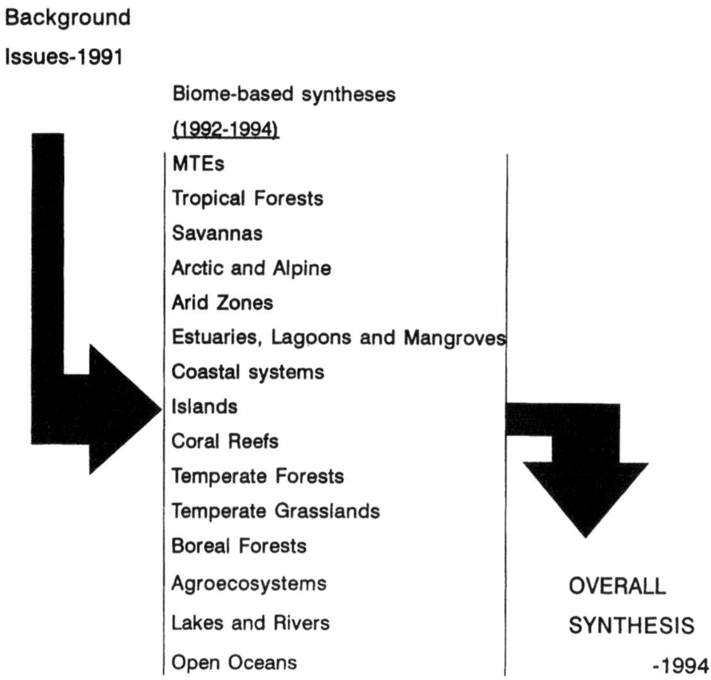

Fig. 7.2. Biodiversity and ecosystem function in mediterranean-type ecosystems is part of a larger programme being conducted under the auspices of IUBS, SCOPE and UNESCO, to answer questions pertaining the potential impacts of global change on complex ecological systems

in which he describes scientific development in the field of ecology as a three stage process: the first being descriptive, the second analytic, and the third synthetic. In the last-mentioned phase it is envisaged that knowledge provided by other disciplines both inside and outside biology will be integrated into a cohesive whole capable of promoting new ecological paradigms. If we refer to these phases as A, B and C respectively, then it is evident that the current search for links between biodiversity and ecosystem function is based on the observations and analyses of phases A and B, and that it is currently groping for answers in C. While the action called for by Ehrlich (1993) must at this stage also rest on knowledge of the first two phases, it is probably only with the anticipated synthesis that sophisticated predictive models will be able to guide (sensu Western 1992) a sustainable management strategy for the long-term safeguard of systems and their diversity, and for the effective rescue and restoration of damaged ones.

Table 7.1. The relationship between theory and methods in ecology and environmental science, and the necessary steps for implementing conservation of biodiversity and ecosystem function (description, conservation and rescue). (After Western 1992)

	Description and classification	Conservation and safeguard	Rescue and restoration
Theory	Classification of biodiversity and ecosystem processes at all levels of scale	Erecting of hypotheses regarding causal links between biodiversity and ecosystem processes	Predictive knowledge of ecosystem resilience and stability, and the functional role of biodiversity
Research methods	Establishment of linked data bases to store information of spatial and temporal diversity in composition, structure and processes[a]	Execution of laboratory, field and natural experiments to test hypotheses[b]	Setting up of system level process experiments to validate predictive restoration models
Conservation methods	Quantification and long-term monitoring of diversity and system processes in natural and disturbed systems	Application of "precautionary principle" and establishement of system process reserves	Establishement of flow paths for essential system processes by preservation – reintroduction – reconstruction of appropriate sytem components

[a] Noss (1990).
[b] Diamond (1986).

7.2 Mediterranean-Type Ecosystems as a Focus

MTEs were targeted for the investigation of ecological complexity, or more specifically, of the relationship between biodiversity and ecosystem function for a number of reasons. Firstly, mediterranean-climate regions, especially those of South Africa and Australia, are biologically diverse (Chaps 1 to 6, this Vol.); secondly, all of the mediterranean-climate regions are subject to intense human pressure, which is leading to fragmentation, transformation and invasion by foreign species (Mooney 1987) with possible loss of ecosystem function; and thirdly, there is a well-established history of cooperative and collaborative research effort between the regions (see Table 1.4 in Chap. 1).

Mediterranean-climate systems have been grouped for ecological study for more than two decades (di Castri and Mooney 1973), and the synergy provided by common purpose has produced a series of important research activities, including a series of international MEDECOS conferences at different mediterranean-climate venues (e.g. Conrad and Oechel 1982; Kruger et al. 1983a; Dell 1984; di Castri et al. 1988; Thanos 1992; Richardson and Cowling 1993) and diverse products (the above and Dell et al. 1986; Tenhunen et al. 1987; and others).

Because MTEs are intrinsically interesting on the one hand, and because they are in great danger of damage through over-exploitation (Johnson 1973) on the other, emphasis has alternated between basic and applied approaches to research and review (Kruger et al. 1983a). Nevertheless, the accumulation of information has strengthened the analytic phase to the point where investment in modelling has become a fairly important thrust, especially in the Californian and South African MTEs (Mooney 1987; Bond et al. 1992). Analytic capabilities, however, are not uniformly well developed for all levels of system organization. At the scale of the sub-organism, organism, and to some extent for key system processes such as nutrient cycling and plant water relations, analysis and predictive understanding are reasonably well advanced. Quantitative analysis at the general ecosystem and landscape levels, however, is only just coming within reach with the rapidly evolving sophistication and accessibility of computational technology. Accurate predictive modelling of whole systems is probably still a way off. The uncertainty which pervades research on the topic of functional value of diversity reflects the struggle that ecologists are experiencing in making progress with this part of the analytic phase, a bottleneck which is hindering both entry into the integrative (C) phase, and the effective application of theory to safeguard and rescue procedures.

Close inspection of definitions and concepts is an important task in the analytic phase. In the reviews presented in this Volume, some underlying concepts have emerged as being particularly relevant to MTE ecology, and which serve as staging points for obtaining clarity on a range of issues. These include:

- ♦ Diversity in ecosystems can be observed and measured not only in terms *composition* (species richness and distribution), but also as variation in *structure* (growth form) and *function* (flowpaths relating to system processes) at all levels of scale (Noss 1990).

- ♦ Any description of system function is relative to the process being observed, a special case of which is the set of system functions supporting human life.

Returning to the narrower mandate of the project – the review and investigation of links between biodiversity and ecosystem function in mediterranean-type ecosystems – it is necessary to reflect on what has been achieved so far. There may not yet be definitive answers to the questions of how biodiversity affects system function, or what the precise relationship between the two might be (Schulze and Mooney 1993b), but there is both experimental and anecdotal support for the notion that fluxes of matter, energy and information in ecosystems are to some extent dependent upon diversity. Success of the MTE portion of the project could therefore be assessed with respect to the following criteria:

- ♦ Do MTEs provide a reasonable testing ground for hypotheses about causative links between diversity and ecosystem function? If the answer is YES, then;
- ♦ Are there opportunities for relevant comparative and collaborative work between the regions?
- ♦ What insights can MTE ecologists offer to policy makers and system managers regarding the functional value of biodiversity in the face of global changes to climate and land use?

In this chapter we attempt to take stock of ecological research in MTEs, in terms of balancing its commitments: (1) between basic science and human well-being; and (2) between global and regional issues. Our retrospective view of the syntheses contained in the pages of this book is based on a semi-quantitative analysis of emphases expressed in those review chapters, results of which will be presented in the next section. At the outset, however, the reader should remember that each of the review chapters is a severe distillation of current regional thinking, and that whole fields of knowledge may not appear for the good reason that the authors regarded them as tangential to the current debate. However, since the current debate is a difficult one, being conducted in territory not yet fully charted, there will be a need to re-evaluate the primary data sets and working perceptions at intervals as the new ecological paradigms take shape.

7.3 Diversity and System Function: a Distillation of MTE Thinking

However successful the MEDECOS alliance might have been at optimising progress in MTE research, activities in each region have inevitably been influenced differently by the local perspectives, needs, and resources of each. Cultural differences have also played their part. This may be a weakness if we are wanting only to improve our ability to test hypotheses through increased replication on a broader geographical basis. On the other hand, diversity of approach allows a more extensive conceptual framework for generalised models of MTE function, and for identifying the critical nodes where research can be deployed to service human needs. The task of supplying useful results to policy makers and managers of ecosystems may thus be negatively affected because specific predictions are less accurate, but the conceptualization of problems and solutions probably ultimately would be more realistic.

Aside from the provision of interim results, a function of this Volume is to assess the profitability of coordinated MTE research to date, and to consider its potential for contributing to the understanding and solution of global ecological problems in the future. Authors based their reviews on an inspection of biodiversity on the one hand, and perceived ecosystem functions on the other, focussing attention on information which they felt would demonstrate or help conceptualize links between the two. The resulting set of regional syntheses embodies both the similarities and differences in approach to the problem. Perceived key concepts which emerged can be divided into five (sometimes overlapping) groups of topics, each referred to here as a *research focus*. These foci have been identified as follows: (1) *Intrinsic system attributes and processes* (IA), which included such topics as hydrology and water relations, nutrient cycling, the role of fire, natural disturbance, productivity and dispersal mechanisms. (2) *Human impacts* (HI) on ecosystems as described by fragmentation, species extinctions, invasions, perturbation and induced shifts in vegetation formations. (3) *Role of diversity* (RD) in ecosystem function as conceptualized in functional groups, keystones and redundancy, the relevance of scale, and the buffering value of biodiversity. (4) *Global change effects* (GC) on biodiversity and ecosystem function in terms of climate and land use change. (5) *Management* (MA) for conserving the functional aspects of ecosystems and providing a means for sustainable utilization. Topics were then subjectively weighted according to the emphasis given to each in the regional syntheses. Although they are an important part of each chapter, descriptions of diversity patterns are not included in this analysis. A comparison of regional emphases within the five research foci is shown in Fig. 7.3, and an ordination of topics against regions in Fig. 7.4.

7.3.1 Similarities and Differences in Approach

In spite of the limitations of an analysis of this type, Fig. 7.3 does indicate a degree of uniformity between the different regional syntheses. It is clear

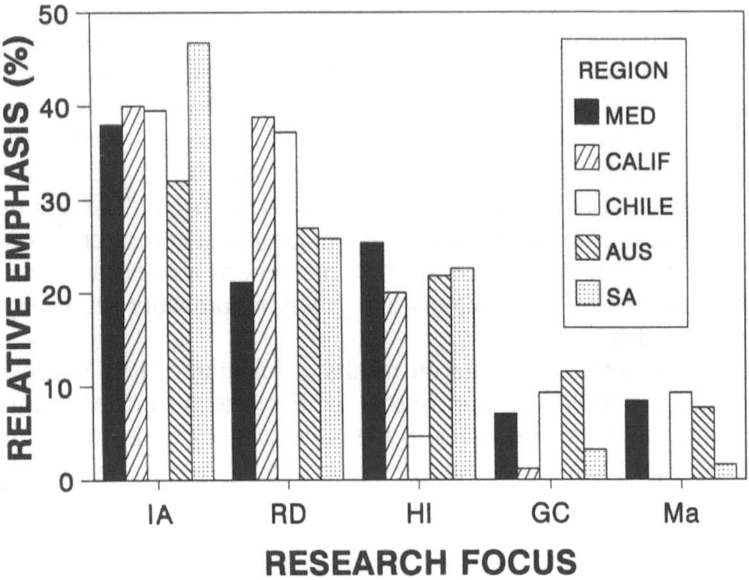

Fig. 7.3. Analysis of attention paid to five key focus areas by MTE research as reflected in the reviews of biodiversity and ecosystem function. *IA* intrinsic attributes and processes of ecosystems; *RD* conceptual roles of biodiversity in ecosystem function; *HI* human impacts on natural ecosystems; *GC* global change in climate and land use; and *Ma* management of ecosystems.

that descriptive component (IA) is dominant in all regions, which correctly reflects the large investment of research effort in providing a sound information base for the later stages of analysis and integration (sensu Grime 1993, as discussed in Sect. 7.1). The analytic phase of ecology is probably best represented here by the "roles of diversity" (RD) category – the next best supported category of reporting, and the area where the conceptual links between diversity and system function are considered. Separation of the human impact (HI = the negative response of systems to human activity) and management issues (Ma = human manipulation of system function) offers the opportunity to gauge the state of applied ecological research in the mediterranean-climate regions. Articulation of human impacts on natural systems forms the basis for conservation management, and in this context is probably reflective of the emerging science of quantitative conservation biology, while management issues are usually raised in the more pragmatic and less academic milieu of land management. The single reported study which makes the most direct headway in building a link between the theoretical issues under consideration and application, is the account of the Chilean honey production industry in which ecological research techniques are applied directly to

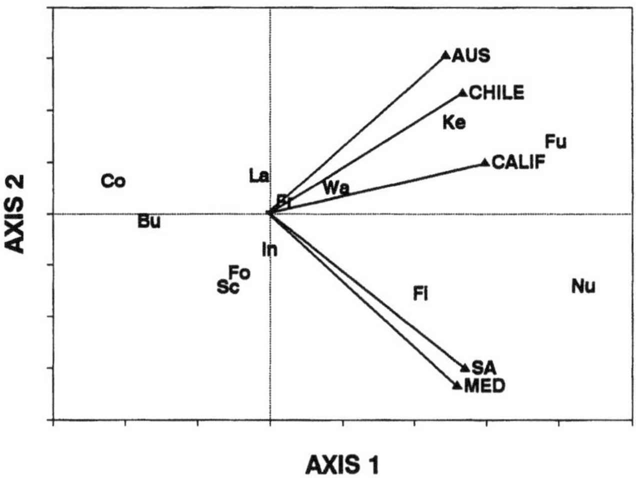

Fig. 7.4. Principle components analysis of attention paid to specific topics against each regional review of biodiversity and ecosystem function in MTEs (Chaps 2 to 6, this Vol.). Topics are as follows, with the broader grouping as described in Fig. 7.2 in parenthesis: *Co* conservation (Ma); *Bu* buffering capacity (RD); *La* land use (GC); *In* invasion (HI); Fo = formation shifts (HI); *Sc* relevance of scale (RD); *Fr* fragmentation (HI); *Wa* Water relations and hydrology (IP); *Fi* fire (IA); *Ke* keystones (RD); *Fu* functional grouping (RD); *Nu* nutrient cycling (IA)

problems encountered in a commercial enterprise exploring the possibilities for sustainable management of a natural, although largely alien, renewable resource (see Fuentes et al., Chap. 4, this Vol.). The review by Blondel and Aronson (Chap. 2, this Vol.) also covers historical accounts of ecomanagement situations in which subjects as diverse as the production of culinary truffles, and charcoal for the glassworks of prerevolutionary France are presented as a demonstration of the long association that humans have had with natural resource management in the Mediterranean Basin. A more profound biological insight into some of the links between biodiversity and ecosystem function, however, is offered by the study of invasions by alien organisms, addressed here under Human Impacts (HI). This topic was addressed most extensively in the syntheses of California, the Mediterranean Basin and South Africa. In the last-mentioned region, work has been done which links the diversity of plant characters (i.e. life history and seed production traits) and the functional consequences of system invasion by preadapted species (Richardson and Cowling 1992).

Research into the effects of global change (GC) on ecosystem function is relatively new, and the rapid growth in this highly quantitative field is probably not adequately represented here. Experimental work into climate change in MTEs is, however, getting under way (e.g. Moreno 1994), while changes associated with land use patterns appear mostly under the human impact (HI) and management (Ma) categories. Conservation, the traditional first line protection of environmental sustainability has also received little direct attention in the regional syntheses, probably because conservation issues are seen as implicit in the investigation of ecosystem function.

While Fig. 7.4 gives some indication of broad similarities between the regions, Fig. 7.5 points to some of the divergences. Australia, California and Chile are grouped around issues of functional groups and keystone species, while South Africa and the Mediterranean Basin paid relatively more attention to nutrient cycling and fire as ecosystem processes. Deeper differences emerge indirectly from the different accounts, and relate to the economic and historical backgrounds of the different regions. Even with regard to base-line data, such as the knowledge of vegetation in each region, we find very different limitations on the quality and quantity of information that is available. Of the five regions, the natural vegetation in South Africa and California is probably both the best conserved, as well as floristically the best documented. The flora of Chile, on the other hand, is not yet fully catalogued due to a shortage of trained botanists working in the region (Fuentes et al., Chap. 4, this Vol.), while western Australia has suffered huge losses of natural vegetation during development of the wheatbelt, and in the Mediterranean Basin all systems have been so heavily impacted by human use over a long period of time, that conventional phytosociological surveys are of questionable value (Blondel and Aronson, Chap. 2, this Vol.). On the other hand, the long duration of human association with ecosystems of the Mediterranean Basin has, according to Naveh (1987), attained the status of co-evolution. Indigenous people in the other regions have all either been displaced, or have been assimilated into the European lifestyle and adopted land use practices similar to those of the colonizers. Systems of the Mediterranean Basin will therefore serve as a control when comparative landscape level studies are undertaken.

Let us now return to the questions posed earlier as a yardstick for measuring the progress made in this project.

- Do MTEs provide a good testing ground? The short answer to this question is probably yes, because of the ever-strengthening comparative information base which covers both patterns of diversity and a quantitative knowledge of many system processes. It could however be argued that an effort needs to be made to make cross-comparisons with biome types which are different from MTEs – those which are not as species-rich, nutrient-poor and seasonally drought-stressed – so that more generalized models can be constructed.
- Is there opportunity for relevant comparative work between the regions? If the operative word here is relevant, then the diverse socio-economic

conditions existing in the five regions provide an excellent opportunity for comparison of ecological issues related to human dependence on diversity and ecosystem function in the face of changes associated with massive urbanization, fragmentation of natural landscapes by agriculture and other development, invasion by alien biota etc. Here again, however, comparison with non-MTEs would be an important step for generalizing the results.

- Where the project has yet to prove itself, is in providing guidance to policy-makers and managers. The answer to this, however, lies more in establishing effective lines of communication between the arena where theoretical and research issues are being considered, and the real world of conservation and land utilization by crisis management.

7.4 The Way Ahead

Biodiversity and ecosystem function are each compelling research topics in their own right. Linking the two and placing an emphasis on the consequences of diversity is an exercise which adds extra dimensions to the topic: some that penetrate deeply into theoretical ecology, and others with immediate implications for environmental management and conservation – the two arrows leading away from the biodiversity compartment in Fig. 7.1. The chapters of this book have not been able to tackle directly the questions implicit in these two arrows because the science of ecology has yet to move fully from the analytic to the synthetic phase (as discussed above; Grime 1993) before quantitative predictions can be made. Feedback and feed-forward loops between elements make it difficult to isolate the direct consequences of biodiversity, and nearly all information has potential for contributing to a predictive understanding of system processes under the influence of biodiversity. Nevertheless, there is an urgent need to sharpen the concepts and tools of research. Here again the "invisible college" of MTE research has an important role to play, both in terms of strengthening its own ability to conduct incisive MTE research, and in contributing to the global ecological research forum. All five MTEs are currently generating sophisticated ecological knowledge, even though they are spread over the divide between the "developed" and the "developing" world: Australia and California are firmly located in the former; Chile and South Africa are embedded in the latter; and the Mediterranean Basin straddles the extremes of the two (Blondel and Aronson, Chap. 2, this Vol.). Connectedness of the collective MTE research community over the wide diversity of this socio-economic spectrum is important in building the much-needed bridges between basic human need and a theoretical understanding of the natural world.

Another important aspect of ecological theory which needs to be addressed and coordinated between mediterranean-climate regions is that of scale and its relevance to the study of ecological complexity. In the regional syntheses, topics were addressed at all scales, from the genetic underpinning of the Mediterranean Basin chapter, through the ecophysiological one espoused by the South African one, to the landscape level evident in the discussion of fragmentation in Western Australia. Holling (1992) has pointed out that there is an aggregation of meaningful concepts in the study of terrestrial systems that lie between certain limits on a temporal scale (from approximately a year to ten decades), and on a spatial scale (from tens of meters to perhaps tens of kilometers). Meaningful research into the links between diversity and system function in MTEs has been, and can be done, at many different scales. What is needed, however, to optimize comparative research effort in MTEs, is to establish suites of projects that are coordinated not only in topic and methodology, but also in scale.

7.4.1 Continuing the Search for a New Paradigm

In moving forward, there is always a need to refer back to tools and ideas current in previous eras, and those developed by different disciplines. Ecology has a rich heritage of eclectic thinking, and there are probably significant advantages to reconsidering some of the alternative paradigms such as those derived from thermodynamics (e.g. in Odum 1975), chaos theory (Green 1991), cellular automata (Phipps 1992) and many others.

There are also more traditional, but less system-oriented concepts that could be considered as we move forward. Throughout this Volume it has been implicitly emphasized that species richness per se is an incomplete measure of biodiversity, but that it is nevertheless a necessary and useful benchmark for analysis of ecological complexity. A case may also be made for the veteran but, in this Volume, less cited measures of equitablity and dominance encapsulated in the Shannon-Weiner and the Simpson indices respectively (Whittaker 1975). In the passionate hunt for keystone species, the functional role of rare species has not been fully explored other than as a long-term genetic buffer. A question which should be asked, especially in systems of the relatively diverse mediterranean-climate regions, is: what is the cumulative role of rare species in system processes? Do they provide system functions that complement those of the dominants, or are they just the functional frayed edge of a community?

The description of a keystone as an element without which some important system process would be radically different (Bond 1993) is appealing, has conceptual value, and is an important marketing icon for conservation. Nevertheless it is one which needs close scrutiny, tighter definition, and much more supporting data before it can be used with confidence in quantitative modelling of ecosystem functions. It is referred to by Mills et al. (1993) as "... a concept ... poorly defined and broadly applied". Synthesis chapters in this book vindicate this call (Blondel and Aronson, Chap. 2, this Vol.), but nevertheless bring us closer to recognizing the nature of a keystone, and directing attention to what processes might be under such control from large herbivores as keystone disturbers (Keeley and Swift, Chap. 3, this Vol.), through keystone resources in trophic webs (Fuentes et al., Chap. 4, this Vol.; Hobbs et al., Chap. 5, this Vol.), to the keystone dispersers of seed for myrmecochorous plant species (Richardson et al., Chap. 6, this Vol.).

Another concept related to keystones is that of the functional group (see Hobbs et al., Chap. 1, this Vol.). While it has been tacitly accepted that a functional group, or guild, is a set of species which perform similar roles in an ecosystem and can substitute for one another (Walker 1992), there is the alternative interpretation that could be indirectly constructive. In organic chemistry a functional group is defined as "... a small number of atoms in an arrangement that determines the chemical properties of the group and of the molecule to which it is attached (e.g. the carboxylic acid group COOH-, or the amine group NH_2). Organic compounds can be considered as structural skeletons with functional groups attached" (Webster's New World Encyclopedia 1992). While this definition is fundamentally different from that of current ecological usage, the idea of a carbon skeleton changing its functional nature in a *quantifiable* fashion through the addition of different appendages is analogous to the change of ecosystem function through presence or absence of particular biotic elements. The idea of an ecosystem as a neutral skeleton with particular bonding sites for attachment of character-providing elements is perhaps an instructive one when we think of nitrogen-fixing legumes invading nutrient-poor fynbos (Richardson et al., Chap. 6, this Vol.), or honey production in central Chile, where both bees and much of their pollen and nectar sources are not endemic (Fuentes et al., Chap. 4, this Vol.).

Ultimately, the combined effort of MTE ecologists involved in the present exercise should contribute to a deeper understanding of the nature and role of biodiversity by extracting what is unique in these species-rich systems. It is appropriate, however, to leave the way open for comparisons with other types of systems – those which are not as climatically temperate, and which are not as speciose – so that the MTE contribution can form part of the larger picture of how natural systems respond to, and influence, global environmental changes.

If the call is for ecologists to provide managers with sound ecological knowledge as a basis for interim decision-making (Younés 1992), then there is a need for researchers to do two things: (1) they must draw more of the existing management knowledge and practice into the ecological research forum to deal with the escalating nature of human-related problems; and (2) they must establish projects that can provide answers to the theoretical issues underlying rescue and restoration of impaired and damaged systems (Western 1992).

Channels must be established to feed forward the benefits of current ecological theory to policy-making structures responsible for the precautionary management of natural systems. Similarly, the tasks of rescue and restoration need to be considered in the forward planning of MTE research as it proceeds further into collaborative research of ecological complexity.

Acknowledgements. We thank Dave Richardson, Ed Witkowski and our colleagues in the Stress Ecology Research Unit of the NBI for input into this chapter.

References

Bond WJ (1993) Keystone species. In: Schulze E-D, Mooney HA (eds) Ecosystem function of biodiversity. Ecological studies 99. Springer, Berlin Heidelberg New York, pp 237-253

Bond WJ, Cowling RM, Richards MB (1992) Competition and coexistence In: Cowling RM (ed) The ecology of fynbos: nutrients, fire and diversity. Oxford University Press, Cape Town, pp 206-225

Conrad CE, Oechel WC (eds)(1982) Dynamics and management of Mediterranean-type ecosystems. Pacific Southwest Forest and Range Experiment Station, Berkeley

Dell, B (ed) (1984) MEDECOS IV: Proceedings of the 4th international conference on mediterranean-type ecosystems. University of Western Australia, Nedlands

Dell B, Hopkins AJM, Lamont BB (eds) (1986) Resilience in mediterranean-type ecosystems. Junk, Dordrecht

Diamond J (1986) Overview: Laboratory experiments, field experiments, and natural experiments. In: Diamond J, Case TJ (eds) Community ecology. Harper and Row, New York, pp 3-22

di Castri F, Mooney HA (eds) (1973) Mediterranean type ecosystems. Origin and structure. Ecological studies 7. Springer, Berlin Heidelberg New York

di Castri F, Floret C, Rambal S, Roy J (eds) (1988) Time scales and water stress: Proc MEDECOS V conference. IUBS, Paris

Ehrlich PR (1993) Biodiversity and ecosystem function: need we know more? Foreword. In: Schulze E-D, Mooney HA (eds) Ecosystem function of biodiversity. Ecological studies 99. Springer, Berlin Heidelberg New York, pp vii-xi

Green DM (1991) Chaos, fractals and nonlinear dynamics in evolution and phylogeny. Trends Ecol Evol 6: 333-337

Grime JP (1993) Ecology sans frontiers. Oikos 68: 385-392

Groves RH, di Castri F (eds) (1991) Biogeography of mediterranean invasions. Cambridge University Press, Cambridge

Holling CS (1992) The role of forest insects in structuring the boreal landscape. In: Shugart HH, Leemans R, Bonan GB (eds) A systems analysis of the global boreal forest, Cambridge University Press, Cambridge, pp 170-191

IUCN (1991) Caring for the earth: a strategy for sustainable living. IUCN/UNEP/WWF, Gland, Switzerland

Johnson AW (1973) Historical view of the concept of ecosystem convergence. In: di Castri F, Mooney HA (eds) (1973) Mediterranean type ecosystems. Origin and structure. Ecological studies 7. Springer, Berlin Heidelberg New York pp 3-10

Kruger FJ, Mitchell DT, Jarvis JUM (eds) (1983a) Mediterranean-type ecosystems. The role of nutrients. Ecological studies 43. Springer, Berlin Heidelberg New York

Kruger FJ, Mitchell DT, Jarvis JUM (1983b) Preface. In: Kruger FJ, Mitchell DT, Jarvis JUM (eds) Mediterranean-type ecosystems. The role of nutrients. Ecological studies 43. Springer, Berlin Heidelberg New York, pp v-viii

Mills LS, Soule ME, Doak DF (1993) The keystone-species concept in ecology and conservation. BioScience 43: 219-224

Mooney HA (1987) The impact of environmental stress on plant performance in mediterranean-climate ecosystems: differing levels of analysis. In: Tenhunen JD, Catarino FM, Lange OL, Oechel WC (eds) Plant response to stress: functional analysis in mediterranean ecosystems. Springer, Berlin Heidelberg New York, pp 661-668

Moreno J, Oechel WC (eds)(1994) Anticipated effects of elevated carbon-dioxide and climate change in mediterranean-type ecosystems. Springer, Berlin Heidelberg New York (in press)

Naveh Z (1987) Landscape ecology, management and conservation of European and Levant mediterranean uplands In: Tenhunen JD, Catarino FM, Lange OL, Oechel WC (eds) Plant responses to stress: functional analysis in mediterranean ecosystems. Springer, Berlin Heidelberg New York, pp 641-657

Noss RF (1990) Indicators for monitoring biodiversity: a hierarchical approach. Conserv Biol 4:355-364

Odum EP (1975) Diversity as a function of energy flow In: Van Dobben WH, Lowe-McConnell RH (eds) Unifying concepts in ecology. Junk, The Hague, pp 11-14

Phipps MJ (1992) From local to global: the lesson of cellular automata In: DeAngelis DL, Gross LJ (eds) Individual-based models and approaches in ecology: populations, communities and ecosystems. Chapman and Hall, New York, pp 165-187

Pitelka LF (1993) Biodiversity and policy decisions. In: Schulze E-D, Mooney HA (eds) Ecosystem function of biodiversity. Ecological studies 99. Springer, Berlin Heidelberg New York, pp 481-493

Richardson DM, Cowling RM (1992) Why is mountain fynbos invasible and which species invade? In: van Wilgen BW, Richardson DM, Kruger FJ, van Hensbergen HJ (eds) Fire in South African mountain fynbos. Ecological studies 93. Springer, Berlin Heidelberg New York, pp 161-181

Richardson DM, Cowling RM (1993) Biodiversity and ecosystem processes: opportunities in mediterranean-type ecosystems. Trends Ecol Evol 8: 79-81

Schulze E-D, Mooney HA (eds) (1993a) Biodiversity and ecosystem function. Springer. Ecological studies 99. Berlin Heidelberg New York

Schulze E-D, Mooney HA (1993b) Ecosystem function of biodiversity: a summary. In: Schulze E-D, Mooney HA (eds) Ecosystem function of biodiversity. Ecological studies 99. Springer, Berlin Heidelberg New York, pp 497-510

Tenhunen JD, Catarino FM, Lange OL, Oechel WC (eds) (1987) Plant responses to stress: Functional analysis in mediterranean ecosystems. Springer, Berlin Heidelberg New York

Thanos CA (ed) (1992) Plant-animal interactions in mediterranean-type ecosystems: Proc MEDECOS VI, 6th Int Conf Mediterranean Climate Ecosystems, Crete 1991. University of Athens, Athens

Walker BH (1992) Biodiversity and ecological redundancy. Conserv Biol 6: 18-23

Webster's new world encyclopedia (1992) Prentice Hall, New York

Western D (1992) The biodiversity crisis: a challenge for biology. Oikos 63: 29-38

Whittaker RH (1975) Communities and ecosystems. 2nd edn. Macmillan, New York

Wilson F, Rhamphele M (1989) Uprooting poverty: the South African challenge. David Philip, Cape Town

Woodmansee RG (1988) Ecosystem processes and global change. In: Rosswall T, Woodmansee RG, Risser PG (eds) Scales and global change. Wiley, London, pp 11-27

Younés T (1992) Ecosystem function of biodiversity: a progress report on the IUBS-SCOPE-UNESCO Programme. Biol Int 24:16-21

Subject Index

aboriginal people 1, 144, 287
  use of fire 246, 257
adaptations 16, 43, 45, 57-59, 76, 97, 203, 249, 256
  to heat and light 248
  to environmental stresses 304
adaptive potential 60, 140
adaptive radiation 77, 134
adaptive responses 10, 45
Afromontane forest 289, 291-294, 320
agriculture 1, 5, 6, 14, 19, 22, 29, 63-65, 70-71, 251, 253, 255, 258-261, 285, 287, 345
agro-ecosystems 65, 101, 104
Agulhas Plain, South Africa 296
Algeria 31, 63, 71
alien oranisms 22, 78, 136-141, 160, 307, 315-322
allelopathy 61, 148
Alps 46, 51
amphibians 80, 131, 186 (see also diversity)
Andes 185
animal husbandry 19
annual plants
  Californian grasslands 147-150
  genetic diversity 135
  invasion of systems by 136-143, 153, 208, 247
  nutrient cycling 82-83, 167-171
  pioneers 52,82,127
  role in ecosystem function 258-260
  stress tolerance 19
Antofagasta, Chile 185
ants 94, 129, 154, 248, 310 (see also diversity)
Argentina 185
Arica, Chile 185
aridity 12
  gradient 83, 221
  aromatic compounds 60
arthropods 244, 288, 302, 308 (see also ants, bees, butterflies, caterpillars, grasshoppers, insects, inverterbrates, mites, moths, scorpions, spiders, termites)
Atacama Desert, Chile 185, 192
Aucó, Chile 209, 211, 213-216
Australia 4, 7, 12, 14, 16, 17, 19, 22, 23, 27, 28, 31, 80, 81, 186, 196, 201, 288, 291, 296, 298, 310, 338, 344-346 (see also Chap. 5)
Avon Valley 247

Balkanic peninsula 56
basil 60
bay leaves 60
Bayreuth workshop 335
bees 61, 191, 217
bioclimatic classification 12
biodiversity (see diversity)
biological control 155
biological invasions (see invasions)
bioturbation, 93-95
birds 61, 87, 141, 308 (see also diversity)
blue tit 96, 97
boxwood 75
Breede River, South Africa 289, 290, 296, 302
butterflies 95, 129, 154, 244, 262

cactus 206
Cade oil 75
California 4, 7, 11-19, 22, 23, 27-29, 31, 61, 63, 68, 80, 82, 83, 188-192, 196, 199-206, 226, 236, 249, 291, 339, 343-345 (see also Chap. 3)
Campanita, Chile 185
Cape Floristic Region 86, 241, 285, 288-300, 307, 309, 315-319, 321-323
Cape folded mountains 13
Cape Hangklip, South Africa 290
carnivory 125, 126, 149, 209, 210
Carpathians 51
Cataluña, Spain 53, 82-84

caterpillars 96
Cenozoic 15
Cerro Campana, Chile 185
CFC 100
chaparral 16, 17, 82, 83, 127-142, 148, 153, 161-171, 173, 196, 199-206, 291, 296, 298
Chile 4, 7, 11, 12, 15-17, 19, 22-29, 31, 61, 62, 131, 132, 236, 249, 342, 344, 345, 347 (see also Chap. 4)
chives 60
climate 11, 24, 185, 191, 233, 234
climate change 50, 122, 124, 125, 134, 140, 223, 254, 343
climate zones 195, 244
  Köppen 2, 11, 233, 234
climatic oscillations 46
CO_2 4, 100, 223
coastal sage 128, 131-133, 137, 138, 146-149, 200
cockatoos 262-263
co-evolution 90, 77, 87
co-existence 51, 158, 295, 300, 301
colonisation 1, 19
  European 287
comparative studies 23-28
competition 72, 79, 135, 143, 146, 150, 158, 172, 196, 208, 209, 213, 249, 298-301, 315
conservation 1, 43, 77, 271, 274, 303, 338, 342-347
convergence 15, 54, 57, 61
  bird communities 61
Cordillera de la Costa, Chile 185, 195, 197, 198, 225
coriander 60
Corsica 56, 64, 65, 75, 80, 92, 93, 96, 97
crassulacean acid metabolism (see CAM under photosynthetic pathways)
crops 29, 66, 146, 258, 259, 263, 308
Cyprus 31, 56, 79, 80

Dalmatia 75
decomposition 3, 4, 72, 77, 94, 145, 193, 203, 248, 260, 264-265, 270, 274, 318
deer 79, 80
deforestation 19, 64-67
degradation 6, 22, 29, 75, 79, 101-103, 168, 224, 259, 271
dehesa (see agro-ecosystems)
density compensation 91, 193, 216, 264, 298
desertification 29, 53, 76, 78, 100
detritivores 73, 129
dioecy 61, 135, 195, 302
disturbance
  biotic responses to 19, 52, 71-74, 87, 102-104, 135, 141-142, 154-156, 161-165, 202, 206, 242, 252, 254, 258, 304
  influence on genetic diversity 96, 131
  influence on habitat diversity 125-126, 136, 154, 246
  regime 71-74, 136-165, 246-248, 255-257, 260-273
  role in changing vegetation boundaries 319-322
  role in invasions 78, 136, 141, 321-322 (see also fire)
diversity
  alpha 5, 17, 46, 71, 122, 131, 152, 189, 242, 286, 295, 297, 298
  amphibians 56-57, 190, 244
  ants 94, 95, 191, 309
  beta 5, 46, 71, 122, 131, 133, 135, 173, 186, 190, 242, 273, 286, 291, 295, 297
  birds 54, 133, 189-190, 244, 297-298
  chemotypic 96
  classification of 104, 266
  climatic 12, 47-49, 51, 185
  community 99
  components of 3, 11, 44, 999, 206, 224, 286, 291, 322
  conservation 99-101, 243, 260, 336
  definition of 3, 7, 44, 104, 286
  delta 46
  determinants of 122
  effects of fragmentation on 140-144
  floristic 17
  functional 295
  fungi 133-134
  gamma 5, 46, 122, 127, 131, 143, 152, 173, 186, 190, 242, 295, 297
  genetic 43, 44, 80, 134-136
  geophyte 206
  growth-form 249, 294, 296, 299

habitat 47-49
hierarchy of 11, 122
impact of alien organisms on 140
influence on stability 44, 94, 139, 149, 196, 224, 300-301, 323, 335
insects 128-131, 244
invertebrates 190-191, 264
landscape 22, 65, 126, 136, 287
landuse impact on 224
loss of 44, 121, 124
mammals 53, 73, 80, 125, 188-189, 243, 298
of approach 341
pheno-morphological 58
physiographic 47
plant 49-53, 122-125, 127-128, 186-188, 194-196, 241-243, 291-297
pollinator 77
postfire patterns 133
reptiles 56-57, 190
rodents 133
root form 250
structural 249, 250, 297-299
threats to 274
topographic 51
vertebrate 133
DNA 43
domesticated organisms 6, 19, 79, 139, 141, 251, 259
"driver" species 10
drought 4, 11, 12, 19, 47, 76, 192, 242
drought-adapted 76, 84, 199, 221
drought-deciduous 202-203, 221
dune vegetation 14

earthworms 92-94
Echo Valley, California 28
ecological equivalence 44, 90, 91, 122, 173, 243 (see also redundancy)
ecosystem
defining function 3-4, 7-8, 44, 121, 193, 273, 286
fluxes 3, 4, 7, 9, 44, 45, 76, 121, 126, 153, 169
transformation 19, 22, 28, 66, 76, 191, 242, 251, 252, 258, 285, 339
ecotone 65, 95, 123, 308
ecotype 59, 137, 158
ecotypic variation 59, 60
efficiency
nitrogen use 168, 202
resource use 153, 249, 258
water use 165, 168
El Roble, Chile 185, 197
elasticity 248
elephant
pygmy 79
endemism 50-59, 74, 83, 123, 127, 186, 187, 195, 241, 243, 244, 290, 309
environmental stresses 76, 304
erosion 15, 94, 101, 258-259, 318
essential oils 60
European Community (EC) 29
evapotranspiration 4, 47, 48, 100, 258
extinction
historical 43-44, 53-54, 59, 78-80, 94, 125-126, 171, 194, 244, 255-257
local 71, 74, 90, 154, 161, 251-252, 309, 315
post-industrial 77-79, 251-253
theory of 141, 223, 255, 262, 273

falcon 210, 213
FAO/UNESCO projects 23
feedback mechanisms 45, 61, 103, 207, 224, 257, 320, 324, 335, 345
fennel 60
fenugreek 60
fire 15, 19, 26, 63, 72, 73, 76, 94, 131-136, 148, 161-168, 242, 256, 260, 263, 269, 286, 291, 295, 298-300, 305-309, 315-317, 319-321, 323, 341, 343, 344
frequency 22, 61, 72, 122, 124, 136, 161-165, 203-206, 246, 247, 291, 321
intensity 168, 318
management 100-102
post-fire biota 196, 198, 251
regime 22, 141, 246, 252-254, 257
firewood 66, 70
flowering 58, 262, 267
flowpath 7, 8, 303
folivory 94, 189, 210, 261
forest
arthropods in 95, 190-191, 216, 310
as bird habitat 73-74, 93, 297-298
clearing 22, 63, 76, 79, 99 (see also deforestation)
edge 95, 291, 319, 321

exploitation 65, 67, 75, 79, 258
fire in 246-247, 269, 317, 321, 323
fragmentation of 253
hydrology 317
nutrient cycling 77, 263-264, 269, 318
pathogens 269
plant-animal interactions in 308, 321
plantations 101
productivity 76, 79, 94, 316, 318
products 75
recovery 65, 70
restoration 270-271
species diversity 190, 247, 295
structure and composition 189, 192, 195, 197, 203-208, 234, 236, 242, 248, 295
fossils 256
foxes 79, 210, 213
fragmentation 22, 96
causes of 21-22, 122, 274, 338
configuration 126, 140
effects on biota 121, 136, 140, 141, 143, 252, 253, 255, 257, 262, 273
effects on physical processes 259-260
theory 141, 142
France 16, 31, 49, 50, 55, 61-88, 94, 95, 101, 343
Fray Jorge, Chile 188, 195, 197, 216
frogs, 244
frugivory 87, 89, 146, 189, 308
fuel load 136, 317
functional equivalence 3, 287, 298, 299, 307, 308
functional groups
annual plants 158-159
birds 91, 308
component of diversity 44-45
defining 4, 10, 81, 144, 193, 266, 286, 323, 347
earthworms 92-94
role in ecosystem processes 7-9, 59, 73, 91, 258-260, 267, 287, 303-307
ecosystem structure 61-62, 85, 269-271, 274, 291-301, 315
flowering phenology 221-223
in chaparral 161-171
in predator/prey systems 213-216
insects 309-310
mycorrhizae 251
rodents 308-309
Fundo Santa Laura, Chile 188, 189, 196, 197, 199, 225
fungi 61, 67-69, 73, 77, 145, 263, 269 (see also mushrooms)
fynbos 16, 17, 196, 289, 291-297, 304, 308, 316, 318, 321

garlic 60
garrigue 17, 20-21, 48, 79
genet 80
genetic variation 77, 134-136
germination 60, 61, 94, 156, 158, 159, 269, 300, 303
glaciation 14, 51-55, 94, 124, 125, 188, 241
Gondwanaland 185-187
gradient studies 82
granivory 144, 145, 147, 149, 188, 210, 213, 297, 308, 310
grasses and grasslands 17, 77, 85, 136-145, 148-150, 153-161, 167, 171, 173, 189, 206, 298, 309, 319, 321
grasshoppers 154, 244
grazing 3, 17, 19, 22, 63-65, 93, 101-104, 126, 139,143, 145, 154, 155, 160, 196, 247, 252, 256, 271, 291, 296, 321
Greece 31, 59, 64, 71
"green slime approach" 286, 288
growth form 52, 81, 82, 84, 86, 139, 196, 236, 300, 302
guild 9 (see also functional groups)

hare 79
heath 234-237, 242, 245, 256, 271, 298, 310
heathlands 17, 242, 248, 272, 291
heavy metals 94
Hérault, France 83-87
herbivory 3, 4, 22, 94, 125-126, 144, 148, 151, 188, 208, 244, 261, 274, 347
hippopotamus
dwarf 54, 79
Holarctic faunal types 57
honey
production industry in Chile 216-24
honeybee (see bees)
hunter-gatherer 19, 22, 287

hunting 53, 78, 79, 125
hypothetico-deductive framework 6

Iberian peninsula 56
IBP (see International Biological Program)
industrialization 224
insects 61, 90, 94, 105, 133, 151, 246, 258, 270, 308, 310, 322 (see also arthropods)
insectivory 188, 210, 213, 297, 308, 309
intelligent tinkering 43
International Biological Program 23, 189, 191, 196
introductions (alien organisms) 22, 52, 53, 80, 121, 137, 139, 247, 252, 255
invasions (see alien organisms, annual plants, disturbance, grasses, introductions)
invertebrates (see diversity)
island biogeography theory 91, 140
Israel 31, 59, 82-86, 291, 296
Italy 30, 31, 63, 65, 79
IUBS 335, 337
IUCN 77, 336

Jurassic 14

Kalbarri, Australia 242
karroid shrublands 289, 292-295
keystone species 8, 10, 45, 59, 79, 87, 91, 92, 125, 141, 142, 144, 148, 194, 196, 207, 209, 221, 223, 255, 266, 287, 310, 322, 344, 346
Knysna, South Africa 291, 316, 318
kwongan 17, 81, 196, 236, 242, 246, 249, 271, 291, 296

lakes 241, 309
landscape heterogeneity 244 (see also mosaic)
Las Chinchillas National Reserve 209
laterites 14
lavender 60
leaching 13-15, 160, 258, 265, 317
legumes 72, 81, 82, 85-88, 102, 158, 251, 266
lichens 133, 145
licorice 60
life history traits 57, 58, 62, 96, 97, 323
limestone 14
lithosols 13, 15
litterfall 248, 319-320
lizards 190

MAB 23
macchia 20-21
mallee 235-236, 245
mammals (see under diversity)
maquis 17, 48, 58, 72, 204, 206
marsupials 188, 256, 263
matorral 17, 54, 56, 73, 131, 188-193, 195-208, 216, 218-225
MEDALUS 29
MEDECOS 23, 339, 340
MEDECS 169
Mediterranean Basin 1, 4, 7, 11-31, 196, 206, 236, 249, 251, 343-346 (see also Chap. 2)
megafauna 53, 125-126, 255, 256, 272
mice 80, 188, 298
Middle Ages 65-66
Middle East 52, 53, 75
migration 1, 61, 70, 94, 95, 123, 143, 185
mining 270
mint 60
Miocene 46, 52, 54, 56, 58, 60, 91, 123
mistletoes 145-146, 260-262
mites 72, 244
models
  conceptual 3, 8, 154, 162, 164, 169, 194, 197, 223, 301, 303, 336
  lottery 300
  patch 143
  predictive 339
  transition matrix 162
mongoose 80
montado (see agro-ecosystems)
Montpellier, France 28
Mont-Ventoux, France 101
Morocco 59
mosaic 5, 14, 17, 19, 21, 61, 73, 74, 95, 97, 105, 125, 136, 137, 201, 218, 224, 241, 246
moths 95, 129, 130, 144, 150, 262
mouflon 80
Mount Lesueur, Australia 242
MTE (see ecosystem)
multiplier effect 10

mushrooms 50, 68, 69, 79, 145 (see also fungi)
mycorrhizae 145, 160, 161, 250, 263
  ecto- 68, 249, 251, 263
  ericoid 303
  sheathing 318
  vesicular-arbuscular 251, 318
myrmecochory 310

nature reserves
  Big Creek Reserve, California 129-130
  Cape Arid National Park, Western Australia 234
  Fitzgerald River National Park, Western Australia 234, 242
  Las Chinchillas National Reserve, Chile 209
  Parque Nacional La Campana, Chile 195
  Stirling Range National Park, Western Australia 234, 242
  Vina Plains Preserve, California 155
Near East 49, 51, 52, 56, 65, 80
nectar 217, 221-223, 267, 270, 297, 299, 302, 347
nectarivores 297
Neolithic 64
New Zealand 186
niche
  regeneration 71, 91, 92, 93, 128, 131, 151, 152, 156, 161, 188, 197, 299-301, 307
  vacant 307
nitrogen 15-16, 72, 78, 81, 83, 87, 101, 102, 154, 160, 168-170, 201-204, 250, 269, 270, 286, 303, 319, 323, 347
  fixation 3, 15, 81, 82, 88, 168, 196, 266, 269, 318, 320
  leaf 167
North Africa 29, 49, 50-53, 56-59, 65, 70, 75, 85
nutrients 15, 27, 32, 101, 121, 160, 167, 168, 196, 198, 201, 203, 249, 250, 258, 259, 273, 274, 287, 288, 302, 318, 319, 323
  cycling 3-6, 62, 75, 76, 81, 82, 100, 104, 125-126, 145, 167-168, 253, 269, 306, 316, 318-320, 323, 339, 341, 343, 344
  "leaks" 44

oak 52, 64, 66, 70, 72, 75, 95, 96, 144, 145, 151, 152, 153
  moth 145 (see also *Quercus*)
  woodlands 68, 104, 173, 189
Oat Mountain Quadrangle, California 138
old fields 104
omnivory 188, 210, 213, 263, 297
onions 60
opposum 188
optimal foraging theory 217
oregano 60
owls 56, 213

Palaeozoic 13, 14
Palearctic region 55, 59, 91
paleobotany 52
palynology 52
Papudo, Chile 188
parasitism 4, 153
parsley 60
pastoralism 1, 19, 287
Patagonia, Chile 185, 189
patchiness 7, 11, 49, 71, 72, 96, 246
pathogen 4, 247, 253, 268, 269
Perth, Australia 234, 275
pH 94, 318, 319
phenology 16, 23, 97, 154, 158, 199, 208, 286
  flowering 217, 221, 248
phosphorus 319
photosynthetic pathway
  C3 154, 248
  C4 154, 248, 321
  CAM 248
phrygana 58
plant-animal interactions (see dispersal, herbivory, pollination etc.)
plant-water relations (see water)
Pleistocene 1, 14, 51-55, 79, 94, 123-126, 134, 136, 188, 255, 257, 261
Pliocene 46, 50, 51, 54, 56, 58, 59, 91, 123
pluviometric quotient 12, 13
podzolization 14
pollen analysis 51, 52
pollen load 217

pollination 4, 61, 77, 130, 131, 150, 207, 266, 303
polyploidy 59
population growth (human) 70, 99, 121, 140, 224
Precambrian 14
precautionary principle 336, 338
predation 4, 95, 154, 305
  larvae 262
  seeds 148
predator/prey systems 209-210
predators 141
primary productivity 76, 94, 168, 286, 303, 316
  net 316
principal components analysis (PCA) 62, 205, 343
process/functional approach 9
productivity 16, 101, 102, 147, 158, 194, 209, 246, 286, 298, 303, 316, 318, 319, 323, 341
Pyrénées 51

Quaternary 12, 14, 15, 52, 90, 241-244

rabbits 79, 142, 148
rainfall
  winter 1, 4, 11, 15, 47, 192, 236, 290, 300
  zones 234, 235, 242, 246
rare and endangered biota 77, 100, 133, 135, 143, 148, 243, 252, 346
rat 80
recruitment 145, 161, 162, 167, 291, 323
redundancy 6, 10, 28, 45, 78, 81, 91, 122, 127, 128, 147, 148, 158, 173, 193, 194, 207, 209, 213, 216, 266, 323, 341 (see also ecological equivalence)
refugia 51, 52, 124, 143
rehabilitation 100, 101, 270
renosterveld 17, 288-292, 295, 309, 319, 321
replacement series 242
reptiles (see diversity, lizards, snakes)
research
  approach 173, 193-194, 224, 303, 341-348
  capacity 29, 31
  global programmes 335-340
  history 23-28
  opportunities 102, 267-272
resilience 4, 9, 101, 248, 255, 258, 272, 273, 315, 338
resource scarcity 242, 248
resource-availability 146
resources 1, 10
  renewable 29, 66
  utilization 146
resprouters 134, 135, 270
restoration 94, 100-103, 137, 260, 268, 271, 273, 338, 348
rivet-popper hypothesis 43
rodents 133, 140, 147, 148, 154, 155, 188, 189, 210
rooting depth 158, 198, 199, 286, 303, 307
roots 61, 64, 72, 79, 145, 167, 197, 249, 250, 258, 299, 304
rosemary 60
rue 60

saffron 60
sage 60, 138
San Diego, California 28
San Dimas, California 28
Santa Barbara, California 128
Santa Cruz, California 128
Santiago, Chile 185, 195, 196, 207, 224-225
scales of magnitude 11, 28, 32, 43
scaling effects 96
sclerophylly 58, 76, 318
SCOPE 23, 335, 337
scorpions 244
sea level 100
sediment transfer 318
seed bank 161-163
seed dispersal 87-91, 261
seeders 134, 135
seedling recruitment 291
serpentines 16
sesame 60
shallots 60
Shannon-Wiener index 5, 218, 346
sheep, 68, 75, 79, 80, 139
shipbuilding 67
Sierra Nevada 122, 124
Sierra Vicuña MacKenna, Chile 185

Simpson index 346
snakes 190
socio-economic factors 31
soil 12-16, 24, 249
bacteria 250
formation 4
nutrients 201, 236
pH 318
parent material 14-16, 201
strata 302, 303
types 240
water 197
Sonoran Desert 83, 192, 204
South Africa 4, 7, 12-19, 23, 27-30, 50, 63, 80, 83, 85, 86, 192, 196, 201, 203-206, 236, 241, 256, 338, 339, 343-346 (see also Chap. 6)
southern Australia 4, 12, 14
Spain 1, 31, 53, 58, 59, 62, 63, 65, 71, 79, 82
speciation 51, 54, 55, 91, 124, 134, 242
birds 55
species equivalence (see redundancy and ecological equivalence)
species interactions 3, 4, 143, 299
species richness 5, 9, 15, 44, 49, 53, 54, 62, 78, 95, 104, 142, 161, 188, 194, 241-243, 248, 259, 271, 290, 291, 296, 297, 298, 300, 302, 307, 308, 309, 323, 339, 346
species turnover 127, 131, 241, 242, 246, 291, 298
species-area relationship 143
spiders 310
stemflow 249
substitutability 10, 28, 266
sub-tropical thicket 294
succession
bird communities 54
post-fire 71
soil biota 73
succulence 79, 191, 203, 204, 206, 296, 297
Swan Coastal Plain, Western Australia 250
Swartboskloof, South Africa 28, 293, 295, 297, 316, 319
sylvo-pastoralism 101

tannins 66, 144
temperature 12, 13
seasonality 11
termites 247
terra rossa 14
Tertiary 12, 14, 15, 90, 123, 185, 241-243, 246
thermal belts 47
thyme 60
Tierra del Fuego, Chile 185
transformation (see ecosystem-transformation)
truffles 67-70
Tunisia 31, 82, 83-86, 102, 103
Turkey 59
Type II error 6

UNESCO 335, 337
Arid Zone Project 23

Valdivian rainforest 190-192, 197
valley bushveld 293
vernal pools 156-158
vertebrates (see under diversity)
vital attributes 103
volatilisation 15, 320

warblers 91, 297
wasps 77, 145, 151
water
balance 164
potential 166, 198, 199, 200
production 316
relations, plant- 4, 199, 258
repellency 317
wheatbelt, western Australian 234, 236, 242, 251, 253, 267, 271
woodpecker 145
World War I 70
World War II 66, 70
wormwood 60

Zapallar, Chile 188, 195

Taxonomic Index

Part I (genera and species)

Abies 49, 124
Abrocoma bennetti 210-215
Acacia 18, 81, 237-242, 291, 307, 315, 318, 320
A. acuminata 237, 240, 261
A. albida (see *Faidherbia albida*)
A. cyclops 319-320
A. dealbata 78
A. extensa 266
A. myrtifolia 266
A. neurophylla 240
A. pulchella 266, 269
A. resinomarginea 240
A. rostellifera 238
A. saligna 316, 319
A. tortilis subsp. *raddiana* 82, 101
Acer 51
Acomys subspinosus 309
Adenanthos 242
Adenostoma ·163
A. fasciculatum 18, 148, 167-172
A. sparsifolium 18
Adesmia 186-187
A. arborea 198, 201-202
A. microphylla 222
Aethomys namaquensis 309
Aextoxicon punctatum 197, 216
Agonis flexuosa 238-239
Agropyron 140
Agrostis microphylla 157
Ailanthus elegantissima 78
Akodon longipilis 188, 210-215
A. olivaceus 188, 210-215
Alectoris 56, 80
A. barbara 80
A. chukar 80
A. graeca 80
A. rufa 80
Allocasuarina 17, 18, 237, 272
A. campestris 238, 272
A. fraserana 238
A. huegeliana 238, 240
Alnus 130
Alsephalus busephalus 78
Alstroemeria 206
Amelanchier 51
Amyema 260
Anagallis minimus 157
Anagasta kuehniella 95
Annona cherimola 222
Anoplolepis custodiens 309-310, 313
A. steingroeveri 309-310
Anthemis cotula 219
Antilocarpra americana 126
Aphaenogaster gibbosa 95
Aphelocoma coerulescens 144
Apis mellifera 216-223
Apodemus 79
A. sylvaticus 73
Arbutus 51, 130
A. menziesii 124, 169
Arctostaphylos 18, 28, 124, 127-128, 130, 134-135, 148, 167
A. glandulosa 135
A. glauca 170-172
Argania spinosa 59
Aristida 140
Aristotelia chilensis 197-198
Artemisia 51, 60, 130, 161
Arthrophora 262
Aspalathus 28, 287, 303, 312, 314
Asphodelus 72, 104, 105
Astragalus 51, 82, 88, 127, 187
Athene cunicularia 210, 213-216
Atriplex 239
Austrocactus 204
Avena 139, 158, 160
A. barbata 155-159
A. fatua 158
Azara dentata 219
A. petiolaris 198

Baccharis 130, 187, 201
B. concava 201
B. linearis 201
Bahia ambrosioides 201, 221-222

Banksia 18, 28, 235, 237, 238, 240, 242, 247
B. attenuata 265
B. menziesii 265, 267
B. prionotes 267
B. tricuspis 262-263
Berberis 187
Bergerocactus emoryi 206
Bettongia penicillata 263
Betula 51
Bidens 222
Blechnum auriculatum 198
Bombycilla 146
Bonjeania 82
B. hirsuta 88
Borrelia burgdorferi 153
Brachypodium 60
B. phoenicoides 88
B. ramosum 72, 88
Bradyrhizobium 251
Brassica 139
B. nigra 163,168
B. rapa 222
Brodiaea orcuttii 157
Bromus 139, 156-158, 160, 163
B. erectus 88
B. hordeaceus 155
B. mollis 157,160
B. rigidus 155
B. rubens 157
Broussonetia papyifera 78
Bubo virginianus 210, 213, 216
Burkea 81
Buteo polyosoma 210, 214-216
Buxus sempervirens 67,75

Cacatua sanguinea 262
Caladenia 242
Calandrinia 186-187
Calceolaria 186-187, 201
C. nudicaulis 219
Callitricha longipedunculata 157
C. marginata 157
Callitris preissii 238
Calluna 51
Calothamnus 242
Calycotome 72, 82
Calyptorhynchus funereus latirostris 262
Canis domesticus 141
C. latrans 141
Capreolus capreolus 79
Carex 127, 187
Carnegia gigantea 204
Carpinus orientalis 75
Carthamus lanatus 219
Casuarina campestris 240
Ceanothus 18, 28, 124, 127-128, 130, 134, 161-170
C. crassifolius 148
C. greggii 148, 170
Cenista pilosa 88
Centaurium venustum 157
Ceratonia 51
C. siliqua 59, 81
Cercis 51
C. silicastrum 81
Cervus 53
C. elaphus 79-80
Chaetanthera ciliata 219
Chamaea fasciata 131
Chamaerops 51
Chinchilla lanigera 210-211
Chirodica wollastoni 313
Chrysomela 155
Cichorium intybus 219
Cirsium vulgare 222
Cissus striata 198
Cistus 28, 51, 63, 72, 104
Citrus limon 219
Clematis 51
Colletia 202
C. spinosissima 222
Colliguaja odorifera 198-201, 217-222
Colophospermum 81
Columba fasciata 146
Conostylis 243
Convolvulus arvensis 219
Coronilla 82
Corylus 51
Crassula aquatica 157
C. erecta 157
Cristaria 187
Crocidura 79
C. russula 73
Crotolaria 82
Cryptantha 127-128
C. intermedia 168
Cryptocarya 189
C. alba 198-201
Cytisus 82, 87

Dactylis glomerata 59-60, 88
Darwinia 242-246
Delias aganippe 262
Deschampsia danthonioides 157
Dicaeum hirundinaceum 261
Didelphis virginianus 141
Dioscorea 187
Diplacus 160
Dipodomys 147-148
D. agilis 147
Diprotodon 256
Distichlis spicata 140
Dorycnium 82
D. hirsutum 87
D. pentaphyllum 88
Downingia cuspidata 157
Drimys winteri 197-198, 216
Drosophila pseudoobscura 136
Dryandra 238, 239, 242, 246

Echinopsis 204
E. chilensis 219
Elatine brachysperma 157
Eleocharis bella 157
E. macrostycha 157
Eliomys 79
Elymus 140
Epacris impressa 272
Ephedra 51
Equinopsis chilensis 217
Equus asinus 78
Erica 28, 72, 287, 302
Eriocyse 204
Eriogonum 127, 160
E. fasciculatum 130, 147
Erithacus rubecula 73, 89
Erodium 139, 155-158
E. botrys 157,160
E. brachcarpum 159
E. cicutarium 219
E. moschatum 157
E. obtusiplicatum 156-159
Eryngium aristulatum 157
Escallonia illinita 198
Eucalyptus 17, 128-129, 235, 239, 241-242
E. accedens 238, 240
E. astringens 238, 240
E. calophylla 18, 234-235, 237-238, 240
E. cornúta 238
E. diversicolor 18, 234-235, 237-238, 240
E. flocktoniae 238
E. globulus 222
E. gomphocephala 234, 237-238
E. lehmannii 238
E. lesouefii 238
E. loxophleba 18, 237, 238, 240
E. marginata 18, 234-235, 237-238, 240, 249
E. obliqua 272
E. oleosa 238
E. platypus 238
E. salmonophloia 18, 237, 238, 240, 245-246, 260-261
E. salubris 18, 237, 238, 240
E. transcontinentalis 238
E. tetragona 239
E. todtiana 265
E. wandoo 18, 237-238, 240, 260
Eulychnia 204
Eupatorium gleachonophyllum 201
E. salvia 201, 219, 222
Euphydryas editha bayensis 154

Fagus 49, 51
F. sylvatica 75
Faidherbia albida 82
Falco sparverius 210, 213-214
Fascicularia 206
Felis domestica 141
Festuca 88
Filago gallica 157
Flourensia thurifera 201
Formica gagates 95
Frankia 251
Fraxinus 51
F. ornus 78
Fringilla coelebs 73
Fuchsia lycioides 201
Fumaria agraria 219

Galega officinalis 217-222
Gastridium ventricosum 157
Gastrolobium 261
G. bilobum 263
Gazella rufina 78
Genetta genetta 80
Genista 82, 87
G. scorpius 88

Geranoaetus melanoleucus 210, 214-216
Glaucidium nanum 210, 213-215
Glis 79
Gnaphalium 130
Greigia 206
Grevillea 242-243
Griselinia scandens 197
Gymnogyps californianus 131

Hakea 28, 242, 291, 307, 312, 315, 319
H. scoparia 267
H. sericea 315-317
Halimium 51
Halosarcia 239
Haplopappus 187-201
Helianthemum 51
Hemizonia fasciculata 157
Herpestes ichneumon 20
Heteromeles arbutifolia 169-170
Hippeastrum 206
Hippocrepis comosa 88
Hirschfeldia incana 217-218
Holcus mollis 88
Holocarpha virgata 158-159
Hordeum 139
Hypericum perforatum 155
Hypochoeris glabra 157
Hystrix 53

Iridomyrmex humilis (see Linpithema humile)
Isoetes howellii 157
I. orcuttii 157
Ixodes neotomae 153
I. pacificus 153

Juncus bufonius 155, 157
Juniperus 52, 75
J. oxycedrus 75

Kageneckia angustifolia 195
K. oblonga 217, 219
Kiggelaria africana 314
Koeleria 139

Lathyrus 82
Laurus 51
L. nobilis 59-60
Lavandula stoechas 72
Lechenaultia 243
Lepechinia calycina 170
L. salviae 201
Leptothorax unifasciatus 95-96
Leucheria 187
Leucopogon 242
Lilaea scilloides 157
Linpithema humile 309, 313
Liolaemus 190
Lithraea caustica 193, 198-202, 217-222
Loasa 187
Lolium 139
L. multiflorum 138
Lonicera implexa 89
Lotu 128
L. corniculatus 88
L. hamatus 157
L. scoparius 168
Loxodonta africana 78
Luma apiculata 216, 222
Lupinus 127, 130, 168
Lysiloma 81
Lythrum hyssopifolia 157

Macaca 53
Macropus 256
Macrozamia riedlei 251
Maculinea 95
Madia sativa 219
Maihuenia 204
Maireana 239
Marmosa elegans 188, 210-213, 215
Maytenus boaria 219
Medicago 82, 87-88
M. sativa 66
Melaleuca 18, 238-239, 242, 272
M. rhaphiophylla 238
M. uncinata 238
Melanerpes formicivorus 145
Melanophila 131
Melilotus 82
Mesophellia 263
Messor capensis 311
Metalasia muricata 314
Microsamia 131
Microseris douglasii 157
Mimulus 127
Monachus monachus 78
Muhlenbergia rigens 140

Mus 79
M. musculus 80
M. spretus 73
Myocastor coypus 80
Myrcengenella chequen 197-198
Myrceugenia corraefolia 197
Myrmica specioides 95
Myrosuris minimus 157
Myrtus communis 89

Navrretia hamata 157
Neochilenia 187
Neoportia 204
Neotoma fuscipes 153
Nolina interrata 135
Nostoc 251
Nothofagus 189, 192
N. dombeyii 216
N. obliqua 192, 216
Nucifraga caryocatactes 90

Ochagavia 206
Octodon degus 189, 207, 210-213, 215
Ogyris 262
Olea europaea 47, 89, 312
Ondatra zibethicus 80
Ononis 82
Ophrys 77
Orphioglossum californicum 157
Opuntia 137, 204
Orthocarpus purpurens 157
Oryctolagus cuniculus 80
Oryx dammah 78
Oryzomys longicaudatus 188, 210-213, 215
Oryzopsis 140
Ostrya 75
Otomys 311
Ovis ammon orientalis 80
O. aries 80
Oxalis 186-187

Palorchestes 256
Panthera leo 78
P. pardus 78
Parabuteo unicinctus 210, 214-216
Parategeticula 150
Parus caeruleus 96-99
Penicillium 134
Penstemon 127
Pentachaeta lyonii 135
Peperomia 197
Perognathus 147-148
P. longimembris 147
P. maniculatus 147
Peromyscus 147
P. californicus 136, 147
P. eremicus 147
Peumus boldus 219
Phacelia 124, 127
P. brachyloba 168
P. cicutaria 168
P. minor 168
Phainopepla nitans 146
Phalaris tuberosa 60
Phascolarctos 256
Phascolonus 256
Pheidole 309-310
P. pallidula 95
Phillyrea 75
P. angustifolia 52
Phryganidia californica 144
Phyllotis darwini 188, 212-213, 215
Phytophthora cinnamomi 247-248, 253, 265, 268-269
Pica nuttalli 131
Picea 124
Pilularia americana 157
Pinus 51, 104-105, 291, 307, 312, 315-316, 319
P. cembra 90
P. halepensis 20-21, 293
P. nigra 75
P. pinaster 264-265, 315
P. radiata 29, 317
Pistacia 75
P. lentiscus 52, 89-90
P. terebinthus 89
Pisum 82
Plagiobotrys undulatus 157
Plantago bigelovii 157
P. erecta 157, 170
P.lanceolata 219
Poa 139, 187
Podanthus mitique 201, 217-219
Pogogyne abramsii 157
Polioptila californica 131
Polygonum 222
Praomys verreauxi 309
Procoptodon 256

Prosopis 81
Protea 310
P. nitida 310
P. repens 310
Protemnodon 256
Proustia cinerea 199
P. pungens 201
P. pyrifolia 198
Prunus 165
P. armeniaca 219
P. ilicifolia 137, 169
P. lyonii 137
Pseudalopex culpaeus 210, 213-216
P. griseus 210, 213-215
Psilocarphus brevissimus 157
P. tenellus 157
Psoralea 82
P. bituminosa 87, 88
P. glandulosa 219
Puya 206
P. berteroniana 198, 219-222

Quercus 51, 130, 144, 161-170
Q. agrifolia 144
Q. calliprinos 52
Q. canariensis 75
Q. coccifera 63-64, 75-76, 104
Q. dumosa 18, 170
Q. ilex 17, 47, 52, 66, 72, 75-76, 92-93, 96-97, 104
Q. petraea 75
Q. pubescens 66, 75, 95-97
Q. pyrenaica 75
Q. rotundifolia 75
Q. sessilifolius 66
Q. suber 72, 75
Quillaja saponaria 198-200, 219-222

Rapanea melanophloeos 312
Raphanus sativus 217-218, 222
Rapistrum rugosum 217-218
Rattus 79
R. rattus 80
Reithrodontomys 147
R. megalotis 147
Retanilla 202
Rhabdomys pumilio 298
Rhamnus 165
R. californica 169
Rhanterium suaveolens 103
Rhaphithamnus spinosus 197
Rhizobium 102
Rhus 312
R. ovata 170
Robinia pseudo-acacia 78
Rosmarinus 104
Rubia peregrina 89-90
Rubus ulmifolius 219

Sacropoterium spinosum 63
Salix 130-197
S. chilensis 197-198
Salmo 126
Salvia mellifera 133, 163
Sarothamnus scoparius 88
Satureja gilliesii 199, 202
Schinus latifolius 222
S. polygamus 201
Schismis barbatus 163
Sclerotheca 94
Selaginella cinerascens 157
Senecio 186-187
S. latifolius 201
Sequoia 124
Sequoiadendron 124
Serinus 308
Silene gallica 157
Silybum marianum 219
Sitanion jubatum 140
Smilax aspera 89-90
Solanum 187
Sophora macrocarpa 202
Spartium junceum 88
Sphyrapicus 145
Spinifex 272
Sporobolus airoides 140
Stachys grandidentata 222
Sthenurus 256
Stipa 76, 139, 187
S. pulchra 140
S. tenacissima 47
Stipagrostis pungens 103
Sturnus vulgaris 139
Stylidium 242
Suncus 79
Sus 59
S. scrofa 79
Sutureja gilliesii 202
Sylvia 56
S. atricapilla 89-90

S. borin 73-89
S. communis 89
S. melanocephala 89, 91
Sylvilagus floridanus 80
Symphyla 72

Talguenea 202
Taraxacum officinale 219
Tecophilea 206
Tegeticula 150
Teline 87
Themeda australis 272
Thylacinus 256
Thylacoleo 256
Thymus 60-61, 95
T. vulgaris 61, 96
Tilia 51
Toxostoma redivivum 131
Trevoa 202
T. trinervis 198-199, 201-202, 218-222
Trichocereus 204
T. litoralis 225
T. chilensis 198, 204
Trichosurus vulpecula 260
Trifolium 82, 88
T. amplectens 157
T. repens 222
Tristerix aphyllus 222
Tuber melanosporum 68
Turdus merula 89
T. philomelos 89
Tyto alba 56, 210, 213-216

Ulex 72
Ulmus 51
Ursus arctos 78
U. horribilis 126
Umbellularia californica 169

Valeriana 187
Verbascum virgatum 219
Veromessor andrei 154
Verticordia 241
Viburnum tinus 89
Vicia 82
Viola 187
Vipera aspis 56
Vitis 68
Vulpia myuros 157

Xanthorrhoea 250

Yucca 150

Zaglossus 256
Zea mays 219
Zylomelum 240
Zygomaturus 256

Part II (families)

Aextoxicaceae 186
Alstroemeriaceae 186
Aphodiidae 290
Apiaceae 60
Arachnitaceae 186
Asteraceae 60, 187, 201, 304-306

Bathyergidae 298
Berberidaceae 187
Brassicaceae 217-219
Bruniaceae 306

Cactaceae 187, 204
Casuarinaceae 251
Chrysomelidae 310
Cynipidae 145
Cyperaceae 187, 239, 270

Dendrobatidae 190
Desfontaineaceae 186
Dioscoriaceae 187
Discoglossidae 56

Ericaceae 301, 304-306, 310

Fabaceae 242, 306
Fagaceae 144
Francoaceae 186

Gomortagaceae 186

Heteromyidae 148, 154
Heterostylacea 186
Histeridae 291
Hydrophilidae 291
Hymenoptera 145

Iguanidae 190
Iridaceae 305

Labiatae 201
Lacertidae 56
Lactoridaceae 186
Lamiaceae 60
Leguminosae 81
Loasaceae 187
Lycaenida 262

Macroscelididae 309
Malesherbiaceae 186
Malvaceae 187
Meliphagidae 261
Mizodendraceae 186
Myrtaceae 239, 242

Nolanaceae 186

Oxalidaceae 187

Papilionaceae 187
Papilionoideae 81
Pieridae 262
Poaceae 187
Portulacaceae 187
Proteaceae 239, 242, 250, 269, 270, 295, 301, 303-306, 310

Restionaceae 239, 298, 301, 304-306
Rhamnaceae 202, 304-306
Rutaceae 306

Salamandridae 56
Saxifragaceae 186
Scarabaeidae 290
Scrophulariaceae 187
Sigmodontinae 189
Solanaceae 187
Staphylinidae 290

Testunididae 56
Thyrsopteridaceae 186
Tortricidae 262
Tropaeolaceae 186

Umbellularia californica 169

Valerianaceae 187
Violaceae 187
Viperidae 56
Vivianiaceae 186

Ecological Studies

Volumes published since 1989

Volume 74
Inorganic Contaminants in the Vadose Zone (1989)
B. Bar-Yosef, N.J. Barrow, and J. Goldshmid (Eds.)

Volume 75
The Grazing Land Ecosystems of the African Sahel (1989)
H.N. Le Houérou

Volume 76
Vascular Plants as Epiphytes: Evolution and Ecophysiology (1989)
U. Lüttge (Ed.)

Volume 77
Air Pollution and Forest Decline: A Study of Spruce (*Picea abies*) on Acid Soils (1989)
E.-D. Schulze, O.L. Lange, and R. Oren (Eds.)

Volume 78
Agroecology: Researching the Ecological Basis for Sustainable Agriculture (1990)
S.R. Gliessman (Ed.)

Volume 79
Remote Sensing of Biosphere Functioning (1990)
R.J. Hobbs and H.A. Mooney (Eds.)

Volume 80
Plant Biology of the Basin and Range (1990)
B. Osmond, G.M. Hidy, and L. Pitelka (Eds.)

Volume 81
Nitrogen in Terrestrial Ecosystem: Questions of Productivity, Vegetational Changes, and Ecosystem Stability (1991)
C.O. Tamm

Volume 82
Quantitative Methods in Landscape Ecology: The Analysis and Interpretation of Landscape Heterogeneity (1990)
M.G. Turner and R.H. Gardner (Eds.)

Volume 83
The Rivers of Florida (1990)
R.J. Livingston (Ed.)

Volume 84
Fire in the Tropical Biota: Ecosystem Processes and Global Challenges (1990)
J.G. Goldammer (Ed.)

Volume 85
The Mosaic-Cycle Concept of Ecosystems (1991)
H. Remmert (Ed.)

Volume 86
Ecological Heterogeneity (1991)
J. Kolasa and S.T.A. Pickett (Eds.)

Volume 87
Horses and Grasses: The Nutritional Ecology of Equids and Their Impact on the Camargue (1992)
P. Duncan

Volume 88
Pinnipeds and El Niño: Responses to Environmental Stress (1992)
F. Trillmich and K.A. Ono (Eds.)

Volume 89
Plantago: A Multidisciplinary Study (1992)
P.J.C. Kuiper and M. Bos (Eds.)

Volume 90
Biogeochemistry of a Subalpine Ecosystem: Loch Vale Watershed (1992)
J. Baron (Ed.)

Volume 91
Atmospheric Deposition and Forest Nutrient Cycling (1992)
D.W. Johnson and S.E. Lindberg (Eds.)

Volume 92
Landscape Boundaries: Consequences for Biotic Diversity and Ecological Flows (1992)
A.J. Hansen and F. di Castri (Eds.)

Ecological Studies

Volumes published since 1989

Volume 93
Fire in South African Mountain Fynbos: Ecosystem, Community, and Species Response at Swartboskloof (1992)
B.W. van Wilgen et al. (Eds.)

Volume 94
The Ecology of Aquatic Hyphomycetes (1992)
F. Bärlocher (Ed.)

Volume 95
Palms in Forest Ecosystems of Amazonia (1992)
F. Kahn and J.-J. DeGranville

Volume 96
Ecology and Decline of Red Spruce in the Eastern United States (1992)
C. Eagar and M.B. Adams (Eds.)

Volume 97
The Response of Western Forests to Air Pollution (1992)
R.K. Olson, D. Binkley, and M. Böhm (Eds.)

Volume 98
Plankton Regulation Dynamics (1993)
N. Walz (Ed.)

Volume 99
Biodiversity and Ecosystem Function (1993)
E.-D. Schulze and H.A. Mooney (Eds.)

Volume 100
Ecophysiology of Photosynthesis (1994)
E.-D. Schulze and M.M. Caldwell (Eds.)

Volume 101
Effects of Land Use Change on Atmospheric CO_2 Concentrations: South and South East Asia as a Case Study (1993)
V.H. Dale (Ed.)

Volume 102
Coral Reef Ecology (1993)
Y.I. Sorokin

Volume 103
Rocky Shores: Exploitation in Chile and South Africa (1993)
W.R. Siegfried (Ed.)

Volume 104
Long-Term Experiments With Acid Rain in Norwegian Forest Ecosystems (1993)
G. Abrahamsen et al. (Eds.)

Volume 105
Microbial Ecology of Lake Plußsee (1993)
J. Overbeck and R.J. Chrost (Eds.)

Volume 106
Minimum Animal Populations (1994)
H. Remmert (Ed.)

Volume 107
The Role of Fire in Mediterranean-Type Ecosystems (1994)
J.M. Moreno, W.C. Oechel

Volume 108
Ecology and Biogeography of Mediterranean Ecosystems in Chile, California and Australia (1994)
M.T.K. Arroyo, P.H. Zedler, and M.D. Fox (Eds.)

Volume 109
Mediterranean-Type Ecosystems. The Function of Biodiversity (1994)
G.W. Davis and D.M. Richardson (Eds.)

Volume 110
Tropical Montane Cloud Forests (1995)
L.S. Hamilton, J.O. Juvik, and F.N. Scatena (Eds.)

Volume 111
Peatland Forestry. Ecology and Principles (1995)
E. Paavilainen and J. Päivänen

Volume 112
Tropical Forests: Management and Ecology (1995)
A.E. Lugo and C. Lowe (Eds.)

Volume 113
Arctic and Alpine Biodiversity. Patterns, Causes and Ecosystem Consequences (1995)
F.S. Chapin III and C. Körner (Eds.)

MIX
Papier aus verantwortungsvollen Quellen
Paper from responsible sources
FSC® C105338

If you have any concerns about our products,
you can contact us on
ProductSafety@springernature.com

In case Publisher is established outside the EU,
the EU authorized representative is:
Springer Nature Customer Service Center GmbH
Europaplatz 3, 69115 Heidelberg, Germany

Printed by Libri Plureos GmbH
in Hamburg, Germany